ELECTRON MICROSCOPY
AND CYTOCHEMISTRY

ELECTRON MICROSCOPY AND CYTOCHEMISTRY

Proceedings of the second International Symposium,
Drienerlo, The Netherlands, June 25–29, 1973

Editors:

E. Wisse

W.Th. Daems

I. Molenaar

P. van Duijn

1974

NORTH-HOLLAND PUBLISHING COMPANY – AMSTERDAM · LONDON

AMERICAN ELSEVIER PUBLISHING COMPANY, INC. – NEW YORK

Library of Congress Catalog Card Number: 73-91448

ISBN North-Holland: 0 7204 4125 0
ISBN American Elsevier: 0 444 10605 7

Publishers:

NORTH-HOLLAND PUBLISHING COMPANY – AMSTERDAM
NORTH-HOLLAND PUBLISHING COMPANY, Ltd. – LONDON

Sole distributors for the U.S.A. and Canada:

AMERICAN ELSEVIER PUBLISHING COMPANY, INC.
52 VANDERBILT AVENUE
NEW YORK, N.Y. 10017

PRINTED IN THE NETHERLANDS

NAT
Ref

The Organizing Committee acknowledges the support
and assistance received from:

Edax Laboratories

European Cell Biology Organization

Janssen Pharmaceutica

Japan Electron Optics Lab. Co. Ltd.

A. de Jong T.H. N.V.

Lameris N.V., representative of Reichert

Leybold – Heraeus N.V.

L K B – Produkten B.V.

Meyvis & Co., N.V.

Ministerie van Onderwijs en Wetenschappen

N.V. Philips' Gloeilampenfabrieken

Siemens

Unilever N.V.

Veco, Zeefplatenfabriek N.V.

PREFACE

The Drienerlo Symposium on Electron Microscopy and Cytochemistry was held from June 25th to 30th, 1973, in The Netherlands. This was the fourth in a series held alternately in England and The Netherlands. The first conference was held in Oxford in 1962, and its Proceedings were published as Parts 3 and 4 of Volume 81 of the Journal of the Royal Microscopical Society. The next symposium took place in 1966 in Leiden; abstracts were published in the *Journal of Histochemistry and Cytochemistry,* Vol. 14, 739 (1966). The third meeting was held in London in 1968.

This book, which is intended to give the reader an impression of the present level of the cytochemical techniques used in electron-microscopy, contains the papers given at the Drienerlo Symposium and covers the various approaches applied within the field of electron-microscopical cytochemistry.

CONTENTS

ENZYME CYTOCHEMISTRY

Electron microscopy and Cytochemistry, eds. E. Wisse, W.Th. Daems, I. Molenaar and P. van Duijn.
© 1973, North-Holland Publishing Company - Amsterdam, The Netherlands.

FUNDAMENTAL ASPECTS OF ENZYME CYTOCHEMISTRY

P. van Duijn
Department of Histochemistry and Cytochemistry
University of Leiden, The Netherlands

SUMMARY

A review is given of the factors that influence the accurate cytochemical localization of enzymes. Theory and practice of fixation procedures are discussed with special regard to the use of fixatives hitherto applied only in immunocytochemistry and to the possibilities of applying the principle of substrate protection. A scheme of the several phases of cytochemical enzyme procedures whose outcome is influenced by the reaction rates of chemical conversions and by the effect of diffusion processes is given. Basic principles of methods based on metal salt reactions are compared with those based on organic dye precipitation reactions. Diffusion coefficients, reaction rate constants of the components of the substrate turnover and the trapping reactions, spot size of the localization site and concentration as well as turnover number of the enzyme involved, are among the factors that determine the final results of the procedure. Study of the influence of variations in fixation and incubation procedures on the reaction as it occurs in the object in combination with studies on models, especially those employing enzymes immobilized in spheres or films, seems necessary to promote further progress of the field.

INTRODUCTION

Since the pioneer-work of Gomori and Takamatsu, many procedures to localize enzyme activities *in situ* have been described.[1] Enzymes can be localized cytochemically, due to their properties as antigens, by immunohistochemical procedures[2] or by autoradiographic methods based on their reactions with labeled inhibitors.[3] However, most of the methods make use of the catalytic activity of enzymes on substrates, which also is the basis of their biochemical characterization.

For enzyme localization at the ultrastructural level, this means that the incubation media should be composed such that substrate conversion will lead to an insoluble, electron-dense, and recognizable final product that is formed as close as possible to the original (*in vivo*) sites of the enzyme molecules. To prevent loss or displacement of the enzyme molecules from their *in vivo* positions and at the same time maintain a recognizable morphology during the incubation procedure, fixation is usually necessary.

In the following, some fundamental aspects of fixation procedures and the requirements of suitable incubation media in terms of their potentials for accurate localization and quantitation of the local enzymatic activities will be discussed. The accuracy of the final results can be influenced by the effect of each separate factor in the sequence of events from tissue preparation and fixation through incubation to observation of the final photographs. For the purposes of the present analysis, several factors involved in fixation and incubation will be discussed separately.

FIXATION

a) Objectives of fixation

A fixation procedure for enzyme cytochemistry should be designed to achieve the following aims as much as possible. It should preserve the morphology of the object adequately, it should stop metabolism so that autolysis is prevented and that the fixed cell and cell organelle skeletons can withstand the incubation procedures that follow. Fixation should also lead to a breakdown of membrane permeability barriers present *in vivo*, such that transmembranic diffusion - at least of low molecular weight substrates and trapping agents - becomes relatively unhampered.

It is becoming more and more clear that aldehyde fixation, especially in the initial phases, does not rapidly lead to this condition, which means that the tonicity of the compounds in the fixation medium that are (temporarily) osmotically active is of importance.[4,5] Finally, an ideal fixation procedure should anchor the relevant enzyme molecules to the cell structure, preferably by a covalent band, without much decrease in their enzymatic activity. In practice it is found that, depending on the type of enzyme and on fixation conditions, aldehyde fixation can result in 90% to 25% inactivation.[6]

b) Theory and practice of fixation

The activity of an enzyme depends on the preservation of the steric conformation at its active site(s) and the presence of co-enzymes. If we define denaturation of a protein as a (gradual) loss of quaternary, tertiary, or secondary structure, it is clear that an enzyme can be partially denaturated but still fully active enzymatically (Fig. 1).

Theoretically, therefore it is not unlikely that for a particular enzyme fixation agents and conditions can be found that, even if they denature the molecule to some extent, preserve its original activity or at least a part of it. The details of how a cross-linking agent will react with a particular enzyme and its influence on the steric conformation at the active site, however, remains largely unpredictable. This despite considerable recent progress in protein chemistry and even for enzymes of which the molecular structure is completely known.

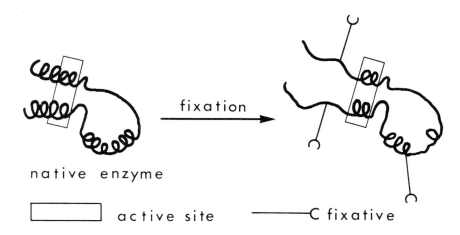

Fig. 1: Theoretical scheme of enzyme fixation. Fixation leads to modification of the original structure of the enzyme molecules: provided that during the fixation procedure the conformation of the active site (sometimes influenced by more remote parts of the molecule) remains relatively intact, loss of conformation (denaturation) caused by the fixation conditions in other regions of the molecule can be compatible with retention of (part of) the enzymatic activity.

It must therefore be stressed that there is still much to expect from trial and error experiments with new fixatives and fixation conditions. The type, ionic strength, and pH of buffers, the presence of neutral salts or hydrophylic compounds such as sucrose and dimethyl sulfoxide, the presence of reducing or oxidative compounds or of heavy metal ions, all can be factors that may prove to be useful for a particular enzyme even if detrimental for the activity of others.

Since there is a strong analogy between the conditions maintaining an active site of an enzyme and that of an immunologically active protein-antigen, enzyme-cytochemistry can learn from experience gained in immunocytochemistry and vice versa. In this respect it may be pointed out that fixatives such as carbodiimide,[7] cyanuric chloride,[8] and dimethylsuberimidate[9,10] successfully used in immunocytochemistry, have so far apparently not been applied for fixation of tissues with the aim of localizing enzymes.

Another field of research from which ideas for new fixatives could be adapted for enzyme cytochemistry are biochemical studies that use cross-linking agents to obtain more information about the steric conformation of reactive groups in proteins. Such agents have been described in great variety,[11,12] and some of them might prove to be suitable fixatives for particular enzymes. On the other hand, fixation media that differentially inactivate one or more of a family of isoenzymes which otherwise are indistinguishable because they have similar substrate specificity,[13] can be of advantage.

c) Substrate protection

In enzymology it has been known for a long time that an enzyme can be protected to some extent against denaturing influences, such as increase in temperature, by the presence of a substrate or a reversible inhibitor. It was found as early as in the last century that sucrose protects invertase to some extent against heat inactivation.[14] Since then, many examples of this effect have become known for both substrates[15] and inhibitors.[16]

Interaction of substrate or inhibitor with the enzymatic site apparently stabilizes the conformation of the enzyme in this region. One could hope that the fixation of enzymes in the presence of a substrate or inhibitor could prevent the fixative from reacting directly with atomic groups in the site or stabilize the conformation of the site in such a way that greater part of its activity remained intact.

This principle of substrate protection (Fig. 2) has been applied successfully in enzyme cytochemistry to the isoenzymes of aspartate aminotransferase.

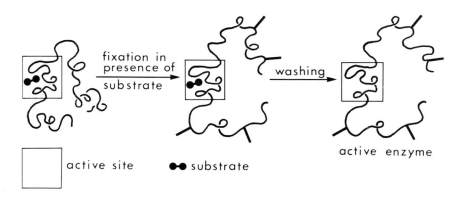

Fig. 2: Active site of the enzyme is protected during the fixation procedure by the presence of substrate or reversible inhibitor.

Fixation with glutaraldehyde or formaldehyde in the presence of ketoglutarate, one of the substrates of the enzyme, decreased the inactivation of the enzyme[17] by the fixatives. The favourable effect of adding ATP to the fixative for demonstration of ATPase activity had already been described independently[18] and has been applied to fix a muscle ATPase.[19,20]

It can be expected that further application of this principle in enzyme cytochemistry might be successful, since for most enzymes a whole range of substrates and reversible inhibitors have been described. Given the sensitivity of protein structure to its chemical environment in terms of pH, salts, etc., it again must be stressed that failure to obtain protection by applying the principle in a certain medium with a certain fixative should not discourage trial and error experiments under other conditions.

d) Quantitative model studies

The effect of a variety of fixation procedures on the accurate cytochemical localization and quantitation of enzymes must finally be judged on the basis of the localization and amount of electron-dense product produced by the action of the enzyme within the biological object. Without detracting from the value of attempting to draw conclusions concerning to the real situation of the enzyme *in vivo* from visual observation or even densitometric quantitation of the pictures obtained, it is clear that both fixation and staining are very complex phenomena, in which a great number of unknown or only partially known factors are involved. The basing of conclusions solely on the final outcome of such complex procedures is a particularly dangerous undertaking. Situations in which a given variation in the procedure is favourable for one step and unfavourable for another, could remain unrecognized if possibilities to study both parameters independently and quantitatively were lacking. Exclusive adherence to such an approach, which is very useful at the start, could in the long run put an end to further development in the field.

A number of quantitative studies, with biochemical,[6] or cytochemical media, [21] on homogenates or sections of the biological object, have been reported. Since in these cases the complexities of the biological object still play a role, models have been developed in which with both cytochemical and biochemical media, quantitative study under more sharply defined conditions, such as the use of purified (iso)enzymes, is possible.

In this respect films of polyacrylamide, which can function as vehicle for purified enzymes preparations as well as for homogenates or for cell organelle suspension, have been used successfully.[22,23] Together with a specially developed film colorimeter (Fig. 3), this method offers an elegant system in which most of the aspects of enzyme cytochemistry can be quantitatively studied in detail, both biochemically and cytochemically. With this system the different sensitivity of the isoenzymes of aspartate aminotransferase for aldehyde fixatives and the effect of substrate protection (Fig. 4), as well as the Michaelis constants for the cytochemical media, have been determined.[17] A similar system has been used to study and improve the conditions that govern light microscopical localization and quantitation of leucocyte alkaline

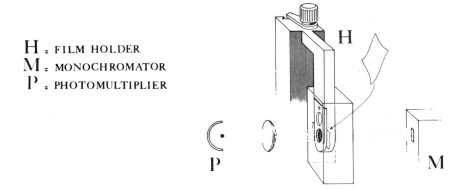

H : FILM HOLDER
M : MONOCHROMATOR
P : PHOTOMULTIPLIER

Fig. 3: Principle of film colorimetry. Films are measured - clamped in holder and covering the lower aperture - in a spectrophotometer cell containing medium of appropriate refractive index. The upper aperture is used as the reference. To correct for differences in film thickness, the measured part of the film is punched out and analysed or dried and weighed.

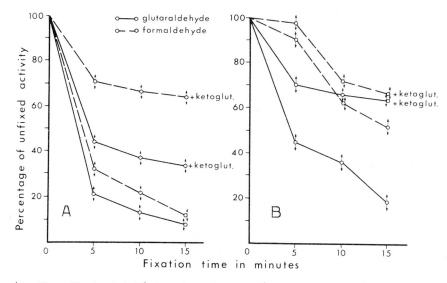

Fig. 4: The effect of 3.7% formaldehyde and 1% glutaraldehyde in 0.05 M imidazole buffer on the activity of the isoenzymes of aspartate aminotransferase. A: Soluble isoenzyme from rat liver incorporated into polyacrylamide films. B: Mitochondrial isoenzyme from rat liver incorporated into polyacrylamide films. Films measured in the film colorimeter after staining with a cytochemical lead medium. (From reference 17)

phosphatase.[24,25]

Films and microscopical beads of cellulose[22] or sepharose[26] also can be used as vehicles to study the quantitative and specificity aspects of both enzyme-cytochemical and immunocytochemical reactions.

LOCALIZATION

The localization of enzymes *in situ* by   formation of light-absorbing or
electron-dense product may at first sight seem to be a problem that can be
discussed without reference to the quantitative aspects of the rate of product
formation. However, in a cytochemical system, contrary to what happens in an
enzyme solution in a test-tube, considerable amounts of final product that
should be deposited at a certain site in the fixed cell can escape by diffusion.
Depending on the reaction kinetics of the system and the diffusion constants,
diffusion can dominate to such an extent that either a false localization or none
at all is obtained, making accurate localization impossible, even in a qualitative
sense.

The processes going on during an enzyme cytochemical reaction in and around
an enzymatic site can be schematically described as in the following Figure 5.

MECHANISM OF CYTOCHEMICAL ENZYME LOCALIZATION

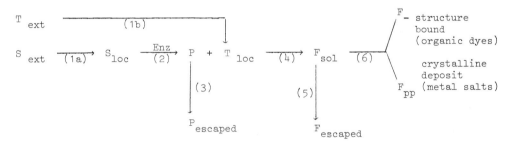

Enz	=	Enzyme
S_{ext} and S_{loc}	=	Concentration of substrate in medium and at the site, resp.
P	=	Primary product
T_{ext} and T_{loc}	=	Concentration of trapping agent in medium and at the site, resp.
F_{sol}	=	Final product when still in solution
F_{pp}	=	Final product precipitated and/or structure-bound

Fig. 5: Scheme of processes occurring during the cytochemical localization of
enzymes. In some cases a separate trapping agent is not present.

The inflow of substrate and trapping agents inside a tissue block, an section,
or a cell is governed by the diffusion processes 1a and 1b. The actual enzymatic
reaction of enzyme with substrate (2) results in the formation of a primary
product (P). This product can either react with a trapping reagent (4) or
diffuse away from the site (3). Since at this stage, too, some time is required
before the final product, even though completely insoluble in a thermodynamic
sense, is actually precipitated, this precipitation reaction (6) also has to

compete with a diffusion process (5) which, if it dominates, can lead to actual deposition of the final product at considerable distance from the enzymatic site.

The end result of the procedure, an amount of final product precipitated near the correct site or at other sites, is determined by a delicate balance between the velocities of all these processes. In cases where diffusion dominates, it could lead to complete false or even negative results. Some of the conditions that govern these processes will now be discussed.

a) Inflow of substrate and trapping agent

The inflow of substrate and trapping reagent from the medium into the object takes some time. The time required to approach a steady-state condition such that substrate and trapping agent concentration are nearly equal to their concentration in the incubating medium, depends on the diffusion coefficient of these molecules as well as on the dimensions of the object. It can be calculated that when substrate and trapping reagents are not used up too rapidly at the sites and for compounds with an effective diffusion coefficient of 3×10^{-6} cm^2 . sec^{-1} (which roughly equals that of the phosphate ion in water), the concentration at the center of a 50-μ section should reach 90% of the external concentration in about 2.5 seconds[27,28] (see Fig. 6).

INFLOW OF SUBSTRATE AND TRAPPING AGENT

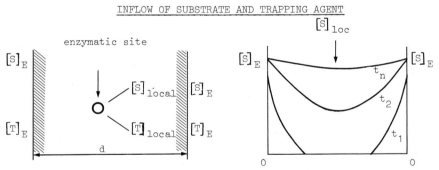

Fig. 6: Substrate gradients of substrates after times $t_1 \longrightarrow t_n$ are shown in the righthand figure. For a diffusion coëfficient of 3×10^{-6} cm^2 . sec^{-1} the time t_n to reach the situation that $[S]_{loc} = \frac{9}{10} [S]_E$ is given by the formula $t_n = 10^{-3} d^2$. For d=10μ this means $t_n = \frac{1}{10}$ sec; for 50μ the formula gives 2.5 sec and for 1 mm a value of 17 min. (After Sundaram et al. 1970)

In practice, both substrate and trapping reagent may not be able to diffuse into fixed tissue as rapidly as phosphate ions free in water. Studies on the penetration of lead ions, the trapping agent in the acid phosphatase reaction, done in tissue blocks treated with sulfide at different times after incubation in lead media, have shown, however, that for a 50-μ section lead ions effectively penetrate to the middle of the object within the first minute.[29] In situations when the exhaustion of substrate is not negligible, however, a considerable diffusion gradient could be maintained, even under steady state conditions.

The conditions determining the steepness of the concentration gradients that can occur when substrate is used up rapidly have been investigated practically and theoretically in enzyme-containing polyacrylamide films[24] Probably, steep gradients in the free space between the active sites in a tissue section are not to be expected provided the local substrate turnover rate within the sites is not too rapid and not too many of them are present.

b) Rate of substrate turnover by the enzyme

For an enzyme that follows Michaelis-Menten kinetics, the rate of substrate turnover (reaction 2) is dependent on the substrate concentration as well as on the enzyme concentration. The course of the reaction can be described as:

$$E + S \xrightleftharpoons[k_{-1}]{k_{+1}} ES \xrightarrow{k_{+2}} E + P \qquad \text{(eq. 1)}$$

where E = enzyme; S = substrate; P = primary product.

It can be calculated that at conditions where the substrate concentration is high compared with the Michaelis-Menten constant (K_m) of the enzyme, a maximum rate of substrate turnover is reached. This can be written as:

$$V_{max} = k_{+2} E_o \qquad \text{(eq. 2)}$$

where k_{+2} = turnover number and E_o is the concentration of the enzyme.

When the enzyme is saturated with substrate, the maximum velocity per gram enzyme is an inherent property of the enzyme and called turnover number. The turnover number can, however, change (in most cases decrease) when something changes in the molecular structure of the enzyme, e.g. by the action of a fixative. Since it seems that aldehyde interaction with some enzymes soon reaches a plateau,[27,30] it might well be that the changed enzyme has a well defined, although lower, turnover number. In addition, it must be pointed out that the turnover number depends on the type of substrate used and on the conditions under which the substrate acts in terms of pH, buffer type, and the presence of activators or inhibitors. In this respect the concentration of trapping agents such as lead ions, which act as inhibitors for many enzymes, can decisively influence the turnover number.

The reaction velocity of the enzymatic reaction is also proportional to the enzyme concentration or, for a given dimension of the site, to a total amount of enzyme in the site. This amount is of course, in a given situation and biological object, the unknown parameter that one should like to estimate.

At lower substrate concentration the reaction velocity is always lower and finally also becomes proportional to the substrate concentration.[24] Since

in general for cytochemical purposes a high substrate turnover is desirable
(*vide infra*) it is necessary to be sure that the substrate concentration at the
enzymatic site is not too low. This means that the substrate that has to diffuse
from the incubating medium into the object should, after a short period, reach
a steady state in which the substrate concentration at the site does not differ
greatly from the external concentration. This can be achieved by choosing the
right thickness for tissue blocks or sections.

c) Trapping in organic dye methods

The rate of formation of the primary product at the surface of the enzyme
depends on the method chosen. In tetrazolium salt cytochemistry the primary
product itself has a low solubility. In this case no trapping agent is needed.
In the case of the diamino-benzidine and dihydroxy-indole methods to localize
peroxidatic activity, the primary product polymerizes to a phenazine[31] or a
melanin[32] polymer, the solubility of which decreases with increasing
polymerization grade. No special trapping reagent is needed in this case either.

Several compounds can act as trapping agents for organic dye methods, such as
potassium ferricyanide to convert indoxyl to indigo[33] or diazonium salts to
convert naphtolic compounds to azo dyes.[24] When the trapping agents can
maintain a constant concentration at the site, the trapping reaction follows
first-order kinetics in respect to the primary product. For the situation in
which it is assumed that the final stain is so rapidly precipitated or
immobilized by binding to cell structures that it has no time to diffuse away to
any extent from the site after it has been formed (reaction (6) being very rapid),

Holt and O'Sullivan[27] calculated the effect of the diffusion process on the
localization sharpness of the final product. They arrived at a localization
factor, defined as the amount of final product actually trapped within the site
divided by the amount of final product trapped when no primary product (P) was
thought to have escaped. This localization factor proved to be independent of
the enzymatic activity in the site but was shown to decrease with decreasing
radius of the site and also decreased when the quotient k/D was lowered (Table 1).
In this quotient k stands for the first-order velocity constant of the trapping
reaction (4) and D is the diffusion coefficient of the primary product P. These
calculations on the effect of process (3) on the localization hold only if
reaction (6) is so rapid that the effect of reaction (5) can be neglected (see
Fig. 5). It must be stressed, however, that for most enzyme-cytochemical methods
based on organic dye methods almost nothing is known about the solubility of the
final products and still less about the velocity with which these products are
actually immobilized. Some final products in polymeric reactions have been found
to react covalently with tissue proteins, so these proteins could actually be
involved in the immobilizing process.[32,34]

It is also possible that the final product has a high affinity (substantivity), by non-covalent binding to relatively omnipresent tissue components such as proteins.[33]

Table 1

EFFECT OF RADIUS SITE AND RATE CONSTANT OF TRAPPING REACTION ON LOCALIZATION FACTOR[a]

radius of site	k[b]		
	10^3	10^3	10
1 μ	95%	57%	4%
0.1 μ	57%	4%	0.1%

a) The localization factor is given as percentage of stain actually trapped compared with trapping of 100% of the primary product. Data from Holt and O'Sullivan (1958)
b) k is the first-order rate constant of the trapping reaction. Values are for a diffusion coefficient of the primary product: $D=10^{-6}$ cm^2 . sec^{-1}

On the other hand, high lipid solubility in the absence of high substantivity can lead to false localizations.[33] Factors such as solubility in water and lipids, as well as substantivity for proteins of the final product, must be taken into consideration. The kinetics of these processes too should be studied, because if these processes are too slow, escape from the site of the final product by diffusion remains possible.

d) Trapping in metal salts methods

Few studies have been carried out on the kinetics of the trapping reactions in metal salt enzyme cytochemistry. The kinetics of the trapping reaction itself (Fig. 5, reaction 4) or of the ensuing precipitation (reaction 6), differs from that occurring with organic dyes. When, for instance, a dilute phosphate solution is added to a medium intended for the localization of acid phosphatase, it depends on the concentration of this solution what happens. At very low phosphate concentrations of the order of 0.1 mM, no visible precipitate is usually formed, even after some hours. Even though the thermodynamically defined solubility product may have been exceeded, the actual precipitation rate may remain very low. An increase in the concentration of the phosphate solution added may bring about rather rapid precipitation, albeit after a certain lag period. At still higher concentrations, precipitation seems to occur immediately. It is therefore clear that for the very rapid precipitation that is essential for avoiding the occurrence of the diffusion processes (3) and (5), attainment of the solubility product will not suffice. A situation of considerable supersaturation is necessary to garantee such a rapid reaction.[35] At such a point the process of the formation of crystal nuclei preceding crystal formation becomes very rapid and becomes independent of accidentally present starting points for crystal formation. This type of nuclei-formation by the precipitation process itself is called homogeneous nucleation to distinguish it from the process of heterogeneous nucleation based on accidently present starting nuclei.

The kinetics of the precipitation of calciumphosphate has been studied extensively to obtain insight into the physiology of bone formation. In such studies the supersaturation level at which homogeneous nucleation and thus very rapid precipitation sets in, has been called the <u>formation product of homogeneous nucleation</u>.[36]

The results of such experiments demonstrate that, depending on the phosphate concentration that can be built up by a phosphatase at its site in the cell, three situations can arise (Fig. 7).

<u>Fig. 7</u>

LOCALIZATION SHARPNESS (INORGANIC SALT METHODS)

	Homogeneous nucleation	Sharp localization
Formation product		
	Zone of heterogeneous nucleation	Possibly wrong localization
Solubility product		
	Undersaturation	No localization

a) Steady-state concentration of primary product inside the site during enzymatic reaction

In the case that the solubility product at the enzymatic site is not reached or is not exceeded to such an extent that precipitation takes place within seconds, the primary product diffuses away. It becomes more and more diluted, and no precipitation takes place in any part of the object. In the case that the building up of the primary product is such that precipitation occurs after a shorter lag period, conditions may be such that a precipitate is formed at another site where the situation for heterogeneous nucleation is more favourable than at the enzymatic site.[35] It is known that this process is accelerated by the presence of crystal nuclei, for instance dust particles, but also as found for calciumphosphate precipitation, macromolecules such as collagen.[36] The artefactual staining of cell nuclei observed with a Gomori medium deficient in lead, could be caused by compounds present in cell nuclei that could act as starting points for lead phosphate precipitation. Whether some particular macromolecular substance present in the cell nuclei is responsible for this phenomenon remains unknown, however.

Only at sites where local phosphatase enzymatic activity has been able to build up a sufficiently high phosphate ion concentration to reach the formation product of homogeneous nucleation can rapid precipitation be expected. This means that in this case reactions (4) and (6) (scheme Fig. 5) will be so rapid that the time-dependent diffusion processes (3) and (5) will have no chance to dislocate either the phosphate ions or the crystal nuclei from their formation site.

Polyacrylamide model films have recently been used to analyse the trapping situation in more detail for the Gomori-lead medium for acid phosphatase, which is notorious for its inconsistent behaviour and artefactual localization. Films impregnated with radioactive phosphate solutions were placed in acetate buffer (pH 5.0) to determine the diffusion constant of phosphate ion diffusing out of the films.[37] The resulting diffusion coefficient of 3.2×10^{-6} $cm^2.sec^{-1}$ decreased by a factor of 1.4 when 14% sucrose was added to the buffer. From these values and the formula given by Holt and O'Sullivan[27] (Table 1) it can be concluded that the effect of adding sucrose to a medium to improve localization sharpness will be minimal.

The system also proved to be useful to obtain data on the relative trapping efficiency of media of different composition used for acid phosphatase localization. A study of the kinetics of the escape of an impregnated P_i^{32} solution when the film was brought into various incubation media, led to an explanation based on the occurrence of a lag period in the precipitation-process at the end of which the rate of the trapping reaction apparently increased to such an extent that no further phosphate escaped. The duration of these lag periods could also be expressed as the value for the mean free path along which phosphate ions could freely move before being precipitated. The shorter this mean free path, the sharper the localization. When the original phosphate concentration in the film was increased, the average free diffusion path was found to be greatly reduced; it reached a value where in this system, further decrease could no longer be measured accurately.

Since the lower limit of accurate measurement of the mean free path for phosphate in the system is about 25μ, it is not completely realistic for diffusion conditions prevailing, for example, around a lysosome, where for accurate localization the free diffusion path should be of the order of 1μ or less. That accurate localization can nevertheless be achieved in these organelles implies that the acid phosphatase activity in a lysosome can build up a higher local concentration of phosphate ions. In the polyacrylamide film system these phosphate concentrations fall in the range where the mean free path is too short to be accurately measurable. Nevertheless, the system is well suited for the study of the relative efficiency of the trapping reactions in media of different composition using phosphate-concentrations that do give measurable intercepts. Such studies[38] have shown that the predominant effect on the free path is caused by the concentration of the lead ions in the medium. The higher the lead ion concentration the more efficient the trapping. It is likely that the concentration of the free lead ion is decisive. Since it is known that several compounds, among them the often-used acetate buffer,[39] form complexes with lead, the presence and concentration of these compounds influences the amount of free lead ion and thus indirectly the trapping efficiency. It was indeed found that increasing the acetate-buffer concentration lengthened the free path. The presence of Cl'ion, possibly involved in the composition of the precipitate was, on the other hand, found to be favourable for rapid trapping. Surprisingly, it was found that in the concentration used in the Gomori medium, glycerophosphate was unfavourable for rapid trapping, which was more efficient when this compound was present in lower concentration or absent.

These results indicate that several factors, probably through different mechanisms,can influence the trapping efficiency of acid phosphatase media. The model system can be used to obtain experimental data on the average free diffusion path of phosphate ions in dependence on the composition of cytochemical phosphatase media in general and, when appropriately modified, for other cytochemical enzyme reactions as well.

e) Importance of local build-up of primary product

Except for pointing to possible improvements in the trapping efficiency of metal salt media, the results of the experiments with models strongly accentuate the importance of the build-up of a certain minimum concentration of phosphate at the site. The local concentrations that can be reached are decisive for reaching the formation product of homogeneous nucleation and thus for the accuracy of the localization.

Recent theoretical treatment in the biochemical literature of the kinetic behaviour of enzymes immobilized in films or spherical particles may shed some light on the factors that determine these levels of primary product. From such theoretical studies it can be derived that the steady-state situation where the internal concentration of primary product P_i in a sphere of diameter R reaches half the value of the concentration of the external substrate concentration is, under certain assumptions, a function of four parameters.[40] It is a function of the diameter R of the sphere, as well as of a factor

$$a = \frac{k_{+2} [E]_{site}}{D_{S.P.}}$$

In this equation k_{+2} is the turnover number of the enzyme for the substrate under the prevailing reaction conditions. $[E]_{site}$ is the enzyme concentration within the site. $D_{S.P.}$ is the diffusion coefficient of substrate and primary product which, to simplify the calculations, have been made equal.[41]

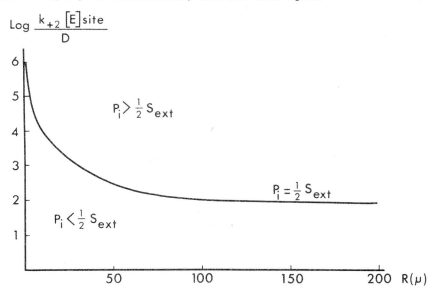

Fig. 8: The curve gives a relation between the radius (R) of the site and the parameter

$$a = \frac{k_{+2} [E]_{site}}{D}$$

for which holds that in the central portion of the spherical site:

$$P_i = \frac{1}{2} S_{ext}$$

The relation is given for $S_{ext} < K_m$ (The curve for $S_{ext} > K_m$ has a similar shape) (calculated from formulas given by Lasch[40]).

From the curve giving the relation between a and R (Fig. 8) for $P_i = \frac{1}{2} S_{ext}$ it can be concluded that to attain a maximal value of P_i it is favourable to have a large diameter of the site. This, in most cases, is a factor determined by the situation in the biological object. It also is favourable to have a high local concentration of enzyme in the site. This again is a factor mostly beyond control. Little control also seems to be possible on the diffusion coefficient of substrate and primary product, especially when both are low molecular compounds, except when D_P could be increased while D_S remains low.

Finally, it is favourable to have a high turnover number of the enzyme. This can be influenced by the choice of substrate and by incubation conditions such as the pH value. It is decreased by the presence of certain type of inhibitors. It is known, for instance, that lead in higher concentration can decrease the turnover number of phosphatases. The turnover number of an enzyme can also be influenced by the fixation conditions. Most fixation procedures irreversibly change the enzyme, leading to a considerable decrease of its turnover number. Prevention of such a decrease would have a favourable influence on $[P]_i$ and thus should improve localization sharpness, especially at sites with low $[E]$.

f) <u>Comparison between trapping mechanism in organic dye and in metal salt methods</u>

From the discussion presented here, it can be concluded that for a metal salt method, such as that for acid phosphatase, it is decisive for obtaining a precipitate at the right site that the phosphate concentration at this site reaches a certain minimum value. In the light of the foregoing theoretical consideration it is clear that especially sites of small dimensions and with a low enzymatic activity could very well be unable to reach such a phosphate concentration. This may be one of the reasons why, for instance, cytochemical localization of acid phosphatase in erythrocytes has not yet been reported, even though the enzyme has been shown biochemically to be present.[42] Too strong inactivation by the usual fixation procedures also may be responsible.

Fixation generally decreases turnover numbers and thus the product $k_{+2}[E]_{loc}$, which is the total enzymatic activity in a site. This means that the extent of inactivation of an enzyme caused by the fixative can determine whether, especially for a small site the localization site of the enzyme shows up at all in the procedure. This may explain why even the Gomori medium, which has long been considered as one of the most fickle in the field of enzyme cytochemistry, has been reported more and more often in recent years to give consistent results when due attention is given to the use of proper fixation procedures.[30] It may also explain why organic dye methods, because of their different trapping mechanism (which has less of an all-or-nothing character), still can demonstrate acid phosphatase sites where metal salts methods fail.[43]

Since enzyme turnover number can be decreased by the presence of inhibitors, a particularly dangerous situation arises in metal salt cytochemistry when the trapping agent is a strong inhibitor of the enzyme. An increase in the concentration of the trapping agents might help to improve localization by decreasing the phosphate concentration necessary to reach the formation product. It also could endanger it, however, since a lower concentration of primary product will be built up when the enzyme is more strongly inhibited. Complexing of the trapping agent with substances from the medium can complicate the situation further. Under certain conditions it may result in a very narrow optimum of trapping agent concentration, such as seems to be the case for some ATPase methods,[21] which under some circumstances work adequately but which also can give rise to artefacts.[44]

CONCLUDING REMARKS

In recent decades enzyme cytochemistry has made considerable progress despite some uncertainties and lack of quantitative data concerning the physico-chemical mechanism on which the methods are based. For many of the methods advocated one can be reasonably sure that positive evidence for the presence of a certain enzyme is reliable. A negative result, however, generally cannot be considered decisive. As this review has attempted to demonstrate, the final result of the cytochemical procedure will be determined by success in finding fixatives giving less inactivation and by further improvement of the incubation media. More quantitative data are needed on all the separate factors that influence the reliability of cytochemical enzyme procedures.

Special efforts should be made to obtain more quantitative data on the rates of reactions (4) and (6) in the trapping reaction (Fig. 5), for both the metal salts and the organic dye methods. If methods can be developed in which these phases in the trapping process of the primary product can be studied quantitatively and subsequently be improved, the disturbing effect of diffusion on the localization of the final product can be minimized.

Since the quality of the end result is always determined by the weakest part of the chain of events, better results with present methods will only be achieved when such a weakest chain has been located and then improved. For this, it is necessary that methods be developed to study each of the separate aspects quantitatively and in isolation.

With such procedures, adequately studied in models, and comparison of the results with evidence from biochemical and diffusion studies, it may be hoped that good localization, also of less active sites, and simultaneous quantitation of enzymes may be attainable on the ultrastructural level.

ACKNOWLEDGEMENTS

The author wishes to thank Drs. C.J. Cornelisse and W.A.L. Duijndam for valuable discussions and Mrs. I.W.L. Planting and Mr. J. Bonnet for help in preparation of the manuscript.

REFERENCES

1. Pearse, A.G.E., 1972: In: Histochemistry: Theoretical and applied, 3rd Ed. vol. 2, Churchill Livingstone, Edinburgh and London

2. Kökfelt, K.F., Goldstein, M., and Hob Joh, T., 1973: Immunohistochemical localization of three catechol synthesizing enzymes: Aspects on Methodology, Histochemie 33, 231-254

3. Ostrowski, K., 1972: The labeled inhibitor method in cytoenzymology, Histochem. J. 4, 467-476

4. Brunk, T.U., and Ericsson, J.L.E., 1972: The demonstration of acid phosphatase in in vitro cultured tissue cells. Studies on the significance of fixation, tonicity and permeability, Histochem. J. 4, 349-363

5. Bone, W., and Ryan, K.P., 1972: Osmolarity of osmium tetroxide and glutaraldehyde fixatives, Histochem. J. 4, 331-347

6. Hopwood, D., 1972: Theoretical and practical aspects of glutaraldehyde fixation, Histochem. J. 4, 267-303

7. Kendall, P.A., Polak, J.M., and Pearse, A.G.E., 1971: Carbodiimide fixation for immunohistochemistry: Observations on the fixation of polypeptide hormones, Experientia 27, 1104-1107

8. Goland, L.Ph., Grand, N.G., Green, F.J., and Booker, B.F., 1969: Immunofluorescence microscopy of cyanurated tissues, Stain Technology 44, 227-233

9. McLean, J.B., and Singer, S.J., 1970: A general method for the specific staining of intracellular antigens with Ferritin-Antibody conjugates, Proc. Nat. Acad. Sci. USA 65, 122-128

10. Davies, G.E., and Stark, G.R., 1970: Use of dimethylsuberimidate, a cross--linking reagent in studying the subunit structure of oligomeric proteins, Proc. Nat. Acad. Sci. USA 66, 651-656

11. Wold, F., 1967: Bifunctional reagents in: Colowick, S.P., and Kaplan, N.O., Methods in Enzymology XI, 617-640

12. Fasold, H., Klappenberger, J., Meyer, C., and Remold, M., 1971: Bifunctional reagents for the crosslinking of proteins, Angew. Chemie (Int. Edition) 10, 795-801

13. Koelle, W.A., Hossaine, K.S., Akbarzadeh, P., and Koelle, G.B., 1970: Histochemical evidence and consequences of the occurrence of isoenzymes of acetylcholinesterase, J. Histochem. Cytochem. 18, 812-819

14. O'Sullivan, C., and Tompson, F.W., 1890: Invertase, a contribution to the history of an enzyme or unorganised ferment, J. Chem. Soc., 834-931

15. Sachar, K., and Sadoff, H.L., 1966: Effect of glucose on the denaturation of glucose dehydrogenase by urea, Nature 211, 983-984

16. London, M., McHugh, R., and Hudson, P.B., 1955: Thermal stabilization of prostatic acid phosphatase by fluoride, Archives of Biochem. & Biophys. 55, 121-125

17. Papadimitriou, J.M., and van Duijn, P., 1970: Effects of fixation and substrate protection on the isoenzymes of aspartate aminotransferase studied in a quantitative cytochemical model system, J. Cell Biol. 47, 71-83

18. Hayashi, M., and Freiman, D.G., 1966: An improved method of fixation for formalin-sensitive enzymes with reference to myosin adenosine triphosphatase, J. Histochem. Cytochem. 14, 577-581

19. Khan, M.A., Papadimitriou, J.M., Holt, P.G., and Kakulas, B.A., 1972: A modified histochemical technique for sarcoplasmic reticular ATPase, Histochemie 30, 329-333

20. Khan, A.M., Papadimitriou, J.M., Holt, P.G., and Kakulas, B.A., 1972: A calcium-citro-phosphate technique for the histochemical localization of myosin ATPase, Stain Technology 47, 277-281

21. Berg, G.G., Lyon, D., and Campbell, M., 1972: Faster histochemical reaction for ATPase in presence of chloride salts, with studies of the mechanism of precipitation, J. Histochem. Cytochem. 20, 39-55

22. Van Duijn, P., and van der Ploeg, M., 1970: Potentialities of cellulose and polyacrylamide films as vehicles in quantitative cytochemical investigations on model substances. In: Introduction to quantitative cytochemistry, vol. 2 (G.L. Wied and G.F. Bahr, eds.) Academic Press, New York and London 1970, 223-262

23. Van Duijn, P., 1972: Conditions for quantitation of histochemical light and electron-microscopical staining methods studied with a model film system, In: Histochemistry and Cytochemistry 1972, p. 7-8, Proc. 4th Internat. Congress Histochem. Cytochem., Kyoto Japan, eds. T. Takeuchi, K. Ogawa, S. Fujita, Summary Repts., Nakanishi, Kyoto 1972

24. Van Duijn, P., Pascoe, E., and van der Ploeg, M., 1967: Theoretical and experimental aspects of enzyme determination in a cytochemical model system of polyacrylamide films containing alkaline phosphatase, J. Histochem. Cytochem. 15, 631-645

25. Van der Ploeg, M., and van Duijn, P., 1968: Cytophotometric determination of alkaline phosphatase activity of individual neutrophilic leukocytes with a biochemically calibrated model system, J. Histochem. Cytochem. 16, 693-706

26. Van Dalen, J.P.R., Knapp, W., and Ploem, J.S., 1973: Microfluorometry on antigen-antibody interaction in immunofluorescence using antigens covalently bound to agarose beads, J. of Immunological Methods 2, 383-392

27. Holt, S.J., and O'Sullivan, D.G., 1958: Studies in enzyme cytochemistry, I. Principles of cytochemical staining methods, Proc. Roy. Soc. B 148, 495-505

28. Sundarum, P.V., Tweedale, A., and Laidler, K.J., 1970: Kinetic laws for solid-supported enzymes, Canad. J. Chem. 48, 1498-1504

29. Holt, S.J., and Hicks, R.M., 1961: The localization of acid phosphatase in rat liver cells as revealed by combined cytochemical staining and electron microscopy, J. Biophys. and Biochem. Cytol. 11, 47-66

30. Holt, S.J., 1959: Factors governing the validity of staining methods for enzymes and their bearing upon the Gomori acid phosphatase technique, Exptl. Cell Res. Suppl. 7, 1-27

31. Seligman, A.M., Karnowsky, M.J., Wasserkrug, H.L., and Hauker, J.S., 1968: Nondroplet ultrastructural demonstration of cytochrome oxidase activity with a polymerizing osmiophylic reagent, diaminobenzidine (DAB), J. Cell Biol. 38, 1-14

32. Van der Ploeg, M., and van Duijn, P., 1964: 5,6-dihydroxy indole as a substrate in a histochemical peroxidase reaction, J. Roy. Micr. Soc. 83, 415-423

33. Holt, S.J., and Withers, R.F.J., 1958: Studies in enzyme cytochemistry, V. An appraisal of indigogenic reaction for esterase localization, Proc. Roy. Soc. 148B, 520-532

34. Fahimi, H.D., and Herzog, V., 1973: A colorimetric method for measurement of the (peroxidase-mediated) oxidation of 3,3'-diaminobenzidine, J. Histochem. Cytochem. 21, 499-502

35. Cornelisse, C.J., and van Duijn, P., 1973: A new method for the investigation of the kinetics of the capture reaction in phosphatase cytochemistry, I. Theoretical aspects of the local formation of crystalline precipitates, J. Histochem. Cytochem. 21, 607-613

36. Robertson, W.G., 1973: Factors affecting the precipitation of calcium phosphate in vitro, Calc. Tiss. Res. 11, 311-322

37. Cornelisse, C.J., and van Duijn, P., 1973: A new method for the investigation of the kinetics of the capture reaction in phosphatase cytochemistry, II. Theoretical and experimental study of phosphate diffusion from thin polyacrylamide films, J. Histochem. Cytochem. 21, 614-622

38. Cornelisse, C.J., and van Duijn, P., 1973: A new method for the investigation of the capture reaction in phosphatase cytochemistry, III. Effects of the composition of the incubation medium on the trapping of phosphate ions in a model system, J. Histochem. Cytochem. in press

39. Burns, E.A., and Hume, D.N., 1956: Acetato complexes of lead in aqueous solution, J. Am. Chem. Soc. 78, 3958-3962

40. Lasch, J., 1972: Theoretical analysis of the kinetics of enzymes immobilized in spherical particles, In: H.C. Hemker and B. Hess, eds., 8th Meeting Fed. Europ. Biochem. Soc, 25, 295-301. North-Holland American Elsevier

41. Katchalski, E., Silman, I., and Goldman, R., 1971: Effect of the micro-environment on the mode of action of immobilized enzymes. In: F.F. Nord, ed. Advances in Enzymology 34, 445

42. Fisher, R.A., and Harris, H., 1969: Studies on the purification and properties of the genetic variants of red cell acid phosphohydrolase in man, In: Fishman, W.H., consulting ed., The phosphohydrolases: Their biology biochemistry and clinical enzymology, Ann. New York Acad. Sci. 166, 380-391

43. Barka, T., and Anderson, P.J., 1962: Histochemical methods for acid
     phosphatase using hexazonium pararosanilin as coupler, J. Histochem.
     Cytochem. 10, 741-753

44. Poelman, R.E., and Daems, W.T., 1973: Problems associated with the
     demonstration by lead methods of adenosine triphosphatase activity in
     resident peritoneal macrophages and exudate monocytes of the guinea pig,
     J. Histochem. Cytochem. 21, 488-498

Electron microscopy and Cytochemistry, eds. E. Wisse, W.Th. Daems, I. Molenaar and P. van Duijn.
© 1973, North-Holland Publishing Company - Amsterdam, The Netherlands.

KINETICS OF THE ACCUMULATION OF LEAD PHOSPHATE
IN ACID PHOSPHATASE STAINING

U. Pfeifer, E. Poehlmann and H. Witschel
Department of Pathology and Department of Forensic Medicine
University of Wuerzburg, German Federal Republic

The efficiency of a cytotopochemical enzyme reaction depends 1) on
the exactness of the localization of the insoluble end product, 2) on
the yield of the reaction. These two criteria are related to each
other, converging to the problem of kinetics. The kinetics of acid
phosphatase staining using the lead salt method were studied by determination of lead as a constituent of the end product.

Methods

Liver tissue of male Sprague Dawley rats, body weight 200-300 g,
was preperfused via abdominal aorta with Ringer-Procain[4] and fixed by
perfusion with a mixture of 2% glutaraldehyde + 3% freshly prepared
formaldehyde in 0.075 M cacodylate, pH 7.2. Slices of 3 mm thickness
were postfixed by immersion for 30-60 min in the same fixans. Frozen
and non-frozen (Oxford Vibratome) sections of 40 micron thickness
were washed overnight in cacodylate buffer containing 10% sucrose.

The most experiments were performed with frozen sections. They
were incubated at 37°C for 10-120 min in a special cuvette[6] allowing
to hang up them into the medium, which contained CMP and lead nitrate
at varying concentrations, 0.05 M acetate buffer, pH 5.0, and 7% sucrose. CMP (cytidine5'-monophosphate disodium salt) was chosen as
substrate[5] because it shows never precipitations which could reduce
incontrollably the concentration of both, substrate and capture ion.

After incubation and washing for 2x1 min in acetate buffer some
sections were processed for light and electron microscopy, the other
were dried, weighed, and wet ashed in 2 ml concentrated nitric acid.
Deionized water was added to a final volume of 10 ml. Lead concentration was measured by atomic absorption spectrometry (Perkin Elmer,
Type 303) and calculated as gamma lead per mg dried tissue[6].

Results and Discussion

The unspecific adsorption of lead by the tissue, as revealed by
incubation without substrate or in a complete medium containing
0.01 M NaF, was independend on the time of incubation and also independend on the concentration of lead nitrate in a range between
1.8 and 7.2 mM. The mean was 6.6 gamma per mg dried tissue, S.D. was
0.8.

Fig. 1 shows that the system is saturated with substrate at a
concentration of 2.8 mM (= 100 mg/100 ml) CMP.

The efficiency of the trapping reaction[3] was studied by varying
the concentration of lead nitrate in the medium between 1.8 and 7.2
mM (= 60 and 240 mg/100 ml). The amount of end product accumulated
after incubating for 120 min decreases with increasing concentration

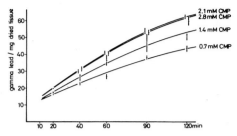

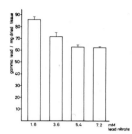

Fig.1: Influence of the substrate concentration on the velocity of the reaction. The 2 S.D. range is given by the vertical bars. The concentration of lead nitrate in the medium was 5.4 mM in all experiments.

Fig.2: Influence of the concentration of the capture ion. The sections were incubated for 120 min at different concentrations of lead nitrate in the medium. The highest amount of end product was found in the low lead experiment.

of lead nitrate (fig.2).

Morphologically in the low lead experiment (1.8 mM) a great deal of the end product is localized outside the lysosomes, especially on the nuclei[1] (fig.3,4). This finding indicates an insufficient trapping capacity, which allows the liberated phosphate to diffuse before being precipitated. In the high lead experiment (5.4 mM) no nuclear staining was observed, and the fine structural localization is almost correct (fig.5,6). The concentration of 5.4 mM needed in our system is high compared to Barka and Anderson[1] probable related to a better preservation of the enzymatic activity by the improved fixation.

The lower amount of end product in the high lead experiments (fig. 2) can be explained by assuming that the end product, when localized exactly, acts as a barrier between the enzyme and the substrate[2,6]. The consequences of such a barrier effect on the kinetics of the reaction are demonstrated by fig.7 which shows that the velocity of the reaction decreases exponentially with the time, when the trapping reaction is sufficient, but takes a more linear course, when it is insufficient.

The results of 3 independend experiments (fig.8) are well in agreement with the model (fig.7). In the 5.4 mM experiment (cf.fig.3,4) the velocity of the reaction decreases exponentially with the time, whereas in the 1.8 mM experiment with its insufficient trapping reaction (cf.fig.5,6) the increase is nearly linear. The initial veloci-

Fig.3-6: Morphological findings after 6o min incubation in a low lead, 1.8mM (fig.3,4), and in a high lead, 5.4mM (fig.5,6) medium. Strong nuclear staining and deposits of end product in the cytoplasm are shown in fig.3,4, whereas in fig.5,6 the localization is almost correct. BC bile canaliculus, N nucleus; no additional staining for electron microscopy.

27

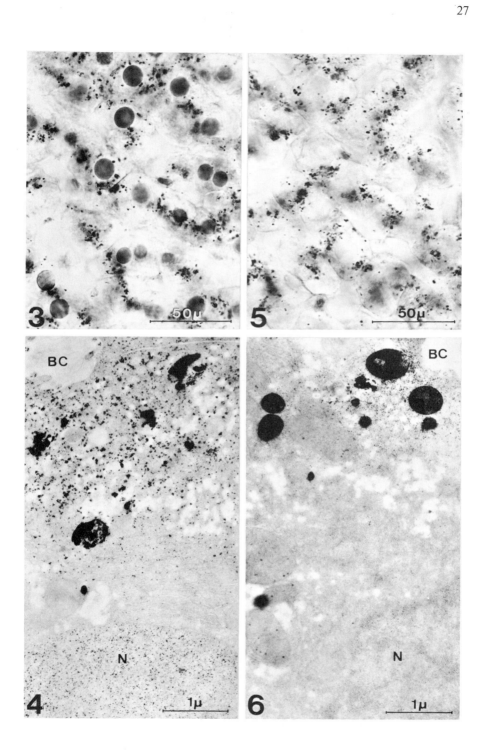

28

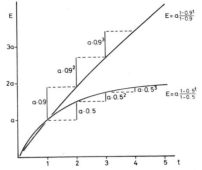

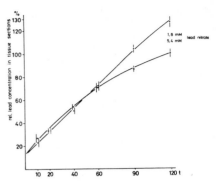

Fig.7: Model of two types of time course of the reaction. 0.5 resp. 0.1 of the active sites are blocked by the end product a at time 1 leading to a clearly exponentially decreasing resp. a more linear time course.

Fig.8: The kinetics of the reaction at 1.8 resp. 5.4 mM is well in agreement with the model. 3 independent experiments. Values from 5.4 mM and 120 min were taken as 100%. 2 S.D. range is given by the vertical bars.

ties show that the higher concentrations of lead ions is not inhibitory. Furthermore the activity is not influenced by the incubation itself, as shown by preincubation.

The reaction is inhibited in frozen sections by DMSO to about 25%. It is further inhibited to about 30% when using non-frozen sections, indicating that the latency is not overcomed by the fixation alone. In non-frozen sections DMSO is inhibitory only to about 15%.

Conclusions

A cytotopochemical enzyme reaction should show an exponentially decreasing velocity due to the barrier effect of the end product, when localized exactly. If the time course is linear, one has to call in question the exactness of the localization. The reaction could be designated as efficient, if the flattened part of the accumulation curve is reached by the incubation time, and if the initial velocity corresponds to the whole activity which can be measured biochemically in the same material, i.e. fixed liver tissue. The latter point needs further studies.

References:
1) Barka,T., Anderson,P.J., J.Histochem.Cytochem. 10, 714 (1962)
2) Barter,R. et al., Proc.Roy.Soc. 144B, 412 (1955/56)
3) Cornelisse,C.J. et al. Acta Histochem., Suppl. 10, 213 (1972)
4) Forssmann,W.G. et al., J.Microscopie 6, 279 (1967)
5) Novikoff,A.B., in Lysosomes, p.36, Churchill, London (1963)
6) Pfeifer,U., Witschel,H., Acta Histochem., Suppl. 11, 111 (1972)

This work was supported by Deutsche Forschungsgemeinschaft. The technical assistance of Miss H. Kreck is gratefully acknowledged.

Electron microscopy and Cytochemistry, eds. E. Wisse, W.Th. Daems, I. Molenaar and P. van Duijn.
© 1973, North-Holland Publishing Company - Amsterdam, The Netherlands.

LEAD SALT METHODS

J.S. Hugon

Department of Pathology,
Medical Center,
School of Medicine,
University of Sherbrooke,
Sherbrooke, P.Q., Canada.

1. Introduction

Ultrastructural cytochemistry permits resolution of the chemical
nature of various intracellular components at the sub-cellular level
and one of the most desirable cellular components to localize are
the enzymes. However, a number of problems arose when the practical
steps were taken to localize these enzymes. Due to the principle of
the electron microscope, based on the electron beam diffraction pattern, it was thought that only methods involving heavy metals could
allow the visualization of the enzymatic activity. The most used
lead salt method in the fifties was the Gomori method for acid phosphatase which was subsequently the basis for many ultrastructural
methods to elicit enzymatic activity. "The early lead taken by the
phosphatases has never been reversed"[57] and up until now the majority of the papers dealing with ultrastructural enzyme localization
are devoted to the phosphatases fine localization.

The bulk of our knowledge on the lead salt methods and their application to enzymatic cytochemistry is due to the pioneer work of
Novikoff's laboratory in the fifties and the early sixties[11,51,52,
54,55]. The first attempt to localize enzymatic activity at the
level of ultrastructure was made by Sheldon et al.[67] in 1955 in a

study of acid phosphatase activity in the intestinal mucosa after brief fixation with osmium tetroxide. Brandes et al.[5] in 1956 showed the activity of alkaline phosphatase on the brush border of the kidney using the calcium salt method of Gomori. With the Wachstein and Meisel techniques[73], Essner et al.[11] in 58 localized adenosine-tri-phosphatase (ATPase) and 5-nucleotidase. In 59,Kaplan and Novikoff[32], using for the first time frozen sections instead of whole tissue blocks, described the ATPase localization in the rat kidney. In these experiments the short fixation with osmium tetroxide was replaced by an overnight fixation in cold 4% formalin with 1% chloride added according to Baker. At the same time Barrnett and Palade[2] localized the acetylcholinesterase activity using thiolacetic acid as substrate. These first attempts using lead salts brought out numerous new facts and allowed a more precise localization in the cells of several enzymatic activities, but the preservation of the fine structure was frequently poor. However, with the use of these techniques derived from light microscopical histochemistry, it was demonstrated that enzyme cytochemistry was possible at the level of ultrastructure.

Although several reviews[66,45,16,68] on the ultrastructural cytochemistry have been published, we think worthwhile to make some comments on the general principles of the lead salt techniques.

2. General principles in lead salt techniques

a) Fixation.

If formalin and osmium were used by the pioneers, the comprehensive study of Sabatini, Bensch and Barrnett[65] of a number of aldehydes has lead to the general use of glutaraldehyde. It allowed a good preservation of the fine structure and did not inhibit too drastically a number of enzymes, especially those involved in the lead salt techniques i.e. the phosphatases, the esterases and the sulfatases. Commercial glutaraldehyde containing polymers and degradation products having an adverse effect on several enzymatic activities, distilled glutaraldehyde was preferred[13]. The perfusion technique appeared also to give more homogenous fixation than the immersion technique for compact animal tissues. Fixation time varied from a few minutes to several hours. Cacodylate buffer 0.1 M has been generally used. Carbonate or phosphate buffers were avoided for their specific action on the lead ions present in the media.

b) Washing.

After the fixation, a thorough washing in the cacodylate buffer with 7.5% sucrose added appeared to be extremely important to wash out the free glutaraldehyde which may interfere with the enzymes at the reactive sites.

c) Sectioning.

Kaplan and Novikoff[32] were the first to promote the use of 20 to 40 μ thick frozen sections. Smith and Farquhar[69] developed a non freezing method of sectioning derived from the Mc Ilvain mechanical tissue chopper. This technique allows excellent preservation of the fine structure. Nevertheless careful use of freezing microtomy avoiding too low sectioning temperatures, and promoting a slow freezing of the tissues, results in a good preservation of the ultra-structures and improves the penetration of substrates and of capturing agents.

d)   Incubation.

The frozen sections are incubated free floating in a stirring
bath.   After the incubation, they are washed several times with cold
buffer, and post-fixed in 2% osmium tetroxide for 30 minutes to one
hour.   During the dehydration some staining may occur with uranyl-
acetate or the Kushida method[34].   Several sections must remain unstained to judge the artefactual deposits.   The staining of the
ultra-thin sections follows the general rules.   However lead or barium sulfate precipitates occurring after an arylsulfatase reaction
are quickly displaced when the lead alkaline citrate stain is used.

e)   Incubation medium.

The cytochemical reaction using lead salts is plagued by problems
involving the solubility of the lead ions in the incubation media,
its concentration in the tissue around the enzymatic sites, the formation of the lead salt precipitate and finally the non-specific
binding of the lead to tissue structures.

1)   Solubility of the lead ions in the incubation media.

Generally the lead nitrate salt is employed but lead citrate and
lead acetate have been preferred by several authors.   Substrates
like p-nitrophenylphosphate form insoluble salts with Pb^{++} at low
pH and should be avoided.   Phosphate or carbonate buffers, interfering with the lead salt, should not be used.   Special care should be
taken with the water component of all the solutions.   Only fresh re-
distilled water boiled and cooled, immediately, before use must be
employed to avoid the formation of lead carbonate precipitates.   If
lead nitrate is fairly soluble at low pH, it forms a precipitate at
pH above 8 unless some chelating agent is employed.   Tris-maleate

buffer seems to be suitable in this case. Certain enzymatic inhibitors like cysteine or the BAL complex may chelate the lead ions preventing any formation of lead salts and should be avoided in the incubating media.

2)   Tissue concentration of lead and formation of lead salt.

The lead salt methods use the cytochemical principle known as the simultaneous coupling reaction. The capture reagent should be in large excess to trap immediately the primary reaction product before it diffuses from the site of reaction.

High concentration of lead in the medium inhibits drastically the enzymatic activity, depresses the concentration of substrate and increases the solubility of the reaction product. Therefore it is of interest to obtain a low lead concentration in the final medium, however rarely less than 2.5 mM. Indeed, around the first microprecipitate occurring at the beginning of the reaction is a layer of unstirred solution in which the amount of lead ions should be sufficient to precipitate all the phosphate ions liberated in the case of a phosphatase reaction. If the enzymatic activity is too high, or lead concentration too low, this layer is quickly depleted and diffusion occurs. In several cases lead ions cannot penetrate intracellular membranes and the precipitate is formed on the external side of the organite as demonstrated by Farquhar et al.[14] for the leukocyte granules, unless some disruptive effect is reached.

3)   Non-specific binding of lead.

This important artefact is dependant on the electrical charge of the lead salt. At high concentration of phosphate ions the net charge is positive, at low concentration it is negative which is generally the case in the histochemical sections. This may explain the fine deposits of lead occurring on red blood cell membranes and other plasma membranes. Moreover the sucrose frequently added to the media may increase the metallophilia of the tissue for the lead. Low concentration of lead has been shown to provoke artefactual nuclear staining probably due to the affinity of several chromatic sites for wandering lead phosphate. Substrate-lead complexes can be formed especially with the di-and tri-phosphate substrates[47] or with thiolacetic acid[20]. These compounds may precipitate on the plasma membranes mimicking enzymatic localizations. Finally high concentration of lead (4 mM) may provoke staining around fat droplets or in tissues participating actively in nutrients transport like the intestinal microvilli.

3. Special methods and results

a) Acid phosphatase.

The original Gomori method was modified for electron microscopy by Holt and Hicks[21] and by Barka and Anderson[1] and remains one of the most used cytochemical methods. However several modifications of substrates have been proposed. Generally β-glycerophosphate is used, but for certain tissues like the lung, γ-naphthyl phosphate should be preferred[12]. Cytidine monophosphate[52], uridine-monophosphate and p-nitrophenylphosphate have been employed. Ericsson and Trump[9] in 65 and Maunsbach[40] in 66 have underlined several requisites for obtaining a good localization, like the ratio of buffer to substrate, the integrity of the substrate, and the methods avoiding diffuse cytoplasmic staining.

Livingstone et al.[39] in 70 described a medium containing lead diazonium chloride. The sections were first incubated in an AS-BI phosphate medium and then coupled with the diazonium salt. The reaction product being soluble in ethanol, they used a water soluble araldite for the embedding.

Poux[60] in her extensive study of enzymatic activities in root meristematic cells used a special technique of prolonged incubation at low temperature (3-4°C) allowing for a better penetration of the compounds. She used also p-nitrophenyl phosphate as a substrate although its solubility was poor at pH 5, but she got a sufficient quantity of substrate and of lead for completing the reaction. She used also a low concentration of sucrose without harmful effects.

The lead method has shown that acid phosphatase is localized in the dense bodies, and in several cisternae of the Golgi apparatus. Some authors have also shown a precipitate in the endoplasmic reticulum. Several reticulo-tubular structures were observed positive in the liver after a total shunt of the portal vein. Acid phosphatase was described on the brush border of the intestinal epithelium of ascaris suum. The presence of an enzymatic activity on nucleoli and nucleoplasm of HeLa cells was shown at the ultrastructural level after a short formaldehyde fixation.

b) Alkaline phosphatase.

Clark[6] used the calcium-lead variant of the Gomori method. However the conversion of calcium phosphate into lead phosphate was an important source of diffusion. A direct lead method was described by Mölbert et al.[46] using a chelating agent but the pH of 7.6 was too low for accurate results. Tranzer[72] described a technique with lead citrate as capturing agent at a very high concentration (12 mM). Mayahara et al.[41] derived a method from the Tranzer medium with a low lead citrate concentration. Hugon and Borgers[23,24,25] used a direct lead method technique at pH 9 with lead nitrate and tris-maleate buffer. These different methods including the Mizutani and Barrnett[44] procedure with cadmium have been compared by Leonard and Provenza[37].

The direct lead salt method at high pH demands extensive care in the preparation of the medium. Freshly boiled distilled water must be used for diluting the compounds. The buffer component has to be prepared from a fresh concentrated solution stocked at a low pH. Intracellular penetration of substrate and lead is slowed down at high pH and frozen sections are preferred. Prolonged fixation and short buffer rinses of the tissues are also advised.

Numerous results have been obtained with these media. In the intestine, it was proved that the activity was present on the brush border, the Golgi apparatus, several dense bodies and smooth membrane profiles. In the salivary glands, renal cell carcinoma, HeLa cells, kidney tubules, amphibian interrenal cells, rabbit leukocytes and bone marow cells, submandibular gland, human liver, and rat foetal thyroid, the activity was observed mainly on the lateral cell membranes and sometimes in the Golgi cisternae and some other intra-cellular sites.

Fig. 1. Glucose-6-phosphatase activity. Absorptive cells of the mouse duodenum. 20 minutes incubation. Activity present in the cisternae of rough and smooth endoplasmic reticulum. Golgi cisternae devoid of precipitate. Remnants of alkaline phosphatase activity on the microvilli. At the arrow, unexplained negativity of several segments of the nuclear envelope. (From D. Menard and J.S. Hugon).

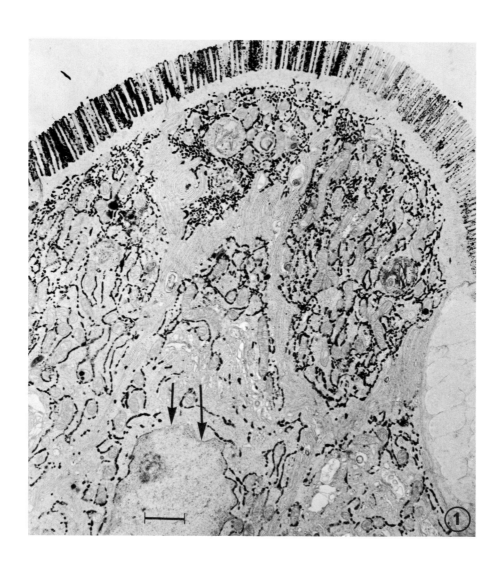

The exact localization of the enzymatic activity on the external
or internal leaflet of the cellular membranes was discussed by Reale
and Luciano[61]. Leduc et al.[35], Iglesias et al.[28] incubating briefly-
fixed sections obtained by ultracryotomy observed fine deposits on
the internal leaflet of kidney cortex microvilli. However the
actual permeability to lead ions of the fixed or frozen sectioned
tissues is poorly understood, and different fixation processes may
change the position of the precipitate on the membranes. Therefore
one should be extremely cautious before asserting that the precipitate has a very definite localization on the membrane leaflets.

c)   Glucose-6-phosphatase.

The medium of Wachstein-Meisel was applied to electron microscopy
by Tice and Barrnett[71]. Ericsson[8] showed the localization of the
enzyme in the liver endoplasmic reticulum. Recently several workers
[26,27,38,58] using a slight modification of the original method at
pH 6.3 have analysed the normal distribution and the response of
glucose-6-phosphatase activity to various stimuli in the intestine,
in the liver and in the adrenal tissue.

d)   Nucleoside phosphatase.

The first demonstration of plasma membrane adenosine tri-phosphatase (ATPase) activity was made in 1958[11]. In 1964, Goldfischer et
al.[18] made a comprehensive review of the methods and the results of
the nucleoside phosphatase techniques. In a series of stimulating
papers, from 66 to 69, Rosenthal, Moses, Tice and Ganote[15,47,48,63,70]

Fig. 2. Glucose-6-phosphatase activity. Mouse liver. The precipitate of lead phosphate fills the stacks of rough endoplasmic cisternae
and the numerous profiles of smooth endoplasmic cisternae embedded
into the glycogen areas. Large lipid droplets partially extracted.
(From D. Menard and J.S. Hugon).

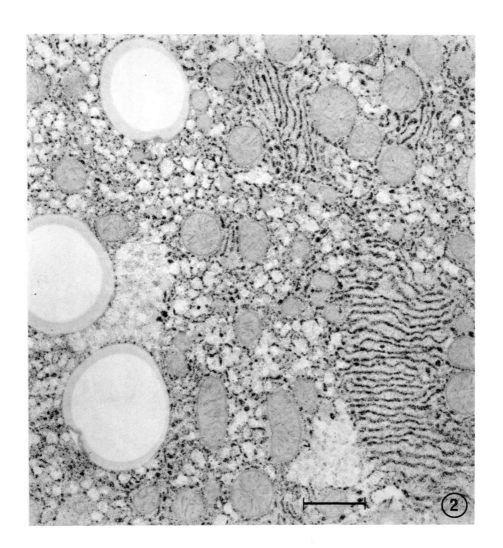

demonstrated the pitfalls of the lead salt methods for the histo-chemical localization of nucleoside phosphatases. They came to the conclusion that the membrane staining after the Wachstein-Meisel procedure was better explained by lead catalyzed hydrolysis of ATP or other nucleoside phosphate than by enzymatic activity. Their conclusions rested on the following data.

1) Pb^{++} strongly inhibits Mg-ATPase and particularly Na-K-ATPase.

2) Pb^{++} catalyzes a non enzymatic hydrolysis of ATP which may contribute to artefactual deposits.

3) The reaction product contains precipitated nucleoside as well as Pb^{++} and phosphate.

4) The components of the incubation medium interact, leading to changes in the pattern of the reaction product deposits.

5) The precipitate due to non enzymatic hydrolysis of ATP occurs at the same sites claimed for the enzymatic activity.

6) The inhibitors of ATPase activity modify the affinity of lead to tissue which is the only cause for the lack of reaction.

In several letters to the editor, Novikoff[50,52] argued with these authors and mentioned a number of facts supporting the thesis that the Wachstein-Meisel procedure does not lead to artefacts. Several authors[3,19,43,29,30,10] have reviewed the histochemical procedure for ATPase and for other nucleoside phosphatase and have gathered the following data in favour of a real visualization of an enzymatic activity with the Wachstein-Meisel method:

Fig. 3. Glucose-6-phosphatase activity. Absorptive cells of an explant of duodenal mucosa of a guinea-pig cultivated 24 hours on Trowell medium. The enzymatic activity is present in all the cisternae of the endoplasmic reticulum. Golgi apparatus is devoid of precipitate. (From M. Kedinger and J.S. Hugon).

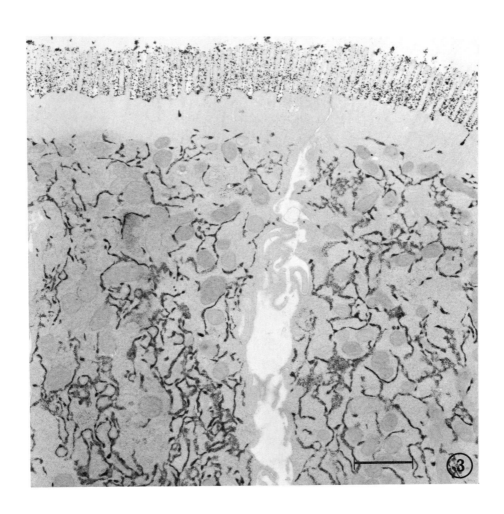

1)   In many instances plasma membrane may stain with ATP and not
with ADP as substrate; however each of these is sensitive to non-
enzymatic hydrolysis by Pb^{++}.

2)   Some nucleoside phosphates used as substrates and sensitive to
non-enzymatic hydrolysis provoke a staining in the endoplasmic reticulum cisternae and not on the plasma membranes.

3)   The histochemical reaction is frequently made at a temperature
and during a time-interval in which non-enzymatic hydrolysis is very
low. However intense staining occurs.

4)   In the presence of specific ATPase inhibitors like NaF or p-chlo-
romecuribenzoate which do not inhibit the non-enzymatic ATP hydrolysis and even accelerate it, the histochemical reaction does not
take place.

5)   Non-enzymatic lead-catalyzed ATP hydrolysis is not significant
in media containing ATP and Pb^{++} at a concentration ratio greater
than 1.

6)   The deposits of lead from non-enzymatic hydrolysis of ATP are
correlated only with the appearance of a coarse precipitate on top
and at the edges of the sections and is proportional to the volume
of the incubating medium.

From all this debate, several data related to the incubation
medium have emerged which modify slightly the original Wachstein-
Meisel method.

Berg et al.[3] have added 2 M Cl^- as a cofactor which provokes a
significant acceleration of the histochemical reaction. This enhan-
cement was obtained by an increase in enzymatic activity due to a
low Pb content (1.4 mM) in the medium and by a better capture of the

reaction product due to a saturation of the medium with respect to lead salts of both ADP and orthophosphate.

Ernst[10] has prepared a new medium for the ultrastructural localization of the ouabain-sensitive potassium-dependent phosphatase activity using K-nitrophenyl phosphate as the substrate and $SrCl_2$ as the capturing agent. The precipitate of Sr Pi was transformed into Pb Pi for the electron microscopic visualization. The relationship of this enzymatic activity to Na-K-activated adenosine triphosphatase was established. This method allows a differentiation between the Mg activated ATPase and the Na-K-activated ATPase and avoids the near total inhibition of the latter enzyme by Pb ions.

Using these new media or the classical incubation in which the substrate to lead ratio is confined between 2.1 and 0.69, using control incubations with enzymatic inhibitors, or with non-enzymatic lead phosphate precipitate, the cytochemist can study with some confidence the nucleoside di-and tri-phosphatase activities.

e) Arylsulfatase.

Techniques for arylsulfatase activity have been developed at the same time by Goldfischer[17] and by Hopsu et al.[22]. The capturing agent was the lead for the former and the barium for the latter. Several substrates have been employed but nitrocatechol sulfate was generally used at pH 5.2. Barium was preferred to lead because its inhibitory effect was less important. Kalimo et al.[31], have discussed the loss of the barium sulfate precipitate from the ultrathin sections during the lead staining at high pH. They recommended uranyl-acetate staining during the dehydratation.

The enzyme visualized is probably an arylsulfatase B. In the animal cells, the localization occurs in the lysosomes, seldom in the Golgi cisternae. In plant meristematic cells, Poux[59] observed a precipitate in the nuclear envelope, in all the profiles of endoplasmic reticulum, and in the peripheral part of the phragmosomes.

f) Esterase.

Lead was employed as capturing agent with thiolacetic acid as substrate by Barrnett and Palade[2] in 59 for the demonstration of specific cholinesterase in the myoneural junction. Complex formation between heavy metals and thiolacetic acid was studied by Haugaard et al.[20]. The same techniques combined with a specific inhibitor was utilized to localize an acid esterase, tentatively considered as a cathepsin. In this case, a precipitate of lead sulfide was observed in the lysosomes and several smooth endoplasmic profiles. However very few workers have used this technique. Kasa and Csillik[33] used an incubation medium with thiolacetic acid as substrate and Pb^{2+} and Cu^{2+} as capturing agents for the localization of cholinesterase. As judged by their micrographies, their results were extremely good. Recently von Deimling and Madreiter[7] have proposed a new method using Fe^{2+} or Bi^{3+} salts for the demonstration of a carboxylesterase activity.

g) Nucleoside phosphorylase activity.

The method of Rubio et al.[64] used the reversibility of the reaction of conversion of inosine to hypoxanthine by nucleoside phosphorylase. The phosphate formed is precipitated by the lead ions present in the incubation medium. Borgers et al.[4] have applied a slightly modified method to formed elements of the blood and to dog coronary blood vessels. The enzyme activity was found in the cyto-

plasmic sap of adventitial cells, adventitial fibrocytes and fibroblasts. Neutrophils, basophils, monocytes and platelets showed a
uniform reaction.

h) Amino transferase.

Aspartate amino transferase has been localized by Papadimitriou
and Van Duijn[56] using the technique of Lee and Torack[36]. The rapid
inactivation of the enzymatic activity by glutaraldehyde fixation was
avoided by the addition of the substrate ketoglutarate to the fixative. The medium contained, $\prec$ -ketoglutarate, 1-aspartate and lead
nitrate. The precipitate of lead oxalo-acetate was observed in mitochondria, plasma membranes, the limiting membrane of peroxisomes, and
in the ground substance between cisternae of endoplasmic reticulum
in intestinal goblet cells.

Ornithine carbamoyl-transferase and aspartate carbamoyl-transferase were localized by Merker and Spors[42] using a lead method.

i) Lipase.

A method for the demonstration of lipase was developed by Murata
et al.[49]. The incubation medium contained tween, taurocholic acid
and calcium chloride. The Ca soap former was subsequently converted
into a lead soap. So far the technique was used only by these
authors.

j) Acetylcoenzyme A caboxylase.

This enzyme was localized by Yates et al.[74] by a reaction in which
hydrolysis of adenosine triphosphate is coupled to the synthesis of
malonyl-Co-A, utilizing biotin, bicarbonate, acetyl-Co-A, ATP and
manganese. The lead phosphate precipitate was located on the outer
surfaces of the endoplasmic reticulum. No deposits were observed on

the nuclei or on the mitochondria. This method was a fair demonstration of a coupled enzyme system. An identical principle was adapted by Reik et al.[62] to the visualization of adenyl-cyclase activity.

4. Conclusion

In conclusion, since the beginning of biological electron microscopy, scientists have tried to relate their morphological observations to an assumed biochemical role. The lead salts methods have helped them to attain this goal. With the works of Barrnett's laboratory on coupled enzymes these methods have reached their final achievement. Unfortunately the high inhibitory effect of lead, its metallophilia for various cell components, and the small number of highly insoluble lead salts do not permit much hope for the future. Other metal techniques with barium, copper, ferrous or cobalt salts are actually being developed with success but the osmium black methods appear even more promising. Nevertheless it seems sound to say that for the near future, lead salts methods, used perhaps with greater care than before, would still provide many interesting observations on the fine structural localization at least of the numerous phosphatases.

Supported by the research grants no MT-3994 of the Medical Research Council of Canada and no 9340-06 of the Defence Research Board of Canada.

The author is greatly indebted to Dr. R. Coté, M.D. for reviewing the manuscript, to Mrs M. Kedinger and Mr. D. Menard for providing the micrographies and to Mr. P. Magny for his helpful technical aid.

References

1. Barka, T. and Anderson, P., 1962, J. Histochem. Cytochem.,10,741.

2. Barrnett, R. and Palade, G., 1959, J. Biophys. Biochem. Cytol.,
    6, 163.

3. Berg, G.G., Lyon, D. and Campbell, M., 1972, J. Histochem.
    Cytochem., 20, 39.

4. Borgers, M., Schaper, J. and Schaper, W., 1972, J. Histochem.
    Cytochem., 20, 1041.

5. Brandes, D., Zetterqvist, H. and Sheldon, H., 1956, Nature,
    177, 382.

6. Clark, S.L. Jr., 1961, Am. J. Anat., 109, 57.

7. Deimling von, O. and Madreiter, H., 1972, Histochemie, 29, 83.

8. Ericsson, J., 1966, J. Histochem. Cytochem., 14, 361.

9. Ericsson, J.L.E. and Trump, B.F., 1965, Histochemie, 4, 470.

10. Ernst, S.A., 1972, J. Histochem. Cytochem., 20, 13.

11. Essner, E., Novikoff, A.B. and Masek, B., 1958, J. Biophys.
    Biochem. Cytol., 4, 711.

12. Etherton, J.E. and Botham, C.M., 1970, Histochemical J., 2, 507.

13. Fahimi, H.D., Drochmans, P., 1965, J. Microscopie, 4, 725.

14. Farquhar, M., Bainton, D., Baggiolini, M. and De Duve, Ch., 1972,
    J. Cell Biol., 54, 141.

15. Ganote, C.E., Rosenthal, A.S., Moses, H.L. and Tice, L.W., 1969,
    J. Histochem. Cytochem., 17, 641.

16. Geyer, G., 1969, Ultrahistochemie, (Fischer Verlag Ed., Jena).

17. Goldfischer, S., 1965, J. Histochem. Cytochem., 13, 520.

18. Goldfischer, S., Essner, E. and Novikoff, A.B., 1964,
    J. Histochem. Cytochem., 12, 72.

19. Goldfischer, S., Essner, E. and Schiller, B., 1971,
    J. Histochem. Cytochem., 19, 349.

20. Haugaard, N., Horn, R.S. and Koelle, G.B., 1965, J. Histochem.
    Cytochem., 13, 566.

21. Holt, S.J. and Hicks, R.M., 1961, J. Biophys. Biochem. Cytol.,
    11, 47.

22. Hopsu, V.K., Arstila, A. and Glenner, G.G., 1965, Ann. Med. Exp. Fenn., 43, 114.

23. Hugon, J. and Borgers, M., 1966, J. Histochem. Cytochem., 14,629.

24. Hugon, J. and Borgers, M., 1967, J. Cell Biol., 33, 212.

25. Hugon, J., Borgers, M. and Loni, M., 1967, J. Histochem. Cytochem., 15, 417.

26. Hugon, J., Maestracci, D. and Menard, D., 1971, J. Histochem. Cytochem., 19, 515.

27. Hugon, J., Maestracci, D. and Menard, D., 1973, J. Histochem. Cytochem., 21, 426.

28. Iglesias, J.R., Bernier, R. and Simard, R., 1971, J. Ultrastructure Res., 36, 271.

29. Jacobsen, N.O., Jørgensen, F. and Thomsen, A.C., 1967, J. Histochem. Cytochem., 15, 456.

30. Jacobsen, N.O. and Jørgensen P.L., 1969, J. Histochem. Cytochem., 17, 443.

31. Kalimo, H.O., Helminen, H.J., Arstila, A.U. and Hopsu-Havu,V.K., 1968, Histochemie, 14, 123.

32. Kaplan, S. and Novikoff, A.B., 1959, J. Histochem. Cytochem., 7, 295.

33. Kasa, P. and Csillik, B., 1966, J. Neurochemistry, 13, 1345.

34. Kushida, H. and Fujita, K., 1966, 6th Int. Cong. E.M., Kyoto,39.

35. Leduc, E., Bernhard, W., Holt, S. and Tranzer, J., 1967, J. Cell Biol., 34, 773.

36. Lee, S.H. and Torack, R., 1968, J. Cell Biol., 39, 716.

37. Leonard, E.P. and Provenza, D.V., 1972, Histochemie, 30, 1.

38. Leskes, A. Siekevitz, P. and Palade, G., 1971, J. Cell Biol., 49, 264.

39. Livingston, D.C., Fisher, S.W., Greenoak, G.E., and Maggi, V., 1970, Histochemie, 24, 159.

40. Maunsbach, A.B., 1966, J. Ultrastructure Res., 16, 197.

41. Mayahara, H. Hirano, H., Saito, T. and Ogawa, K., 1967, Histochemie, 11, 88.

42. Merker, H.J. and Spors, S., 1969, Histochemie, 17, 83.

43. Mietkiewski, K., Domka, F., Malendowicz, L. and Malendowicz, J., 1970, Histochemie, 24, 343.

44. Mizutani, A. and Barrnett, R., 1965, Nature, 206, 1001.

45. Mölbert, E.R., 1967, in: Methods and Achievements in Experimental Pathology, vol.2, Year Book Medical Publisher, Chicago, p. 1.

46. Mölbert, E.R., Duspiva, F. and von Deimling, O., 1960, J. Biophys. Biochem. Cytol., 7, 387.

47. Moses, H.L. and Rosenthal, A.S., 1968, J. Histochem. Cytochem., 16, 530.

48. Moses, H.L., Rosenthal, A.S., Beaver, D.L. and Schuffman, S.S., 1966, J. Histochem. Cytochem., 14, 702.

49. Murata, F., Yokota, S. and Nagata, T., 1968, Histochemie, 13, 215.

50. Novikoff, A.B., 1967, J. Histochem. Cytochem., 15, 353.

51. Novikoff, A.B., 1959, J. Biophys. Biochem. Cytol., 6, 136.

52. Novikoff, A.B., 1963, Ciba Found. Symp. Lysosomes, p. 36.

53. Novikoff, A.B., 1970, J. Histochem. Cytochem., 18, 916.

54. Novikoff, A.B., Essner, E., Goldfischer, S. and Heus, M., 1961, in: Interpretation of Ultrastructure, vol. 1, ed. Harris, R.J., (Academic Press, New York) p. 149.

55. Novikoff, A.B. and Goldfischer, S., 1961, Proc. Natl. Acad. Sci. U.S., 47, 802.

56. Papadimitriou, J.M. and Van Duijn, P., 1970, J. Cell Biol.,47,71.

57. Pearse, A.G., 1972, in: Histochemistry, vol. 2, ed. Churchill Livingstone, London.

58. Penasse, W. and Fruhling, J., 1973, Histochemie, 34, 117.

59. Poux, N., 1967, J. Histochem. Cytochem., 14, 932.

60. Poux, N., 1970, J. Microscopie, 9, 407.

61. Reale, E. and Luciano, L., 1967, Histochemie, 8, 302.

62. Reik, L., Petzold, G. Higgins, J., Greengard, P. and Barrnett, R., 1970, Science, 168, 382.

63. Rosenthal, A.S., Moses, H.L., Beaver, D.L. and Schuffman, S.S., 1966, J. Histochem. Cytochem., 14, 698.

64. Rubio, V.R., Wiedmeier, T. and Berne, R.M., 1972,
    Am. J. Physiol., 222, 550.

65. Sabatini, D.D., Bensch, K. and Barrnett, R.J., 1963,
    J. Cell Biol., 17, 19.

66. Scarpelli, D.G. and Kanczak, N.M., 1965, Int. Rec. Exp. Path.,
    4, 55.

67. Sheldon, H., Zetterqvist, H. and Brandes, D., 1955, Exp. Cell
    Res., 9, 592.

68. Shnitka, Th. and Seligman, A., 1971, Ann. Rev. of Biochem.,
    40, 375.

69. Smith, R. and Farquhar, M., 1963, Nature, 200, 4907.

70. Tice, L.W., 1969, J. Histochem. Cytochem., 17,85.

71. Tice, L.W., and Barrnett, R., 1962, J. Histochem. Cytochem.,
    10, 754.

72. Tranzer, J.P., 1965, J. Microscopie, 4, 409.

73. Wachstein, M. and Meisel, E., 1957, Ann. J. Clin. Path., 27, 13.

74. Yates, R.D., Higgins, J.A. and Barrnett, R.J., 1969,
    J. Histochem. Cytochem., 17, 379.

Electron microscopy and Cytochemistry, eds. E. Wisse, W.Th. Daems, I. Molenaar and P. van Duijn.
© 1973, North-Holland Publishing Company - Amsterdam, The Netherlands.

LEVAMISOLE AND ITS ANALOGUE R 8231 : NEW POTENT INHIBITORS OF ALKALINE PHOSPHATASE.

M. Borgers

Department of Cell Biology, Janssen Pharmaceutica,
Research Laboratories, Beerse (Belgium)

Summary

Low concentrations, respectively 0.5 mM levamisole and 0.1 mM R 8231, sufficed to achieve complete inhibition of AlPase of various mammalian tissues (kidney, liver, heart, lung, brain, spleen, lymph nodes, thymus, adrenals, blood vessels, pancreas, thyroid, testes, ovaria, uterus). The intestinal variant was not influenced. Specific phosphatase activities as ATPase, TPPase, NDiPase, AMPase, G-6-Pase and ACPase were not affected.

Introduction

The inhibitory properties of L-tetramisole (levamisole)* and of its analogue R 8231** for alkaline phosphatases have been recently described in a biochemical report[5].

The present study deals with the cytochemical application of these AlPase inhibitors and their influence on the localization of specific phosphatases.

Materials and methods

Adult rats, mice and guinea-pigs of both sexes were used. After perfusion with cacodylate buffered 2 % glutaraldehyde for 5 min small tissue samples from various organs were excised, rinsed in buffer for 2 hours, sectionned to 30 microns with a freeze-microtome and incubated in the appropriate media for AlPase[2] and for specific phosphatases (ATPase, NDiPase, TPPase, AMPase, G-6-Pase and ACPase)[1].

Variations in preparation procedures included 1) different fixatives, fixative concentrations and fixation times; 2) different substrates and substrate concentrations, incubation time and pH changes.

---

*L-tetramisole : (-) -2, 3, 5, 6-tetrahydro-6-phenylimidazo (2, I-b) thiazole
        hydrochloride

**R 8231 : (±) -6(m-bromophenyl)-5, 6-dihydroimidazo (2, I-b) thiazole oxalate

The inhibitor concentration added to the incubating solutions varied from 0.01 mM to 2,5 mM levamisole or R 8231. In parallel, the dextro-isomer of tetramisole (D-tetramisole) was added in the same concentration.

Results

Complete inhibition of AlPase activity was obtained after incubation in the AlPase solution containing 0.5 mM levamisole or 0.1 mM R 8231 at the following sites : proximal convoluted tubules and glomeruli of the kidney (Fig 1 A-B); myocardial cells and blood vessels of the heart; bile canaliculi and blood vessels of the liver; bronchioli, type II alveolar cells and vascular endothelium of the lung (Fig 1 C-D); vascular endothelium of the pancreas, thyroid and brain; adventitial cells of large blood vessels; lymphocytes and vascular endothelium of thymus, spleen and lymph nodes; seminiferous tubules of the testes; stroma of the theca folliculi and follicular cells of the ovaria; endometrium and submucosa of the uterus; lamina propria of the small intestine; blood sinusoids and cortical cells of the adrenals.

Sites of activity not influenced by either compound were : absorptive epithelium of small intestine; some cytoplasmic granules in sinusoidal endothelium and hepatocytes of the liver.

Variations in fixation and incubation procedures did not affect the degree of inhibition obtained. Addition of D-tetramisole had no influence upon the distribution of AlPase activities.

The distribution of other phosphatase activities remained unaltered in the presence of the inhibitors, however, the interfering action of non-specific AlPase on the specific substrates was totally suppressed.

No difference in response towards the inhibitors was noted between the species under investigation.

Fig 1 A-B : <u>Kidney.</u> AlPase, incubation for 15 min at 37°C in the medium without (A) and with 0.1 mM R 8231 (B). The precipitate is visible in (A) on the brush border (bb), in apical vacuoles (v) and a secondary lysosome (ly) of a proximal convoluted tubule cell whereas in (B) the above mentionned sites are unreactive.

C-D : <u>Lung.</u> AlPase, incubation for 60 min at 37°C in the presence of 0.5 mM D-tetramisole (C) and 0.5 mM L-tetramisole (D).

The reaction product is present in (C) at the apical plasma membrane (pm), in an apical granule (gr) and in the Golgi apparatus (g) of a bronchiolar Clara cell. No precipitate is seen in (D), an area comparable to (C).

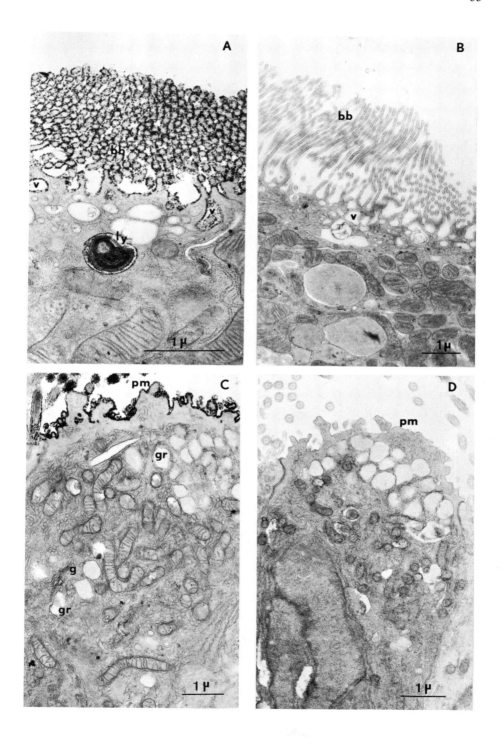

54

Discussion

These cytochemical observations on AlPase inhibition in various tissues of rodents correlate well with the biochemical data of VAN BELLE on dog AlPases [5], i. e. the potency of the inhibitors, the substrate and pH independency, the organ selectivity, the nature of the inhibition and the stereospecificity of tetramisole.

Levamisole and R 8231 are comparable to 1-homoarginine [3,4] as far as the nature and the organ selectivity of the inhibition are concerned but they vary considerably in potency and in substrate preference.

The fact that specific phosphatases are apparently not influenced by these AlPase inhibitors implies that they can be used to differentiate specific phosphatase activities from non-specific alkaline phosphatase activity, especially in tissues where both types of enzymes are abundantly present.

References

1. Borgers, M., Schaper, J., Schaper, W., 1971, J. Histochem. Cytochem. 19, 526.
2. Hugon, J., Borgers, M., 1966, J. Histochem. Cytochem. 14, 429.
3. Lin, C. W., Fishman, W. H., 1972, J. Biol. Chem. 247, 3082.
4. Rufo, M. B., Fishman, W. H., 1972, J. Histochem. Cytochem. 20, 336.
5. Van Belle, H., 1972, Biochim. Biophys. Acta, 89, 158.

Electron microscopy and Cytochemistry, eds. E. Wisse, W.Th. Daems, I. Molenaar and P. van Duijn.

COMPARISON OF ALKALINE PHOSPHATASE REACTIONS
ON SEA URCHIN EMBRYOS

Giuseppe Millonig and Jutta Millonig

Laboratory of Molecular Embryology, C. N. R.

Arco Felice, Naples - Italy.

1. Introduction

The Gomori Takamatsu (1) reaction with conversion of the calcium phosphate into lead phosphate is a very sensitive method but usually produces unspecific lead precipitates in the cytoplasm, over the chromatin, and the nucleolus. Several electron histochemical methods were suggested during the last years, using incubation media at high pH with lead nitrate. The formula of Tranzer (2) includes in the Gomori formula sodium citrate as a complexing agent for the lead ions; the formula of Hugon and Borgers (3) uses Tris buffer instead of Veronal to chelate the lead; the Oledzka et al. (4) formula is identical to the latter, substituting beta-glycerophosphate with cytidine monophosphate as substrate. To compare the sensitivity of the different methods and some modifications, and to determine the parameters of fixation, rinsing, and Os postfixation, sea urchin embryos have been used as test object. These larvae (plutei) have a primitive digestive tract with an oral cavity, an oesophagus, and an intestine in which alkaline phosphatases have been localized by light histochemistry (5). Several reasons suggested the use of this test object: different methods can be tested simultaneously on samples taken from a large homogenous population of embryos; since they are less than 70 μ large and the intestine is open, the epithelial cells are immediately exposed to the different solutions allowing rather exact timing; the solutions are quickly substituted in the test tube; a preliminary quantitative evaluation of the enzyme content and of the method can be done on the intact larvae without sectioning and embedding with the light microscope, since the lead sulfide deposits are observed in the intestine of the almost translucent embryos.

2. Materials and Methods

Embryos of Arbacia lixula at the pluteus stage (about 48 hrs after fertilization) were fixed in 1,5 % glutaraldehyde in 0,1 M phosphate or cacodylate buffer with 2 % NaCl for 30' at 20°C, rinsed for 30' with 2 % NaCl, and incubated in the different media suggested in the literature. For LM, samples were developed briefly with ammonium sulfide and observed "in toto", for EM rinsed for 30' in 2 % NaCl and postfixed in 1,5 % OsO_4 in 0,1 M phosphate buffer pH 6,8 + 2 % NaCl and embedded in ERL (6). For comparison, mouse duodenum was treated in a similar way: the glutaraldehyde concentration was 3 % in buffer + 1 % NaCl, the rinsing solution 1 % NaCl, and the osmium solution contained 1 % NaCl.

A modified Tranzer medium was prepared in the following way:

0,2 M Tris buffer at pH	9.2	3,0ml
Na-glycerophosphate, beta	1 %	0,2ml
Na - citrate	3 %	0,3ml
lead nitrate	3 %	0,3ml

The lead solution is added under stirring. If a weak opalescence is formed, one or two drops of citrate will dissolve it. The solution is completely clear and stable for several hours. To the medium for the sea urchin embryos 0,2 ml of 20 % NaCl was added. Incubation is carried out at 38°-40°C for 1 to 2 hrs. KCN and 1-phenylalanine (to be dissolved in the buffer) are used as inhibitors. Incubation media without substrate were used as controls.

3. Results

With the light microscope it can be observed that the intensity of the different reactions decreases gradually from the Gomori method (GM) (Fig. 1) to the Tranzer method (TM) (Fig. 2) and to the Oledzka et al. method (OM) (Fig. 3). The Hugon-Borgers mixture (HBM) produces a very weak reaction (Fig. 4) when it is used, as suggested, at 4°C. When the temperature is raised to 37°-40°C, the intensity appears to be between the TM and the OM. The modified TM (MTM) (Fig. 5) can be located between the GM and the TM. The incubation time was 7 min.

With the electron microscope the reaction product is found with all methods on small membrane bound vesicles within the intestinal cavity and on the luminar cell membrane. Also here the lead particles decrease in number according to the method used, which parallels the LM observations. The membranes at the intercellular spaces appear less positive with the HBM and OM. Inside the cell, with the GM, the majority, but not all, Golgi cisternae of the intestinal cells are filled with reaction product (Fig. 6). The intensity is slightly reduced with the MTM (Figs. 7 and 8), but is higher than with the TM; the HBM and the OM are even lower in sensitivity. Here only few Golgi vesicles are positive. In the apical part of the cell, vesicles with lead precipitates are observed with the GM, the MTM and the TM. Unspecific precipitates are very strong in the GM, much less in the HBM and OM, even if the solutions are very carefully prepared with freshly distilled water. The MTM and the TM do not need such precautions and only in few cells fine unspecific dense deposits can be visualized.

4. Discussion

The results indicate that the Gomori technique with conversion of calcium to lead phosphate is more sensitive than the direct lead methods, confirming data obtained on leucocytes (7). Unfortunately, all attempts to avoid the unspecific precipitates in the tissue, by careful rinsing or by other modifications, failed. Also the poor tissue preservation could not be avoided. By replacing in the Tranzer medium the Veronal with Tris buffer, which is known to enhance the enzyme activity, and by increasing the concentration of substrate and lead, the sensitivity could be improved. When glycerophosphate is replaced by 10 mg CMP, the reaction is apparently slightly less efficient. A number of tests on mice duodenum and kidney gave with this method promising results.

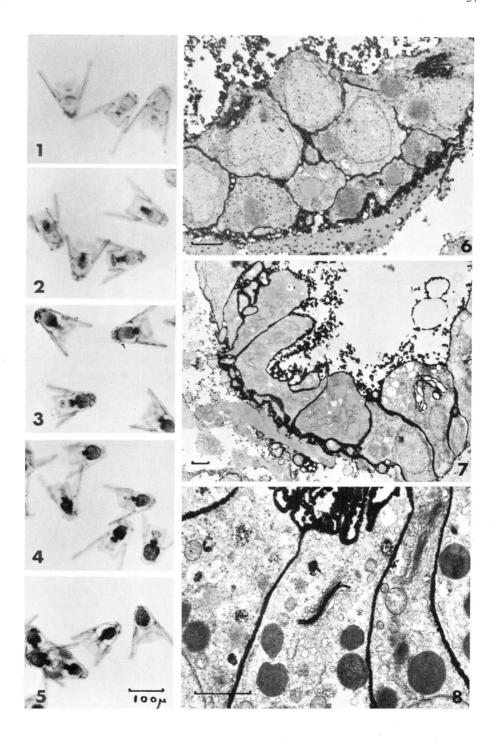

The authors wish to thank Mrs. R. Parisi Pisa for excellent technical assistance.

5. References

1. Pearse, A. G. E. , 1960, Histochemistry (Little, Brown & Co. -Boston).
2. Tranzer, J. P. , 1966, VI Int. Cong. Elect. Micr. Kyoto, p. 91.
3. Hugon, J. and Borgers, M. , 1966, J. Histochem. and Cytochem. 14, 629.
4. Oledzka-Slotwinska, H. , Creemers, J. and Desmet, V. , 1967, Histochemie 9, 320.
5. Evola-Maltese, C. , 1957, Acta Embryol. et Morphol. Exp. 1, 99.
6. Spurr, A. R. , 1969, J. Ultr. Res. 26, 31.
7. Ford Bainton, D. and Farquhar, M. G. , 1968, J. Cell Biol. 39, 286.

Electron microscopy and Cytochemistry, eds. E. Wisse, W.Th. Daems, I. Molenaar and P. van Duijn.
© 1973, North-Holland Publishing Company - Amsterdam, The Netherlands.

ULTRASTRUCTURAL STUDIES ON REGAN AND NON-REGAN ISOENZYMES

OF ALKALINE PHOSPHATASE IN HUMAN TUMOURS

Mitsui Sasaki   and   William H. Fishman

Tufts Cancer Research Center, Tufts University,
School of Medicine, Boston, Mass.

1. Summary

Regan isoenzyme (placental type) alkaline phosphatase can be distinguished from non-Regan (non-placental type) alkaline phosphatase cytochemically by use of specific amino acid inhibitors. L-Phenylalanine inhibits
Regan isoenzyme while L-homoarginine inhibits non-Regan isoenzyme.
Ovarian cancer patients were first typed by light microscopy on ascites
cell smears stained by Kaplow's reaction in the presence separately of
L-phenylalanine and L-homoarginine. Ultrastructural localization of alkaline
phosphatase was done on glutaraldehyde-fixed cells by the Hugon-Borgers
technique without aminoacids. In general, the cells with only Regan isoenzyme exhibited mitochondrial membrane sites, those with non-Regan isoenzyme showed localization on cell membrane whereas the cells producing
both isoenzymes exhibited mitochondrial <u>and</u> cell membrane sites.

2. Introduction

The data presented in this paper are based only on cases of ovarian carcinoma of the adenocarcinoma type.
L-Phenylalanine sensitivity of the Regan alkaline phosphatase (an isoenzyme
of the placental type first discovered in the serum and lung cancer tissue
of a patient named Regan) has been known as one of the main characteristics
by which it is differentiable from other alkaline phosphatase isoenzymes[1,2].
This is the property which is most suitable for histochemical use at the
present time.

Analogously, L-homoarginine has been incorporated into a histochemical
technique for evaluating non-L-phenylalanine-sensitive isoenzymes of alkaline phosphatase[3].

Accordingly, a number of ascites cancer cells from individual patients
with ovarian cancer were evaluated with Kaplow's staining technique[4] in the
presence separately of L-phenylalanine and L-homoarginine. These were classified as Regan type (L-phenylalanine sensitive), non-Regan type (L-homo-
arginine-sensitive) and mixed Regan and non-Regan type (sensitivity to both

L-phenylalanine and L-homoarginine). Here appears to be an ultrastructural specificity in the sites of Regan and non-Regan isoenzymes and the evidence for this suggestion is presented in this paper.

3. Materials and Methods

For light microscopy, smears and tissues were fixed for 2 minutes in (1:3) acetone-formolcalcium on ice, washed with distilled water and the incubation carried out in two stages, first for 2 hours at 0°C (pre-incubation) and then for 30 minutes in fresh solutions at room temperature. The pre-incubation solutions all contained 0.05 M propanediol buffer, pH 9.6 and 50 mM concentrations of D-phenylalanine, L-phenylalanine and L-homoarginine for the individual separate tests; no substrate was present. The incubation solutions also at pH 9.6 all contained 44 mM sodium L-naphthyl phosphate and 1.0 mg Fast Blue BBN per ml incubation solution and in the test experiments 12.5 mM concentrations of D-phenylalanine, L-phenylalanine and L-homoarginine. Substrate and aminoacid free controls were used.

For electron microscopy, fixation of cells and tissue was carried out with cold 2.5% cacodylate-buffered glutaraldehyde for 2 hours and washing with 0.05 M cacodylate buffer pH 7.4 containing 7.5% sucrose. The histochemical reaction was carried out in the Hugon-Borgers medium[5] for 45 minutes at room temperature. Afterwards, standard osmium tetroxide treatment, ethanol dehydration, propylene oxide rinsing and Epon 812 mixture embedding were done. Ultrathin sections were made (LKB 111 Ultramicrotome), mounted on mesh grids, double-stained by uranyl acetate and lead nitrate and observed on the JEOLCO 100-B electron microscope.

4. Results

Three representative cases are presented.

The first (Regan-isoenzyme) showed positive cytoplasmic granules but no nuclear or cell membrane stains in the cancer cells at the light level whereas under the electron microscope, most of the deposits were localized in the spaces between the outer and inner mitochondrial membranes but not on the cell membrane.

The second (non-Regan isoenzyme) which was L-homoarginine sensitive exhibited cell membrane reaction but not any for intracytoplasmic organelles.

The third (mixed Regan and non-Regan isoenzymes) showed a more generalized intracytoplasmic granule membrane distribution (mitochondria, Golgi, endoplasmic reticulum and a pronounced cell surface membrane reaction including microvilli.

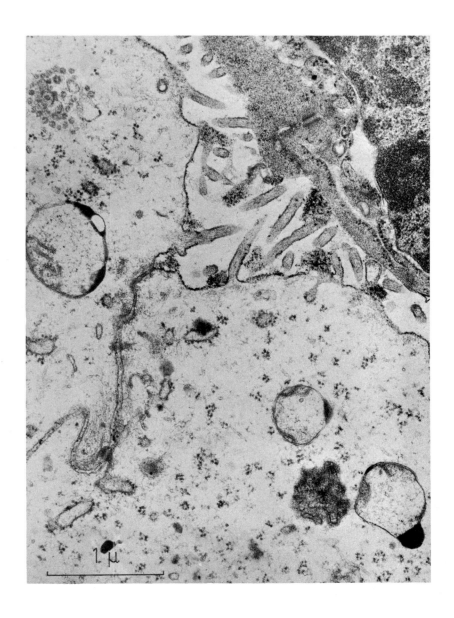

Mitochondria, cell border and microvilli with positive
alkaline phosphatase activity. CASE 3

5. Discussion

The interest in Regan isoenzyme is directed to the possibility that it is a phenotypic product of the activation of a placental gene in cancer cells. The present data suggest that in ovarian cancer, some individuals will express the Regan phenotype, others the non-Regan phenotype and still others will show both phenotypes. The possibility that Regan and non-Regan isoenzymes may each invest specific membrane sites at the ultrastructural level is suggested by the present results and warrants further investigation.

Acknowledgements

This research has been supported in part by Grants-in-aid from the National Cancer Institute, National Institutes of Health (CA-12924; CA-13332; K_6-CA-18453) and from the John A. Hartford Foundation, New York.

References

1. Fishman, W.H., Inglis, N.R., Stollbach, L.L., and Krant, M.J., 1968 Cancer Res. 28, 150.

2. Fishman, W.H., 1973, Carcinoplacental isoenzyme antigens in Advances in Enzyme Regulation, volume 11, George Weber, Editor (John Wiley, Publisher).

3. Rufo, M.B., and Fishman, W.H., 1972, J.Histochem.Cytochem. 20, 336.

4. Kaplow, L.S., 1955, Blood 10, 1023.

5. Hugon,J.S., and Borgers, M., 1966, J.Histochem.Cytochem. 14, 629.

Electron microscopy and Cytochemistry, eds. E. Wisse, W.Th. Daems, I. Molenaar and P. van Duijn.
© 1973, North-Holland Publishing Company - Amsterdam, The Netherlands.

FINE-STRUCTURAL LOCALIZATION OF ARYLSULPHATASES IN RAT ADRENAL
CORTEX

Bácsy, E. and Rappay, Gy.
Institute of Experimental Medicine,
Hungarian Academy of Sciences, Budapest, Hungary

1. Introduction

It has been shown or suggested for a number of endocrine cells
that lysosomes participate in the secretory process or in its
regulation. Szabó et al.[1] have demonstrated in the cells of the
adrenal zona fasciculata that the number and size of lysosomes
with acid phosphatase activity changes after hypophysectomy or
ACTH treatment. The role of lysosomes in corticoid hormone secretion is still not understood. The fine-structural localization of
the activity of acid hydrolases other than phosphatases may
elucidate the problem. In this work arylsulphatases have been
studied. Preliminary results have already been published[2].

2. Material and methods

Adrenals of male CFE rats weighing ca. 150 g were fixed either
by perfusion with 2 % formaldehyde and 2.5 % glutaraldehyde in
0.1 M cacodylate buffer, pH 7.2, containing 5 % saccharose, for
15 minutes, or by immersion in 5 % glutaraldehyde in the same
buffer for 2 hours. Perfused adrenals were immersed in this latter
fixative for 2 hours as well. 50 μ slices were prepared with a
TC-2 Sorvall chopper and incubated for arylsulphatase activity
according to the barium procedure of Goldfischer[3]. In some cases
the pH of the medium was modified from 5.5 to 6.0. After washing
and postfixation with 1 % osmium tetroxide for 1.5 hours the
slices were embedded in Durcupan ACM. Ultrathin sections were
briefly stained with lead citrate and uranyl acetate, and examined
in a JEM 6-AS electron microscope. 0.5 μ sections of the same
material were viewed unstained in the electron microscope with
100 kV accelerating voltage.

3. Results

Primary and secondary lysosomes exhibited strong barium deposit
in the cells of the zona glomerulosa and fasciculata. Sulphatase
active lysosomes were often found to be in close contact with
lipid droplets, some of the lipid droplets seemed to be eroded
near the interconnection with the lysosome. Reaction deposit was
regularly found in the Golgi region, the outer Golgi saccules
remaining free of barium precipitate. Some cysternae of the Golgi
region were continuous with smooth endoplasmic reticulum cysternae
surrounding mitochondria, and contained reaction product as well.
Endoplasmic reticulum cysternae, especially those being near to
mitochondria, were often seen to contain barium precipitate.

Barium deposit was seen, in addition, in small, round areas of mitochondria, and as a faint, diffuse precipitate in the marginal zone of some lipid droplets in specimens incubated both with and without substrate. In some areas, especially in angles between lipid and mitochondrion, precipitate was seen diffusely in the cytoplasm. This precipitate was never found in preparations incubated without substrate; diffusion of the enzyme during tissue preparation could not be excluded.

Using lead instead of barium in the incubation medium resulted in heavy staining of primary and secondary lysosomes, without any precipitate in other localization.

4. Discussion

The cysternae containing arylsulphatase activity in the Golgi region probably belong to the GERL. The activity of arylsulphatase in the smooth endoplasmic reticulum seems not to be restricted to the Golgi region but it is found in cysternae near to or between mitochondria. We do not know definitely whether this is the case in the living cell or the enzyme is displaced in the smooth ER continuous with GERL post mortem.

The role of sulphation-desulphation processes in the secretory activity of the adrenal cortex is not clear. The presence of sulphated steroids and of the non-specific sulphate donor compound, PAPS in the adrenal has been demonstrated. Fry and Koritz[4] have shown the participation of lysosomes in the degradation of PAPS. The close connection between lipids, mitochondria, and arylsulphatase-containing lysosomes and ER suggests an immediate intervention of this enzyme in steroid metabolism or transport.

5. References

1. Szabó,D., Stark,E. and Varga,B., Histochemie 10, 321 /1967/
2. Rappay,Gy., Kondics,L. and Bácsy,E., Histochemie 34, 271 /1973/
3. Goldfischer,S., J.Histochem. Cytochem. 13, 520 /1965/
4. Fry,J.M. and Koritz,S.B., Proc.Soc.Exp.Biol.Med.140,1275 /1972/

Fig.1. Arylsulphatase activity in two zona fasciculata cells. Well-localized reaction product in primary and secondary lysosomes. Some of the lysosomes are in close contact with a lipid droplet. In the cell on the right a part of the GERL is seen to contain reaction product and to be continuous with a sulphatase-containing ER cysterna surrounding a mitochondrion.

Fig.2. Arylsulphatase activity in the GERL of a zona fasciculata cell, the outer Golgi saccule being free of reaction product.

Fig.3. Electron micrograph of a 0.5 μ section of adrenal cortex incubated for arylsulphatase activity. Reaction deposit in parts of the Golgi region.

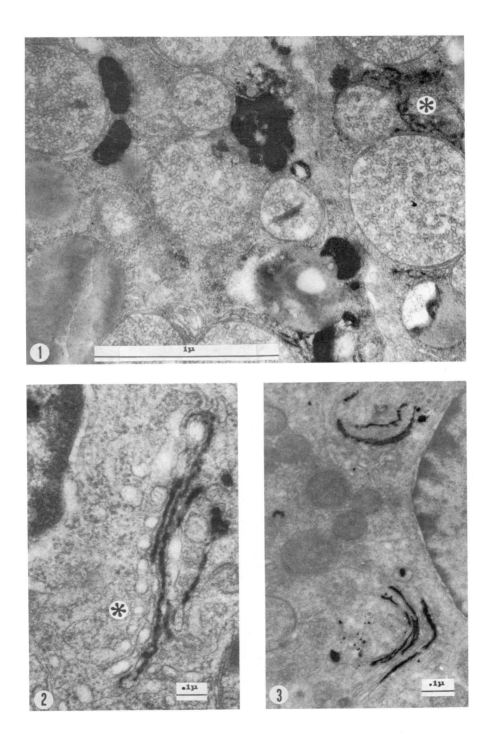

Electron microscopy and Cytochemistry, eds. E. Wisse, W.Th. Daems, I. Molenaar and P. van Duijn.
© 1973, North-Holland Publishing Company - Amsterdam, The Netherlands.

DEMONSTRATION OF NAD-PYROPHOSPHORYLASE ON ULTRASTRUCTURAL LEVEL

Unger, E., Buchwalow, I.B.[*], Hartmann, M.

Dept. of Exp. Cell Res. and Dept. of Biochem., Central Inst. of
Microbiol. and Exp. Therapy, Acad. of Sci. DDR, Jena;[*] Lab. of
Histochem. and Cytochem., Inst. of Exp. and Clin. Oncology,
USSR Acad. of Med. Sci., Moscow

NAD-pyrophosphorylase(EC No.2.7.7.1) is a typical nuclear enzyme
catalizing the reaction: ATP+NMN $\rightleftharpoons$ NAD +PP. To detect it histochemically   it is possible to make experiments on the basis
of a NAD-depending reaction or a pyrophosphate precipitation method with heavy metal ions.

The NAD-formation by isolated NAD-phosphorylase prepeared from
pig liver or by a suspension of mouse liver nuclei was investigated
under addition of lead ions, ATPase inhibitors or aldehyds in two
different buffers(see table1).

Table 1:Inhibition of NAD-formation

glycylglycine buffer pH 7.4, 0.1M; incubation 60min, 24°C
or 38°C(+);conc.ATP 1.2 mM,conc.NMN 2.5mM; liver nuclei

agent	concentration	inhibition	
		mouse liver nuclei	pig liver enzyme
lead acetate	3.8mM	17%(+)	–
sodium fluoride	15mM	19%(+)	–
N-ethylmaleimide	10mM	20%(+)	–
potassium cyanide	10mM	16%	13%
formaldehyde	0.45M	96%	96%
FA prefixation	1%	79%	17%
GA prefixation	2.5%	99%	99%
tris-maleate buffer	0.06M	3%	-1%

(The formed NAD was measured after reduction with alcohol
dehydrogenase) (1)

It can be concluded that a lead method is possible. An ATPase
inhibitor in the incubation medium after (or without) prefixation
with formaldehyde can be used.

Table 2: Incubation mediums

agent	medium	I	II	III
glycylglycine buffer pH 7.4		0.1M	–	–
tris-maleate buffer pH 7.6		–	0.06M	0.06M
NMN		2mM	2.5mM	2.5mM
ATP		2mM	2mM	2mM
magnesium acetate		–	0.01M	0.01M
magnesium chloride		0.01M	–	–
nicotinamide		–	0.25M	0.25M
lead nitrate		2mM	–	–
lead acetate		–	2mM	–
manganese chloride		–	–	4M
saccharose		0.2M	0.3M	0.3M

incubation time: 30min lead method, 60min manganese method
incubation temperature: 37-38°C

Control reactions are:
- incubation without ATP or (and) NMN or lead ions
- prefixation with glutaraldehyde
- incubation with addition of an ATPase inhibitor (med
- washings with acetate buffer pH 5.0 (med.III)

Ultrastructural investigations of isolated mouse liver nuclei:
With the help of medium I and II NAD-pyrophosphorylase activity is localized in parts of the nucleolus and in chromatine. This conclusion is derived from the following facts. An unspecific alkaline monophosphatase hasn't been found in these nuclei. An acid monophosphotase is only positive in a range near pH 6.4 in the nuclear envelope. ATPase remains as a disturbing factor because it isn't to inhibit completely by any inhibitor. However, the decrease of the intranuclear ATPase activity after incubation with sodiumfluoride is great enough to differentiate this enzyme from NAD-pyrophosphorylase. Medium III forms an crystalline precipitate with a localization like in medium I or II. The precipitates are insoluble in acetate buffer pH 5.0. After incubation in a medium without ATP there isn't any precipitation. Therefore the medium III precipitate must be manganese pyrophosphate and not the result of a nucleotide pyrophosphatase action.

NAD-pyrophosphorylase can be demonstrated by both a lead method and a manganese method. Each method has its restrictions. But taking these facts into considerations each method is usable.

The NAD-pyrophosphorylase localization is in coincidence with the biochemical works of SIEBERT et al. [2] who mainly found the NAD-pyrophosphorylase activity in the necleolus and KAUFMANN et al. [3] who suggested an association of the enzyme in the desoxyribonucleoprotein.

(1) A. Kornberg, 1950, J.Biol.Chem., 182,779
(2) G. Siebert et al. 1966, J.Biol.Chem. 241,71
(3) E. Kaufmann et al. 1972, Exp.Cell.Res. 71,209

fig.1, fig.2 - medium I, 0.01M NaF added, without prefixation whole medium, precipitates are in parts of nucleolus and chromatine, fig.2 contrast in these regions is less, medium without NMN and nicotinamide.
fig.3 like fig.1, nucleolus part, filament arrangement of contrast see(↑).
fig.4,5,6 - medium II,formaldehyde prefixation.
fig.4 whole medium, precipitates are in nucleolus and chromatine.
fig.5 medium without NMN and nicotinamide, precipitation is lower than in fig.4, fig.6 medium without NMN and ATP, no important unspecific lead absorption is to be observed.
fig.7,8 - medium III, formaldehyde prefixation.
fig.7 crystalline precipitates are mainly in the nucleolus but also in other parts of nucleus, fig.8 medium without NMN and nicotinamide, no crystals are formed.
In all cases it followed glutaraldehyde and osmiumtetroxide fixation after incubation.

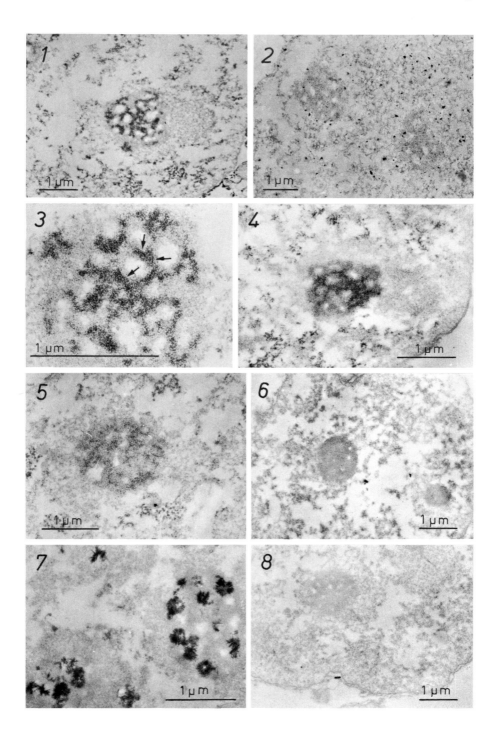

Electron microscopy and Cytochemistry, eds. E. Wisse, W.Th. Daems, I. Molenaar and P. van Duijn.
© 1973, North-Holland Publishing Company - Amsterdam, The Netherlands.

ULTRASTRUCTURAL LOCALIZATION OF ENZYMES USING OSMIOPHILIC PRINCIPLES

S.J. Holt

Department of Cytochemical Research
Courtauld Institute of Biochemistry
Middlesex Hospital Medical School
London W1P 5PR, England

With the exception of procedures that rely on immunochemical or radioautographic principles, most methods originally developed for the localization of enzymes using the light microscope depended upon the precipitation of either insoluble metal salts or organic dyes formed by capture of one of the products of enzymically catalysed substrate decomposition with a suitable reagent included in the incubation medium. It is not surprising, therefore, that the development of enzyme localization methods to be used with the electron microscope first relied on metal salt reaction products because of their known electron scattering properties, thus ensuring good contrast in the electron image. Such methods have had wide and successful applications, but in many cases are limited in the spacial resolution attainable because of the discrete granular nature of the reaction product, although, of course, such products have an advantage in that they can unambiguously be recognized in or on the ultrastructural components of cells. This is illustrated in fig. 1, in which the crystalline nature of the deposit of lead phosphate produced by acid phosphatase hydrolysis of β-glycerophosphate in the presence of lead ions is clearly seen. However, it would be rash to conclude that the distribution of the reaction product in this case faithfully represents that of the enzyme in the stained kidney lysosomes. This is not only because of the size of the grains of the crystalline deposit (an extreme case here), but also because of the lack of precise knowledge of the factors that govern the sites of precipitation of insoluble metal salts from their supersaturated solutions produced locally, in this case, by capture of enzymically produced phosphate ions by lead ions.

The exploitation for electron microscopy of the alternative organic dye methods, which by careful choice of substrates and capture reactions can be made to yield substantive amorphous dyes, with a higher potential for better localization than that achievable with metal salt methods, has been limited by the very low electron scattering power of such deposits and their solubility in the various solvents and embedding media used to prepare stained tissue for electron microscopy. However, in the last few years this limitation has been overcome to a large extent by the utilization of organic visualization products which react with osmium tetroxide to produce insoluble osmium containing derivatives. This reaction usually accompanies post-fixation of the cytochemically stained aldehyde-fixed tissue currently used in enzyme localization methods, and results in increased electron scattering due to the presence of osmium, an element of high atomic number (76) in the final reaction product.

An early application of this principle was the successful use by Hanker, et al. of osmiophilic thiosemicarbazide $[H_2NNHCSNH_2]$ (TSC) to stain periodate oxidized polysaccharides[1]. The aldehyde groups formed by oxidation of polysaccharides condensed with the hydrazino group $[H_2NNH-]$ of TSC and, after rinsing to remove excess TSC, the sections were exposed to osmium tetroxide vapour which reacted with the thiocarbamyl moiety $[-CSNH_2]$ of the aldehyde semicarbazone to produce "osmium black", a visible, insoluble, electron scattering reduction product of osmium tetroxide. This modification of the periodic acid-Schiff reaction made possible the demonstration of macromolecular carbohydrates in ultrastructural preparations of a variety of tissues[1]. In later procedures of Seligman et al.[2,3], thiosemicarbazide was replaced by thiocarbohydrazide $[H_2NNHCSNHNH_2]$, which is a more effective reducing agent for osmium tetroxide.

Although these procedures were not for the localization of enzymes, they did establish the practical applicability of the osmiophilic principle in ultrastructural cytochemistry. Concurrently, other studies by the same group of investigators[4] showed that other sulphur-containing radicals were also effective in reducing osmium tetroxide to osmium mercaptides, lower oxides of osmium and their hydrates, or

Fig. 1. Part of glutaraldehyde fixed rat kidney proximal tubule cell after incubation in β-glycerophosphate at pH 5.0 in the presence of lead ions, followed by post-fixation in osmium tetroxide. The lead phosphate reaction product of acid phosphatase activity forms scattered crystalline clumps in the stained lysosomes.

Fig.2. Proximal tubular cell of mouse kidney aldehyde fixed 6 min. after intravenous injection of cytochrome c and incubated with DAB and hydrogen peroxide at pH 3.9, followed by post-fixation in osmium tetroxide. The amorphous osmium-containing reaction product is clearly visible in the brush border at the lower left of the field, in pinocytotic invaginations and in apical vesicles, showing that the marker has been reabsorbed from the urinary space. No counterstain. By courtesy of Dr. M.J. Karnovsky.

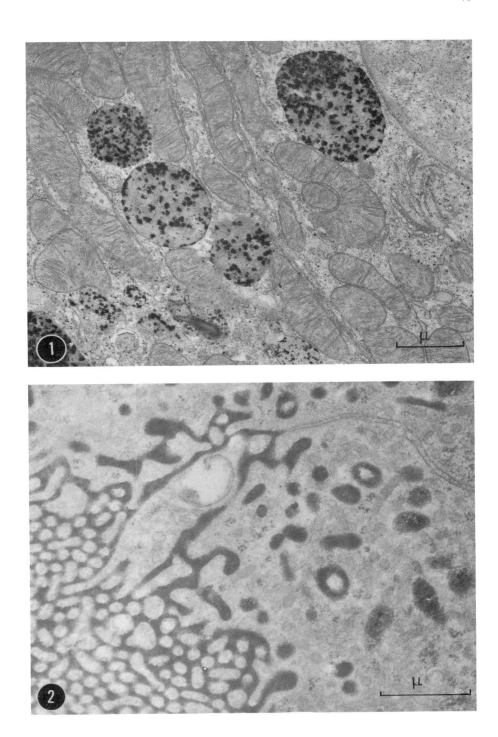

possibly to osmium itself; all black lipid insoluble pigments capable
of being used as localization agents. These findings were applied in
early methods for the localization of esterases, phosphatases and
cytochrome oxidase[1], and although the precision of localization
obtained in these and other enzymic methods examined by the Seligman
group[5,6,7] was not good at that time, a basis was nevertheless established for the later development of more sophisticated and satisfactory osmiophilic methods for the localization of enzymes which will be
discussed below.

During this same period, a method was developed by Graham and
Karnovsky[8] for the ultrastructural localization of exogenous peroxidase which depended upon the oxidative polymerization of 3,3'-diamino-
benzidine (DAB) catalysed by the enzyme in the presence of hydrogen
peroxide. The brown insoluble reaction product can be detected in
the electron microscope without treatment with osmium tetroxide, but
considerably greater contrast follows post-osmication. Subsequent
in vitro experiments showed that the initial polymer is, in fact,
osmiophilic and gains about 38% in weight when treated with osmium
tetroxide[9]. Other peroxidatic markers, such as cytochrome c were
later used as ultrastructural tracers, using the same staining
principle, by Karnovsky and Rice[10], and fig. 2 illustrates a typical
result of using this method.

The amorphous non-droplet character of the DAB reaction product
allowed the development by Seligman, et al.[9,11] of a staining method
for cytochrome oxidase which gave a much higher spacial resolution
of enzymic activity (fig. 3) than the osmiophilic Nadi type reagents
used earlier by the same group[1,7]. At high magnification, the DAB
reaction product was seen to be confined to the intercristate spaces
and the outer compartment of the mitochondria, a finding which
suggested that the enzyme was located on the outer surface of the
inner mitochondrial membrane, and which is consistent with the

---

Fig. 3. Formaldehyde fixed rat liver stained by the DAB method of
Seligman et al. for cytochrome oxidase[9,11]. The osmium-containing
polymeric oxidation product of DAB has high contrast and is seen to
lie in the mitochondrial intercristate spaces and in the outer
compartments. From ref. 11 by courtesy of Dr. A.M.
Seligman and Springer-Verlag. No counterstain.

Fig. 4. Osmiophilic reaction product produced in rat heart mitochondria when incubated with succinate and the monotetrazole BSPT[18] is
present on most of the cristae and on the inner and outer mitochondrial membranes. No counterstain. From ref. 21 by courtesy
of Dr. A.M. Seligman and the Histochemical Society, Inc.

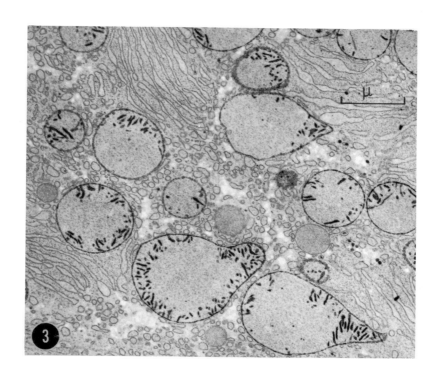

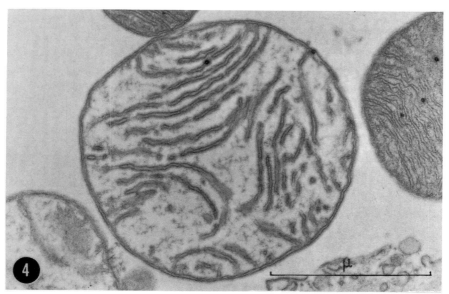

chemi-osmotic hypothesis for oxidative phosphorylation in the mitochondrion put forward by Mitchell[12].

The improvement in resolution of oxidase activity associated with
the use of DAB as an electron donor in ultrastructural cytochemistry
revealed the superiority of methods that yield polymeric osmiophilic
reaction products over those that do not, and led the Seligman group
to search for other bis-phenylenediamines to compare with DAB[13,14].
Two of these, N,N'-bis(4-aminophenyl)-N,N'-dimethylethylenediamine
(BED) and N,N'-bis(4-aminophenyl)-1,3,-xylylenediamine (BAXD) behaved
differently from DAB[11]. Whereas DAB and BAXD were both oxidized by
mitochondrial cytochrome oxidase, BED was oxidized by a terminal
oxidase in the endoplasmic and sarcoplasmic reticulum. Although the
exact nature of the terminal oxidases involved in the oxidation of
BED and BAXD has not yet been elucidated[15], it appears that there is
great future scope for the use of these and other bis-phenylenediamines in the study of oxidase systems by high resolution cytochemical
methods. One must not forget, however, that in spite of the high
inherent resolution theoretically possible with the amorphous
osmium derivatives of such oxidized diamines, some controversy has
arisen concerning the existence of diffusion artefacts in DAB cytochemistry[16,17].

A similar course of developmental events to those associated with
the demonstration of cytochrome oxidase activity took place with
methods designed for the high resolution localization of succinic
dehydrogenase (SDH) using osmiophilic principles. The inclusion of
two of the previously studied osmiophilic thiocarbamyl groups in
nitro-blue tetrazolium allowed some advances to be made in the ultrastructural localization of SDH in mitochondria[9,18], but difficulties
of interpretation and possible artefactual staining[19], together with
poor penetration of the acceptor tetrazole into tissue and its tendency to lose some of its osmiophilic sulphur[20], prompted a search for
better acceptors. This was attempted in two ways. First, by the
synthesis of a distyryl derivative of nitro-blue tetrazolium, which
depended on the unsaturation of the styryl radicals $[C_6H_5CH=CH-]$ for
its osmiophilia[20]. Although this yielded an osmiophilic formazan
on reduction by SDH in the presence of succinate, it also possessed
several drawbacks[21]. One was that the tetrazolium salt was as
osmiophilic as the derived formazan. Another was the long exposure
to osmium tetroxide at high temperature that was necessary to achieve
a useful degree of osmication. A third was poor penetration into
tissue caused by the presence of the two positively charged

quaternary nitrogen atoms of the ditetrazole, and a fourth was the
tendency of the diformazan to coalesce into droplets and granules
before osmication, thus lowering the accuracy of spacial delineation
of the enzyme. These limitations led to a second approach by the
Seligman group to the improvement of osmiophilic SDH localization
methods, in particular, the preparation of a series of monotetrazoles
containing benz thiazolyl and phthalhydrazidyl radicals[21], the latter
having already been shown in earlier work to confer lipophobic
properties upon osmiophilic dyes used for the ultrastructural localization of aminopeptidase and transpeptidase activities[22], and thereby discourage droplet formation. These modifications resulted in
tetrazoles that were not themselves osmiophilic under normal conditions of exposure to osmium tetroxide, but gave formazans in the SDH
reaction that were readily osmicated under the same conditions.
Unexpectedly, these monotetrazoles did not lead to a significant
improvement in their penetration into tissues, but of the ten studied,
2-(2'-benzthiazolyl)-5-styryl-3-(4-phthalhydrazidyl) tetrazolium
chloride (BSPT) was selected as having properties worthy of further
investigation for the localization of SDH[21].

The results of staining rat heart muscle, very briefly fixed in
2% formaldehyde, in buffered succinate containing BSPT are shown in
fig. 4 in which it is seen that most of the mitochondrial membranes
show an almost amorphous deposit of the osmicated reaction product
of the cytochemical procedure. However, this is not in accord with
the findings of biochemical studies on isolated inner and outer mitochondrial membranes, which show that SDH is not present in the outer
membrane,[23] a discrepancy that may be due to the participation of other
enzymes or carriers in the final pathway of reduction of BSPT in the
cytochemical procedure. However, the existence of such discrepancies
is unfortunate, for, ideally, biochemical and cytochemical methods
should augment each other's findings, the former providing quantitative evidence which is not yet available from ultrastructural cytochemistry, and the latter providing the spacial information that
cannot be obtained at present by fractionation techniques. Nevertheless, it must be recognized that an immense effort has been made by
Seligman and his numerous associates in their tenacious studies of
osmiophilic principles for the localization of SDH and the other
enzymes discussed above. Such efforts will undoubtedly continue now
that knowledge of how to achieve precise deposition of cytochemical
markers for enzyme activity has reached the present encouraging
state, and one looks forward to an eventual satisfactory integration

of these naturally complementary cytochemical and biochemical techniques.

Our own investigations into methods for localizing enzymes using osmiophilic principles has taken a different course from those described above, and have been based upon the use of substrates derived from indoxyl for the localization of hydrolytic enzymes. In spite of the apparent potentialities for high resolution reflected by the results of light microscope methods using this principle for the localization of esterases[24], the indigoid dyes produced as visualizing agents do not resist the dehydration and embedding procedures commonly used for electron microscopy, nor, as might be expected, do they afford more than feeble contrast in the electron microscope[25]. We have therefore subsequently been almost wholly concerned with investigating the principle that indoxyl, liberated from some appropriate substrate by enzymic activity, would couple with diazonium salts to give azoindoxyl dyes which would be osmiophilic, not by producing osmium black on treatment with osmium tetroxide, but by forming an insoluble osmium chelate[26]. This expectation was based on the fact that some of the most stable osmium chelates known are derived from 2,2'-dipyridyl[27] and 1,10-phenanthroline[28], in which osmium is chelated to nitrogen atoms having approximately the same spacing as those in azoindoxyl dyes[26]. In preparing such chelates from high valence osmium compounds, such as osmium tetroxide, it is obligatory to react them with the nitrogenous ligand under reducing conditions[27,28], such as those obtaining when stained tissues are post-fixed in osmium tetroxide. Of the many azoindoxyl dyes produced from a variety of indoxyl acetate substrates for esterases[29] in the presence of various simple and complex diazonium salts[30], only one, that derived from hexazotized pararosaniline[31], met the requirements of satisfactory osmiophilia for ultrastructural cytochemistry. When

---

Fig. 5. Glutaraldehyde fixed rat kidney incubated in a medium containing indoxyl acetate and hexazotized pararosaniline, followed by post-osmication[25]. The osmium-containing reaction product of esterase activity is clearly seen in the large lysosome at the lower right, in the endoplasmic reticulum, and in the perinuclear cisterna of the nucleus in the upper right of the field. No counterstain.

Fig. 6. Rat liver after brief formalin fixation incubated in the indoxyl acetate/hexazotized pararosaniline medium[25]. The membranes of this tract of rough surfaced endoplasmic reticulum are sharply delineated by the osmiophilic reaction product of esterase activity and the associated ribosomes also appear to be lightly stained, although this may represent their natural density. No counterstain.

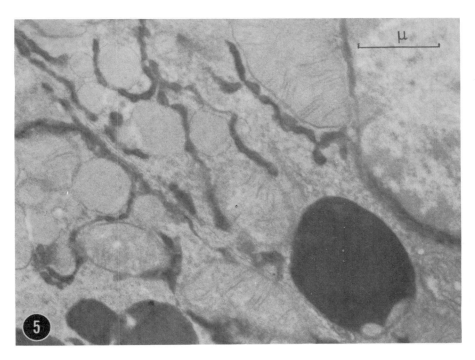

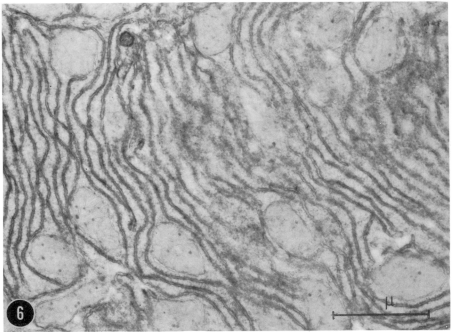

fixed tissue sections were incubated in indoxyl acetate in the presence of this hexazonium salt, followed by treatment with osmium tetroxide, sometimes with the addition of low concentrations of reducing
agents such as ethanol or formaldehyde[32], an amorphous insoluble
osmium complex was formed which gave satisfactory localization of
esterase in the lysosomes and endoplasmic reticulum of rat kidney
proximal tubule cells[26] (fig. 5). Azoindoxyl dyes produced from the
same substrate, but with various other coupling agents, even after
treatment with osmium tetroxide, did not resist the rigours of dehydration and embedding procedures, and it was concluded that this difference in behaviour depended upon the unique structure of the osmicated product of the dye derived from indoxyl and hexazotized pararosaniline. Metal chelates are often more soluble in organic solvents
than the dyes from which they are derived, largely because they are
less polar than the free dyes. It appears, however, from considerations of the structure of an osmium chelate of the azodye produced
from indoxyl and hexazotized pararosaniline, that in this case, the
complex would be an osmium linked polymer having an alternating
system of positive and negative charges throughout its structure,
thus giving it some of the characteristics of a metal salt[26,30].
This is considered to be the basis of the insolubility of this
complex. Theoretical calculations indicated that the complex should
contain about 70% by weight of osmium and subsequent analysis of
X-rays generated from stained areas in the electron beam confirmed
that osmium was present to a high degree compared with that in unstained areas[33].

One of the most controversial problems as far as the localization
of esterases is concerned has been the previously conflicting results
given for rat liver by biochemical and by cytochemical techniques[34,35].
The former have shown that most of the esterase activity against
indoxyl acetate is present in the microsome fraction of the tissue,
whereas the latter have indicated that the activity is largely lysosomal. This discrepancy appears to be due to the unique sensitivity
of rat liver esterase to the effects of fixatives. Only when fixation
is minimal (e.g. in 1% formaldehyde for 1 hour) does sufficient activity survive in the endoplasmic reticulum for it to be demonstrated
by the azoindoxyl staining technique (fig. 6). Curiously, there is
no similar problem in the case of the endoplasmic reticulum of the
livers of mouse, Guinea-pig and rabbit, its esterase activity being
easily demonstrable after much more drastic fixation[36]. Such findings illustrate the fact that the problems associated with the ultra-

structural localization of enzymes are not only those related to the
nature of the enzymic reaction involved, and that of the visualizing
product, but also to the effects of fixation, a process which at the
moment appears to be required for most ultrastructural investigations.
In some cases, lack of fixation and sectioning or chopping of sections in preparation for staining can lead to extensive extraction of
the enzyme under investigation, as in the case of liver esterases[35].

From what has been described above, it appears that one of the
prerequisites for enzyme localization using osmiophilic techniques
is that the reaction product should be polymeric, whether it be
osmium black or other types of osmium linked polymer. Recently, an
ingeneous combination of these principles has been achieved by the
use of a single agent as both substrate and capture reagent for the
localization of cholinesterase in the motor end plate. In this work,
Davis et al.[37] used a substrate containing an esterase susceptible
thiolester group and a diazonium group in the same molecule. This
yielded an osmiophilic polymer on hydrolysis in which the units were
linked by diazothioether groups. They also synthesized an indoxyl
acetate containing a diazonium group as an alternative substrate for
the esterase. In this case the polymer formed on hydrolysis was
composed of indoxyl moieties linked by azo groups and, as expected,
was osmiophilic. The thiol substrate was the S-acetylthiol-3-toluene-
diazonium ion and the indoxyl substrate was 3-acetoxy-5-indolediazon-
ium. Because of the strong positive charges on these diazonium ions,
neither substrate was hydrolyzed by aliesterases, as tested against
rat kidney. In each case the osmicated polymeric reaction product
was somewhat granular in nature, particularly with the thiol substrate.
Nevertheless, the practicability of this new principle has been established and one looks forward to the results of further investigations
along these lines which have been promised by these authors[37].

The solution to the various problems still remaining in the design
of methods for the ultrastructural localization of enzymes, whether
by osmiophilic or by new principles yet to be formulated, appears
to be, as in the past evolution of cytochemical staining methods,
that each method must be thoroughly investigated from all points of
view - the effects of fixation, the specificity of the enzymic reaction with the substrate (designed to give an electron scattering
product), the solubility and contrast properties of the stain, the
presence of diffusion or adsorption artefacts, and the kinetic aspects
of the enzymic and capture reactions employed[38]. These requirements
impose far greater constraints upon cytochemical methods for enzyme

localization than those based on biochemical principles. However, the need for further knowledge of the precise localization of enzymes in cells, and of the variations that occur during developmental and pathological (particularly neoplastic) changes, makes every effort directed towards the perfection of high resolution cytochemical techniques one of great importance in cell biology. It is therefore hoped that the substantial progress seen during the relatively brief history of cytochemical techniques for the ultrastructural localization of enzymes will be vigorously maintained during the coming years.

References

1. Hanker, J.S., Seaman, A.R., Weiss, L.P., Ueno, H., Bergman, R.A. and Seligman, A.M., 1964, Science 146, 1039.

2. Seligman, A.M., Hanker, J.S., Wasserkrug, H.L., Dmochowski, H. and Katzoff,L., 1965, J. Histochem. Cytochem. 13, 629.

3. Seligman,A.M., Wasserkrug, H.L. and Hanker, J.S., 1966, J. Cell Biol. 30, 424.

4. Hanker, J.S., Weiss, L., Dmochowski, H., Katzoff, L. and Seligman, A.M., 1965, J. Histochem. Cytochem. 13, 3.

5. Seligman, A.M., Kawashima, T., Ueno, H.,Katzoff, L. and Hanker, J.S., 1970, Acta Histochem. Cytochem. 3, 29.

6. Rutenburg, A.M., Kim, H., Fischbein, J.W., Hanker, J.S., Wasserkrug, H.L. and Seligman, A.M., 1969, J. Histochem. Cytochem. 17, 517.

7.Seligman, A.M., Plapinger, R.E., Wasserkrug, H.L., Deb, C. and Hanker, J.S., 1967, J. Cell Biol. 34, 787.

8. Graham, R.C. and Karnovsky, M.J., 1966, J. Histochem. Cytochem. 14, 291.

9. Seligman, A.M., Karnovsky, M.J., Wasserkrug, H.L. and Hanker, J.S. 1968, J. Cell Biol. 38, 1.

10. Karnovsky, M.J. and Rice, D.F., 1969, J. Histochem. Cytochem. 17, 751.

11. Seligman, A.M., Wasserkrug, H.L. and Plapinger, R.E., 1970, Histochemie, 23, 63.

12. Mitchell, P., 1967, Fedn Proc. Fedn Am. Socs exp. Biol. 26, 1370.

13. Plapinger, R.E., Linas, S., Kawashima, T., Deb, C. and Seligman, A.M., 1968, Histochemie, 14, 1.

14. Seligman, A.M., Seito, T. and Plapinger, R.E., 1970, Histochemie 22, 85.

15. Holtzman, J.L. and Seligman, A.M., 1973, Archs Biochem. Biophys. 155, 237.

16. Novikoff, A.B., Novikoff, P.M., Quintana, N. and Davis, C., 1972, J. Histochem. Cytochem. 20, 745.

17. Seligman, A.M., Shannon, W.A., Hoshino,Y. and Plapinger, R.E., 1973, J. Histochem. Cytochem. 21, 756.

18. Seligman, A.M., Ueno, H., Morizono, Y., Wasserkrug, H.L., Katzoff, L. and Hanker, J.S., 1967, J. Histochem. Cytochem. 15, 1.

19. Haydon, G.B., Smith, S.O. and Seligman, A.M., 1967, J. Histochem. Cytochem. 15, 752.

20. Seligman, A.M., Nir, I. and Plapinger, R.E., 1971, J. Histochem. Cytochem. 19, 273.

21. Kalina, M., Plapinger, R.E., Hoshino, Y. and Seligman, A.M., 1972, J. Histochem. Cytochem. 20, 685.

22. Seligman, A.M., Wasserkrug, H.L., Plapinger, R.E., Seito, T. and Hanker, J.S., 1970, J. Histochem. Cytochem. 18, 542.

23. Schnaitman, C. and Greenawalt, J.W., 1968, J. Cell Biol. 38, 158.

24. Holt, S.J. and Withers, R.F.J., 1958, Proc. R. Soc. B148, 520.

25. Holt, S.J. and Hicks, R.M., 1962, Br. med. Bull. 18, 214.

26. Holt, S.J. and Hicks, R.M., 1966, J. Cell Biol. 29, 361.

27. Burstall, F.H., Dwyer, F.P. and Gyarfas, E.C., 1950, J. chem. Soc. p. 953.

28. Dwyer, F.P., Gibson, N.A. and Gyarfas, E.C., 1950, J. Proc. R. Soc. N.S.W. 84, 68.

29. Holt, S.J. and Sadler, P.W., 1958, Proc. R. Soc. B148, 481.

30. Holt, S.J. and Hicks, R.M., to be published.

31. Davis, B.J. and Ornstein, L., 1959, J. Histochem. Cytochem. 7, 297.

32. Leduc, E.H., Bernhard, W., Holt, S.J. and Tranzer, J.P., 1967, J. Cell Biol. 34, 773.

33. Holt, S.J. and Sheldon, F., 1972, Histochemistry and Cytochemistry, Proc. 4th. Internat. Congress Histochem. Cytochem., Kyoto. p. 279.

34. Underhay, E., Holt, S.J., Beaufay, H. and de Duve, C., 1956, J. biophys. biochem. Cytol. 2, 635.

35. Barrow, P.C. and Holt, S.J., 1971, Biochem. J. 125, 545.

36. Holt, S.J. and Barrow, P.C., to be published.

37. Davis, D.A., Wasserkrug, H.L., Heyman, I.A., Padmanabhan, K.C., Seligman, G.A., Plapinger, R.E. and Seligman, A.M., 1972, J. Histochem. Cytochem. 20, 161.

38 Holt, S.J. and O'Sullivan, D.G., 1958, Proc. R. Soc. B148, 465.

Electron microscopy and Cytochemistry, eds. E. Wisse, W.Th. Daems, I. Molenaar and P. van Duijn.
© 1973, North-Holland Publishing Company - Amsterdam, The Netherlands.

THE AVIDITY OF HEAVY METAL DIAZOTATES FOR ANIMAL LYSOSOMES AND PLANT VACUOLES DURING THE ULTRASTRUCTURAL LOCALISATION OF ACID HYDROLASES

Beadle, D.J., Dawson, A.L., James, D.J., Fisher, S.W., and Livingston, D.C.

School of Biological Sciences, Thames Polytechnic, London SE18
and Department of Chemistry, Imperial Cancer Research Fund,
Lincoln's Inn Field, London WC2

Introduction

Over a period of several years we have been interested in the use of naphthol esters and lead-containing diazonium chlorides for the ultrastructural local-isation of acid hydrolases in a number of plant and animal tissues. Acid phosphatase, non-specific esterase and β-glucuronidase have been studied in insect and mouse tissues, baby hamster kidney (BHK/21) and Chinese hamster ovary (CHO) cultured cells, strawberry receptacles and root cells of Vicia faba . Both the diazotates of lead phthalocyanin[1] and triphenyl-p-aminophenethyl lead (LPED)[2,3] have been used as coupling salts and the appropriate naphthol ester has been used as substrate. Our observations have shown that these heavy metal diazotates have an avidity for lysosomal structures when fixed in glutaraldehyde-cacodylate mixtures, an effect first noticed by Smith and Fishman[4] using a mercury-containing diazotate.

Methods

Small pieces of mouse kidney, midgut from various insects, strawberry receptacles and roots of Vicia faba were fixed at 4°C for 3h. in 2.5% glutaraldehyde buffered with O.1M cacodylate at pH7.2. BHK/21 and CHO cells were spun into pellets and similarly fixed for 2O min. All tissues were then washed in several changes of cold cacodylate buffer before demonstrating enzyme activity. Acid phosphatase was demonstrated in all animal tissues by incubating for 3O min. at 37°C in a medium containing naphthol AS Bl phosphate (2O mg. %) as substrate and LPED (2.5 mg. %) as coupling salt in O.2M acetate buffer at pH 5.O. β-Glucuronidase was similarly demonstrated using naphthol AS- β-D-glucuronide as substrate. Non-specific esterase was demonstrated in plant material using naphthol AS D acetate as substrate in tris-HCl buffer at pH 7.1 and incubating for

8 - 20 min. Acid phosphatase was also demonstrated in insect midgut and non-specific esterase in strawberry receptacles using lead phthalocyanin diazotate (10 mg/ml) as coupling salt. Both simultaneous and post-coupling techniques were used. Control material was reacted in the absence of substrate with an additional sodium fluoride control (1×10^{-3} M) for acid phosphatase.

After incubation all tissues were washed in buffer, post-fixed in 1% osmium tetroxide, dehydrated in Durcapan-water mixtures and embedded in Araldite. All sections were examined unstained.

Results

The early experiments using LPED as a coupling salt showed end-product in lysosomes of kidney cells[2] and insect midgut cells[3] in test sections only. No-substrate controls were negative. The end-product was electron opaque and easily identifiable (Fig. 1). However, later work using a wider range of tissues showed LPED deposits in control as well as test material. This proved to be the case with BHK/21 and CHO cells as well as kidney (Fig. 2) and insect tissues. The LPED in control material appeared to be associated only with lysosomes and particularly secondary lysosomes. Any membranous material within the lysosome such as myelin figures showed particularly heavy deposits. No other subcellular organelles or membrane systems showed deposits of LPED. In plant material test sections showed the deposition of LPED in vacuolar structures and endoplasmic reticulum (Fig. 3), although in the latter case it was confined to the membranes aligned with the plane of the newly formed wall. Control sections appeared to be negative.

Fig. 1. Chinese hamster ovary cells incubated in medium to demonstrate β-glucuronidase with LPED. End-product is in secondary lysosomes.

Fig. 2. Mouse kidney incubated in LPED without enzyme substrate. The LPED is bound in a lysosome.

Fig. 3. Epidermal cells of V. faba incubated for non-specific esterase using LPED. End-product is in vacuoles and endoplasmic reticulum.

Fig. 4. Midgut cells of Carausius morosus incubated with lead phthalocyanin and without enzyme substrate showing deposits in the endoplasmic reticulum and nuclear membrane.

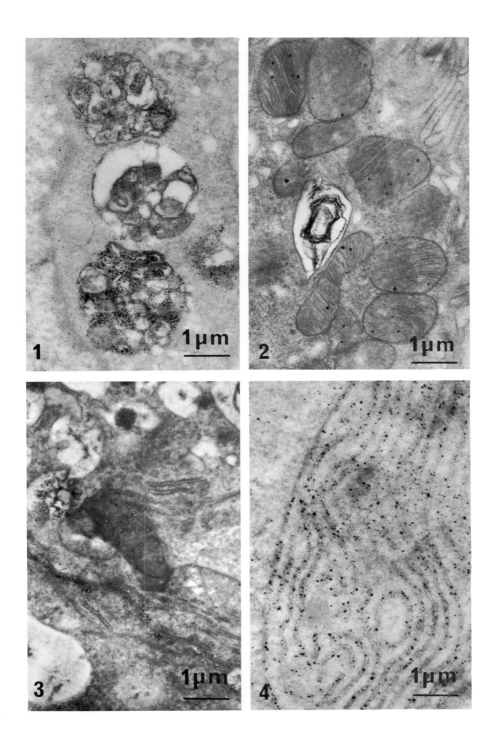

When lead phthalocyanin was used as a coupling salt in ripe fruits it was found in the Golgi apparatus and a variety of different sized vacuoles in both test and control material. In the insect midgut lead phthalocyanin was found in lysosomes, Golgi apparatus, endoplasmic reticulum and the nuclear membrane in both test and control sections (Fig. 4).

Discussion

These observations endorse the remarks made by Smith and Fishman that the electron density of metal deposits in subcellular organelles, that are unrelated to released naphthols, present a serious limitation to the use of heavy metal diazotates in ultrastructural enzyme cytochemistry. This is particularly true when such compounds are employed in conjunction with glutaraldehyde-cacodylate fixation. With compounds such as lead phthalo-cyanin that bind to a number of membranous structures it is likely that some component of the membrane, such as sulphydryl groups, might be respon-sible for the binding but it is more difficult to explain the avidity of LPED for lysosomes. It is possible that when the cause of this binding becomes known some modification of the procedures employed may yet allow these techniques to be used with some confidence.

References

1. Livingston, D.C., Coombs, M.M., Franks, L.M., Maggi, V. and Gahan, P.B., 1969, Histochemie 18, 48.

2. Livingston, D.C., Fisher, S.W., Greenoak, G.E. and Maggi, V., 1970, Histochemie 24, 159.

3. Beadle, D.J., Livingston, C.D. and Read, S., 1971, Histochemie 28, 243.

4. Smith, R.E. and Fishman, W.H., 1969, J. Histochem. Cytochem., 17, 1.

Electron microscopy and Cytochemistry, eds. E. Wisse, W.Th. Daems, I. Molenaar and P. van Duijn.
© 1973, North-Holland Publishing Company - Amsterdam, The Netherlands.

STUDIES ON THE STRUCTURE AND FUNCTION OF CELL ORGANELLES:
3,3'-DIAMINOBENZIDINE CYTOCHEMISTRY

Alex B. Novikoff

Department of Pathology,
Albert Einstein College of Medicine, Yeshiva University,
Bronx, New York

1. Introduction

In our laboratory cytochemistry has been used chiefly for identifying cell organelles and for helping to elucidate their structures and functions.

Often simply visualizing cell organelles can be of considerable use in histological studies. Mesangial cells are dramatically identified if a frozen section of glutaraldehyde-fixed rat kidney is incubated in the Wachstein and Meisel phosphatase medium (1) with nucleoside tri- and di-phosphates such as ATP or ADP. The plasma membranes of these cells become visible because they possess high levels of phosphatases which hydrolyze these substrates (2). Similarly, the macula densa cells are vividly stained when formaldehyde-fixed sections of rat kidney are incubated for NADH-NBT reductase activity (Fig. 4 in ref. 3). These cells have plump mitochondria, with high levels of this reductase activity. Eosinophils can be visualized against a colorless background, under specific conditions of DAB incubation (Fig. 3) [1].

By making possible the visualization of organelles in situ, cytochemistry has supplemented (a) biochemical analyses of subcellular fractions isolated from homogenates, and (b) electron microscopic observations (4). The most striking instances concern cytoplasmic particles not readily recognized by their fine structure, the lysosomes and the microperoxisomes. Thus, it has added significantly to our knowledge of the types and biological behavior of lysosomes (5-7) and peroxisomes (8).

Cytochemistry can also be useful in identifying organelles, and even fragments of organelles, in subcellular fractions as shown recently by Farquhar and her collaborators (9).

It became apparent long ago that unfixed tissue, even when sectioned in improved cryostats, was ill-suited for cytochemical studies because of inadequate preservation of intracellular

---

[1] This may prove of value in automated enumeration of eosinophils in blood samples.

morphology. In 1958 it was reported that lactic dehydrogenase and NADH-tetrazolium reductase activities survived formaldehyde fixation and that mitochondria could be visualized in frozen sections of tissues fixed overnight in cold 4% formaldehyde-1% $CaCl_2$ (10); also see ref. 11 [2] [3].

Our studies therefore centered on enzyme activities which visualize cytoplasmic organelles in well-fixed tissue (14-16). In general longer fixation improves the preservation of organelles; therefore fixation for as long as the reactive constituent (such as an enzyme) permits is desirable.

There are many unresolved questions in regard to enzyme activities involved in the cytochemical visualization of organelles. Is it really an ATPase which is demonstrated in the plasma membrane (15) by incubating sections of fixed tissues in the lead medium of Wachstein and Meisel (1) or in the strontium medium of Ernst (17)? If so, is it the Na^+, K^+-stimulated, ouabain-inhibited ATPase? Is the same enzyme responsible for both the thiamine pyrophosphatase activity which visualizes the Golgi apparatus (18) and the nucleoside diphosphatase activity which, in addition to the Golgi apparatus, visualizes the ER of some cells (19)?; see refs. 20 and 21. Does fixation lead to loss of "latency" of the acid phosphatase activity which permits visualization of lysosomes, as has been suggested (22)? Is the high alkalinity of DAB media (23-26) used to visualize peroxisomes and microperoxisomes responsible for the peroxidatic activity of the catalase?; does aldehyde fixation also favor such peroxidatic activity (27, 28)? Can diffusion of reaction product, or enzymes and other metalloproteins give false reactions with DAB?

---

[2] In these studies the tetrazolium salt, NBT, was used. A later tetrazolium, tetra-nitro BT, does not form artifactual formazan deposits at the surfaces of lipids as does NBT (11, 3). However, in unusual instances, there is a redistribution, sometimes rather quickly, of the TNBT formazan so that it no longer visualizes mitochondria. Another tetrazolium salt, MTT, was found to be inadequate for mitochondrial visualization (11).

[3] Pearse (12) in the 1960 edition of his book, wrote, "There is thus no need to use formalin fixation and I consider that its use prior to the demonstration of any dehydrogenase system, even if it is tough enough to survive such a procedure, is a retrograde step in dehydrogenase histochemistry which must be utterly condemned" (p. 581). In the 1972 edition he states (13), "My agreement with the principle of minimum fixation constitutes an absolute withdrawal from the position which I adopted in 1960" (p. 923).

However, our central concern is the visualization of organelles, as well-preserved as possible. Thus, shorter incubations are always desirable even when the mechanisms by which a reagent accelerates the reaction are unknown. This is why manganese ions are used in DAB techniques for visualizing mitochondria and ER (29, 30). Seligman et al. (31) consider manganese "an unnecessary feature". When I chose manganese from the various activating ions tested I was aware of possible complicating factors, but these did not interfere with the DAB reactions. Nor did the manganese alter the intracellular or intra-organelle localizations. Whatever the mechanism, manganese ions shorten the incubation time, thus helping to preserve cell organelles.

Cytochemistry has made significant contributions to cell biology, but its capacities should not be overestimated. For example, it is unlikely that present DAB techniques can test Mitchell's chemosmotic hypothesis of energy coupling in mitochondria, as has been attempted (32). Nor can a tetrazolium salt such as MTT (see ref. 11) reveal a single functionally-abnormal mitochondrion in a cell whose remaining mitochondria appear normal, as has been proposed (33).

It is well known that negative results of cytochemical procedures must be interpreted cautiously. Fixatives or other ingredients of the incubation media may inactivate enzymes; inadequate trapping agents may permit escape of enzyme reaction products. Positive results for enzyme activity may also be misleading if they result from diffusion and adsorption, or from non-enzymatic breakdown of substrate. Still another possibility needs to be borne in mind: enzymes revealed cytochemically may not be active in vivo. For example, DAB procedures reveal endogenous peroxidase in the ER of some cells. However, this does not prove the in vivo activity of this enzyme within the ER. Inside the living cell, enzymes may be "latent" or "masked". For instance, tyrosinase activity can be demonstrated in mouse melanoma sections using the old "DOPA-oxidase" technique (34). Melanin is deposited at the sites of tyrosinase: melanosomes, premelanosomes and GERL. In the absence of cytochemical treatment only melanosomes show melanin. From the cytochemical results we can conclude that tyrosinase is also present in premelanosomes and GERL. Yet this enzyme is inactive in vivo.

2. DAB cytochemical procedures and tracer studies

It was in 1966 that Graham and Karnovsky published the first cytochemical results with DAB (35). The fate of injected horse-radish peroxidase was followed in the kidney of the mouse. Sections of aldehyde-fixed tissue were incubated at room temperature in a DAB-H_2O_2 medium at pH 7.6 and were then treated with OsO_4. Sites of peroxidase were visualized as a homogeneous electron-opaque material (osmium black) readily seen in the electron microscope (Fig. 1).

A large literature has accumulated since 1966 in which DAB cyto-chemical methods were used to trace exogenously administered horse-radish peroxidase and other metalloproteins in many cells and tissues. The early literature is listed in ref. 29; for some recent work see refs. 36-43. The wide variety of pinocytic mechanisms and the fusion of pinocytic vesicles with lysosomes (Fig. 2) have been elucidated by these studies. Especially significant have been the observations bearing on the pore systems in blood capillary endo-thelia (36-38, 41) and on membrane recirculation in nerve cells and other cells (42, 43).

DAB cytochemical techniques have contributed significantly to two other areas of cell biology, the intracellular localization of antigens and antibodies in situ and the visualization of cytoplasmic organelles.

3. Diffusion artifacts in DAB cytochemistry

I agree with Seligman et al. (31) that DAB is "a remarkably useful reagent for ultrastructural cytochemistry"; I had earlier spoken of it as a "remarkably reliable cytochemical reagent" (44). The reaction product is a homogeneous polymer, readily visible by light microscopy and electron microscopy. The large-scale, readily apparent, diffusion artifacts such as often occur in phosphatase cytochemistry are not evident when well-fixed tissue is incubated in DAB media. Yet diffusion artifacts on a smaller scale do occur, and these can be troublesome (45). Thus, membranes and ribosomes can stain artifactually (Fig. 5). Such artifacts may be observed in well-fixed cells if there is excessive accumulation of reaction product, and they are probably due to diffusion and adsorption of oxidized DAB (45). Adsorption artifacts might also result from diffusion of metalloprotein to adjacent structures. This is more likely to occur in inadequately-fixed tissue, particularly if kept in buffer for a prolonged period prior to incubation (46). In assessing the relative significance of various sources of artifact

(Fig. 4) the observations of Archer and Hirsch (47) may be relevant. They found that repeated freezing and thawing released most lyso-somal enzymes from the eosinophilic granules, but that peroxidase remained granule-bound.

Fahimi (46) now considers the staining of ribosomes adjacent to peroxisomes in rat liver, as previously reported by him and others (see ref. 45), to be an artifact due to diffusion of catalase from peroxisomes and its adsorption by adjacent ribosomes. Prolonged rinsing of the tissue in buffer is thought to cause the diffusion.

My views on the possibility of DAB diffusion artifact drew a reply from Seligman et al. (31) which contains unsupported assertions [4]. However, the authors do consider it possible that some DAB polymers produced during incubation may be small enough to diffuse and to "attach to acidic structures such as ribosomes". They state that unreacted DAB can also be adsorbed by such structures. Finally, they believe that "some hemoprotein may diffuse and bind to certain sites in spite of fixation".

4. DAB and the intracellular localization of antibodies, antigens, and receptor sites in situ

This section is brief because Nakane, Bernhard and other contri-butors to this volume will make evident the significant contribu-tions to these important areas of biological research.

Horseradish peroxidase was used by Leduc, Avrameas and Bouteille (48) to hyperimmunize rabbits and then to follow the appearance of antibody (anti-peroxidase) in immunocompetent cells. Sections of spleen were treated with peroxidase and then incubated in the DAB medium of Graham and Karnovsky.

---

[4] These are: (1) that en bloc staining of tissue with uranyl acetate and staining of thin sections with lead "should be avoided in evaluating the question of artifacts in enzyme ultrastructural cytochemistry"; (2) that "good morphology" is not retained when incubating tissue at high pH (9.0-9.7); see Fig. 12; (3) that Fahimi reported "ribosomal staining for peroxidase (our under-lining); (4) that the use of $MnCl_2$ is a "dangerous procedure"; (5) that "the numerous electron opaque materials" used in our methods account for the staining we attribute to oxidized DAB; see Fig. 8; (6) that glutaraldehyde is not superior to formaldehyde for "demonstrating cytochrome oxidase activity"; see Fig. 8; and (7) that we have confused the literature "by attributing the oxidative activity by organelles under one set of conditions to all condi-tions, when using DAB". Two of our statements are quoted out of context: (1) that "simply increase in polymer length may be ruled out as the cause of staining of unreactive sites", and (2) that Karnovsky had apparently withdrawn the claim that the Seligman et al. procedure was due to cytochrome oxidase; I was referring to his statement about the controls, not the animals injected with cytochrome c.

Fig. 1. Reproduction of figure 3 of Graham and Karnovsky (35).
Portion of a cell in the proximal convoluted tubule of the mouse
kidney, 3 min after intravenous injection of horseradish
peroxidase. Reaction product is seen in tubular invaginations (T)
and in an apical vacuole (AV). BB shows the brush border.
(Courtesy of Dr. Morris J. Karnovsky)

Fig. 2. Hepatocytes 16 hr after intravenous injection of bovine
hemoglobin into a rat (400 mg/100 g); incubated for acid
phosphatase activity followed by incubation in DAB, pH 7.6; from
(94). There is much less DAB reaction product, due to hemoglobin,
in the large droplet than in sections not previously incubated in
the acid phosphatase medium. The dense deposits (lead phosphate)
result from acid phosphatase activity. The hemoglobin droplets
have been transformed into digestive vacuoles by fusion with
lysosomes (L), probably dense bodies. The arrow indicates a
portion of GERL. A bile canaliculus is seen at BC.

Fig. 3. Frozen 10 μ sections of rat duodenum fixed 90 min, cold
3% glutaraldehyde-0.1 $\underline{M}$ cacodylate, pH 7.4. (a) Incubated 60 min,
37°, DAB, pH 8, 0.001% H_2O_2, 5 x 10 $^{-3}$ $\underline{M}$ $MnCl_2$ (ref. 29); (b)
Incubated 60 min, 37°, DAB, pH 9.7, 0.05% H_2O_2, 1 x 10 $^{-3}$ $\underline{M}$ KCN;
and (c) Incubated 90 min, 37°, in medium prepared as follows:
dissolve 20 mg DAB-tetra-HCl in 9.0 ml 0.05 $\underline{M}$ propanediol buffer,
pH 9.0; adjust to pH 9.7; and add 1.0 ml 0.01 $\underline{M}$ KCN. In (a)
arrows indicate regions of stained mitochondria. Such staining is
inhibited by the KCN in (b) and (c). In (a) and (b) erythrocytes
as well as eosinophils are intensely stained. In (b) some of the
erythrocytes are indicated by arrows. In (c), the omission of
H_2O_2, together with the presence of KCN, inhibits erythrocyte
staining. Thus, only the eosinophils are stained.

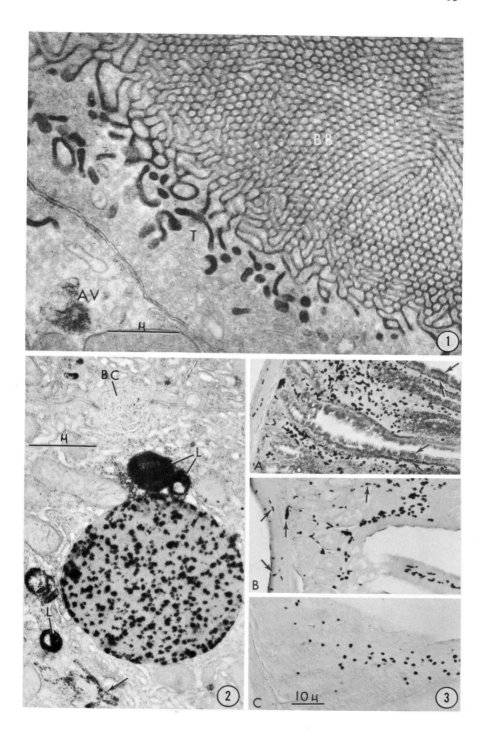

Fig. 4. Frozen 10 μ sections of rat liver fixed 3 hr, cold 2.5%
glutaraldehyde-0.1 M cacodylate buffer, pH 7.4. Upper two sections
incubated in the Burstone cytochrome oxidase medium (Sigma Technical
Bulletin No. 185); middle two in the Seligman et al. DAB medium,
and the lower two in the DAB, pH 6 medium. Incubations were at room
temperature for 30 min. The sections at the left were incubated
without cytochrome c additions. Only the DAB pH 6 shows light
staining of mitochondria; also intense staining of erythrocytes.
With added cytochrome c (right panel) staining due to mitochondria
is evident in both Burstone and Seligman et al. media; see
enlargements in Figs. 6 and 7. The mitochondrial staining in the
DAB, pH 6 medium is enhanced by cytochrome c; erythrocyte staining
is not.

Fig. 5. Portion of an eosinophil in rat liver, incubated in DAB,
pH 9.7; from (45). The regions marked by black X's do not have
reaction product whereas the regions shown by the white X's have
much reaction product. Either DAB reaction product or peroxidase
has diffused from the eosinophilic granules and has been adsorbed
to ribosomes. Adsorption is also seen in areas of the plasma
membrane (short arrows) and nuclear membrane (long arrow).

Fig. 6. A higher magnification of Fig. 4, Burstone procedure with
added cytochrome c. A portal area is marked by P and a central
vein by C.

Fig. 7. A higher magnification of Fig. 4, Seligman et al. procedure
with added cytochrome c. P and C, as in Fig. 6.

Fig. 8. Rat kidney, fixed in 4% glutaraldehyde-0.1 M cacodylate
buffer, pH 7.4, for 5 hr. Non-frozen section (25 μ) prepared
according to Smith and Farquhar (95) and incubated 50 min at 37° in
the medium of Seligman et al. with added cytochrome c (Beard and
Novikoff, unpublished). Tissue was not rinsed in uranyl acetate
and section was unstained. Note that all mitochondria show reaction
product (A) except for one which is negative (B); cf. ref. 96. The
indicated area is enlarged 1½ X in the Inset. Reaction product is
in the outer membrane (short arrow) and inner membrane (long arrow)
which is continuous with the cristae (C); cf. Fig. 9 of ref. 54.

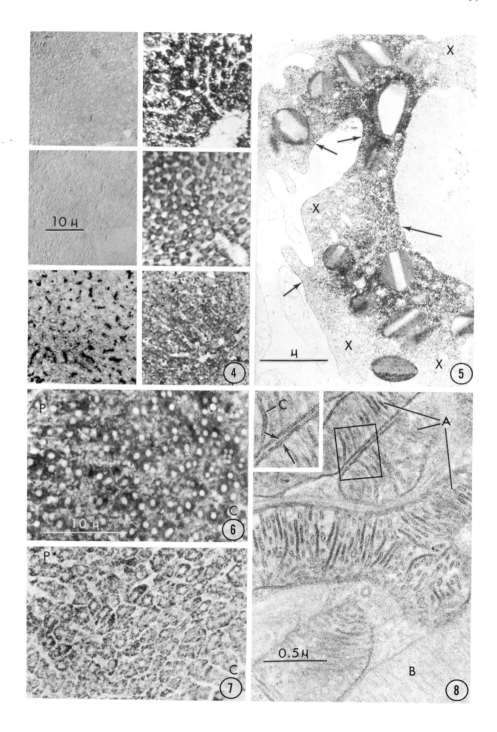

10 μ

X
X
X
X
μ
4
5

P
C
10 μ
6

P
C
7

C
A
0.5 μ
B
8

The coupling of horseradish peroxidase to specific antibody in order to detect antigenic sites was developed independently in the laboratories of Nakane and Avrameas. The literature is reviewed by Avrameas (49). Other contributions came from the laboratories of Spicer (50) and Sternberger (50a). A volume on immunocytochemistry deals extensively with this subject (51).

Recently the DAB procedure has been used to detect receptor sites for concanavalin A by Bernhard and Avrameas (52).

5. DAB and the visualization of cytoplasmic organelles

By varying the incubation conditions (concentration of DAB, temperature, pH, H_2O_2 levels and addition of activating ions) different organelles of animal and plant cells may be visualized (23, 29, 30, 16); see Fig. 3.

Mitochondria. Because of the insolubility of the reaction product in cell lipid and the absence of large-scale diffusion artifacts, DAB methods which visualize mitochondria in fixed tissues are much superior to tetrazolium or other cytochemical techniques, for light microscopy as well as for electron microscopy. DAB procedures have contributed to the demonstration of the functional heterogeneity of mitochondria within a single cell (Fig. 8); and to the confirmation (32, 30) of the "sidedness" of cytochrome c as suggested by Racker and his colleagues (53).

The DAB procedures for mitochondria have engendered controversy. In both title and text, Seligman et al. (32) gave the clear impression that it was cytochrome oxidase that was being localized (see, e.g., ref. 53a). In our laboratory it was found that the Seligman et al. medium stains mitochondria in glutaraldehyde-fixed tissues (Figs. 7 and 8), essentially in the same manner as our pH 6 medium (23, 54). We concluded that it is cytochrome c that is being localized.

My experience with cytochemical staining began at a time when the distinction between the stable Nadi M reaction and the labile Nadi G reaction was still being stressed. It was believed that cytochrome oxidase, the enzyme system which Nadi G proved to be, would not resist fixation. Gomori, in his 1952 book, wrote "cytochrome oxidase is a rather sensitive enzyme, readily destroyed even by drying and intolerant to formalin" (55).

However, in 1972 Reith and Schüler (56) demonstrated that oxygen consumption by isolated mitochondria was stimulated by the addition of DAB, and also by cytochrome c, thus indicating that DAB can function in the respiratory chain that includes cytochrome c and cytochrome oxidase. This has been confirmed by extensive studies by Cammer and Moore (57) and in unpublished work referred to by Seligman et al. (31). It therefore seemed desirable to test the assumption that cytochrome oxidase is sensitive to fixation.

Figs. 4, 6 and 7 show some results of experiments designed for this purpose. Rat liver and kidney were fixed as follows: cold 4% formaldehyde-1% $CaCl_2$ (prepared from paraformaldehyde) for 5 min, 10 min, 20 min and overnight; cold 2.5% glutaraldehyde (Ladd Co., Burlington, Vt.)-0.1 M cacodylate buffer, pH 7.4 (58), 3 hr; and cold 2.5% glutaraldehyde-2% formaldehyde-0.05% $CaCl_2$, 3 hr. Frozen sections were incubated according to the Burstone method for cyto-chrome oxidase (59), the Seligman et al. medium (32) and the pH 6 medium (23, 29). With formaldehyde up to 20 min mitochondrial staining occurs with all three media, even without added cytochrome c. Cytochrome c enhances the staining. However, these tissues, like unfixed tissue, are unsuitable for electron microscopy. Formaldehyde-glutaraldehyde and glutaraldehyde-fixed tissues, both adequate for electron microscopy (Fig. 8), are unstained by the Burstone "indophenol oxidase" and the Seligman et al. DAB procedures. With the addition of cytochrome c(1 mg/ml medium) mitochondrial activity is evident in rat liver and kidney sections, even at room temperature (Figs. 4, 6-8). Of the three procedures, the pH 6 procedure is the least sensitive to fixation, but it too gives deeper mitochondrial staining upon the addition of cytochrome c. After overnight formaldehyde fixation, mitochondrial activity is lost with all three procedures in the absence of cytochrome c. With added cytochrome c mitochondrial activity is evident in the thick limb of Henle and other regions of the kidney, but not in liver at the incubation periods tested (up to 90 min at room temperature). The added cytochrome c may have more ready access to the electron transport chain in the renal cells.

The mechanisms by which fixation makes cytochrome c unavailable for the oxidative reactions are unknown. Also unknown is the manner by which added cytochrome c gains access to the respiratory chain in the membrane, while itself not oxidizing the indophenol or DAB in the medium. It is relevant to note that the addition of cytochrome c which enhances mitochondrial staining in the pH 6

medium has no such effect upon the simultaneous (sub-optimal) staining of peroxisomes in the epithelial cells of rat liver and kidney.
The possibility that incubation at temperatures above 30° may affect
the availability of cytochrome c is raised by the results of Cammer
and Moore (57).

Because of crystal formation and dye solubility the indophenol
oxidase procedure is unsuitable for electron microscopy. The ultrastructural localizations with the Seligman et al. (Fig. 8) and pH 6
procedures (54, 23) are the same. Not only are the internal membrane
and its cristae stained but also the outer mitochondrial membrane is
stained.

In evaluating the extent to which either the Seligman et al. or
the pH 6 procedure requires cytochrome oxidase activity in tissue
sections, it should be stressed that DAB oxidation by cytochromes
occurs in membranes which lack cytochrome oxidase. One instance is
the outer mitochondrial membrane staining (Fig. 8) where probably
the cytochrome b_5-like component of the retinone-insensitive cytochrome c reductase and/or monamine oxidase (also a metalloprotein)
oxidize(s) the DAB. Another example is the DAB staining of various
cytochromes of the electron transport chain in chloroplasts (see
below).

Chloroplasts. In 1970 Nir and Seligman (60) showed that when
spinach or Elodea leaves were illuminated in a DAB medium at pH 7.0
the membranes of the grana and stroma were intensely stained. In
contrast, the outer chloroplast membrane was unstained. The authors
considered the observations insufficient to determine which of the
photosystems interacted with DAB.

Nir and Pease (61) used essentially the same DAB medium, with
spinach leaves, but the tissue was embedded in glutaraldehyde-urea
resin rather than Epon. According to the authors, such embedding
permits precise localization of the sites of photooxidation. DAB
oxidation is demonstrated to be on the inner side of the thylakoid
membrane, due to photosystem I. This coincides with the suggestion
of Arntzen et al. (62) that photosystem I is localized on the inner
side of the membrane and photosystem II on the opposite side.

An abstract by Vigil et al. (63) suggests that DAB can also be
oxidized by photosystem II.

Endoplasmic reticulum. A variety of mammalian cells, but a small minority, possess endogenous peroxidase activity in the ER which can be visualized with the original Graham-Karnovsky medium, with a more optimal medium (29, 30), or with other modifications. The cell types reported to 1971 are listed in ref. 30. To this list may be added the rat exorbital lacrimal gland (64, 65).

Perhaps the most interesting of the known ER peroxidases is that in the thyroid epithelium. It is generally assumed that the peroxidase thus revealed is the iodine peroxidase, responsible for incorporation of iodine into thyroglobin. ^{125}I autoradiography argues strongly that this incorporation occurs outside the cell, at its luminal surface (66). This would suggest, that except under unusual circumstances (67), the peroxidase is inactive in the ER (cf. discussion, above, of inactive tyrosinase). In our laboratory we have distinguished two types of granules, peroxidase-positive "A granules" and peroxidase-negative "B granules" (Fig. 10). It has been suggested (68, 69) that the A granules carry the peroxidase directly from the apical ER to the lumen, bypassing the Golgi apparatus and that the B granules carry uniodinated thyroglobulin from GERL to the lumen.

Recently we have extended the observations of Venkatachalam et al. (70) on ER peroxidase in the mucus cells in the crypts of rat colon (71). As is true with other cytochemical studies [e.g., see Max Wachstein's studies on kidney (72)] marked species differences exist. ER peroxidase activity is present in colon, caecum and rectum of the rat (Fig. 9) and mouse, and in the colon of the hamster. It is absent from the mucus cells of the caecum and rectum of the hamster, and all three regions of the large intestine in rabbit, dog, cat, guinea pig and also from a limited number of biopsies of human colon and rectum (71). This (also the absence of activity from the mucus-secreting cells in the surface epithelium) complicates speculations regarding the anti-bactericidal function of peroxidases in the large intestine (70) and elsewhere.

Peroxisomes and microperoxisomes. Elsewhere (8) we have discussed the remarkably parallel role cytochemistry has played in advancing our knowledge of lysosomes (staining for acid phosphatase activity) and of peroxisomes (alkaline DAB reactions for catalase). Both lysosomes and peroxisomes began as biochemical concepts of Christian de Duve and his collaborators. Lysosomes are now known to have important diverse roles in the physiology and pathology of cells; cytochemistry played a significant part in attaining this in-

Fig. 9. Frozen 10 μ section of rat rectum fixed 30 min, cold 1%
glutaraldehyde-2% formaldehyde, 0.1 $\underline{M}$ cacodylate buffer, pH 7.4;
incubated 30 min in DAB, pH 8, 0.001% H_2O_2, with 1 x 10^{-3} $\underline{M}$ KCN.
Mucous cells in crypts show intense ER staining. The lumen is
at the upper left (L). The stained cells in the lamina propria
are eosinophils (cf. Fig. 3).

Fig. 10. Apical region of an epithelial cell of rat thyroid
gland, fixed in 3% glutaraldehyde-0.1 $\underline{M}$ cacodylate buffer, 20
min perfusion and 80 min immersion. Non-frozen section
incubated for 120 min in DAB, pH 8, 0.001% H_2O_2; from (68).
Reaction product is present in "A granules" (A), ER including
GERL (GE). The "B granules" (B), nucleus (N) and Golgi stack
(to the left of GE) are negative.

Fig. 11. Frozen 10 μ section of guinea pig kidney, fixed 3 hr,
cold 2.5% glutaraldehyde-0.1 $\underline{M}$ cacodylate buffer, pH 7.4.
Incubated 90 min, DAB pH 9.7 medium; from (16). Numerous
microperoxisomes (short arrows) are evident in some cells of
the convoluted portion of the distal tubule (D). The
peroxisomes (long arrows) in the proximal convolutions (P) are
also seen more clearly when KCN eliminates the mitochondrial
staining (23).

Fig. 12. Portion of a hepatocyte in normal human liver; fixed
90 min, cold 2.5% glutaraldehyde-0.1 $\underline{M}$ phosphate buffer, pH 7.4.
Incubated 60 min, DAB pH 9.7 medium (16). A portion of a large
lipofuscin granule and an adjacent DAB-positive microperoxisome
are seen. Of the three constituents of the lipofuscin granule
(A, B, C) part C is DAB-positive (79). Note that the delimiting
membrane of the lipofuscin granule (arrow) is continuous with
the ER membrane (upper arrowhead); the ER membrane is also
continuous with the delimiting membrane of the microperoxisome
(lower arrowhead).

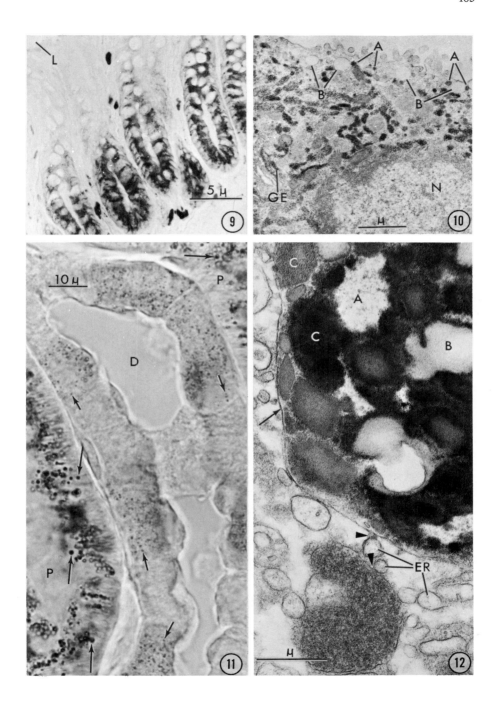

formation which is on the threshold of playing a role in medical
practice. The functions of peroxisomes in animal cells are beginning
to be revealed - both for hepatic peroxisomes, the original member
of the "family" (73) and the microperoxisomes, the recent addition
to the family $\overset{5}{\vee}$. It was largely the alkaline DAB media developed
in our laboratory (23, 16) which showed microperoxisomes to be
ubiquitous in mammalian cells and which helped: (a) to establish
the basis of identifying these organelles by their ultrastructure
alone and (b) to reveal the intimate relation of the organelles to
the endoplasmic reticulum, of which they may be considered to be
localized dilatations (8, 74-76). It is of interest that the ER
itself shows no alkaline DAB reactivity, even where the peroxisomes
are attached - by wide connections in the case of hepatic peroxisomes and numerous slender attachments in the case of microperoxisomes. The work of Lazarow and de Duve (77, 73) may provide an
answer: in rat hepatocytes, the catalytically-active catalase
appears to be formed from precursors within the peroxisome itself.
The heme moiety, essential for DAB reactivity, is apparently
attached only inside the peroxisome.

The DAB medium we use currently (16) contains 5×10^{-3} $\underline{M}$ KCN in
order to eliminate the background staining due to the mitochondrial
reaction, at suboptimal pH, obtained with earlier media. This makes
possible the light microscopic studies of many microperoxisomes
(Fig. 11). In the electron microscope, microperoxisomes have a
moderately opaque matrix, without a nucleoid, and are surrounded by
a tripartite membrane that has numerous slender continuities with
smooth ER.

Cytochemical studies have revealed the intimate structural rela-
tionships of microperoxisomes to lipids of various kinds (Fig. 12)
(16, 76, 78, 79, 80). These observations, together with suggestive
biochemical findings (81), indicate that these organelles may play a
role in lipid metabolism (see also refs. 82-84).

Other cell structures. The non-enzymatic staining with DAB of
structures containing nucleic acids (85) is of much potential value.

Other non-enzymatic DAB oxidations have been described for
melanin and norepinephrine (86). Similarly, portions of lysosomes
are sometimes stained during DAB incubations. Unusually intense
DAB staining of lysosomes at pH 8 and with a high H_2O_2 concentration
occurs in thyroid epithelial cells (30).

---

[5] Summations of the current knowledge of peroxisomes may be found in
the November 1973 J. Histochem. Cytochem.

Lysosomes of considerable interest are the lipofuscin granules which contain special lipids and which accumulate products of cell metabolism, including molecules with iron and other metallic ions. It is the iron-containing (87), acid phosphatase-rich (88) regions [which we refer to as the C component (79)] which oxidize DAB. In human hepatocytes the smooth ER is continuous with the delimiting membranes of both lipofuscin granules and microperoxisomes found adjacent to the granules (79) (Fig. 12).

6. Roles of the endoplasmic reticulum

As we were led by our acid phosphatase studies on lysosomes to emphasize the role of the ER, including GERL (89, 7), so DAB cytochemistry has also focussed our attention on the remarkably diverse roles of the ER in animal cells. As seen above, it has been suggested that thyroid peroxidase bypasses the Golgi apparatus and goes directly from apical ER to the follicular lumen, via A granules (Fig. 10).

From biochemistry it is known that microsomes, largely derived from ER, contain many enzymes of lipid metabolism. This is reflected morphologically in diverse ways. In mammalian hepatocytes small lipid droplets are moved through the ER or in ER-bound vesicles to the Golgi apparatus (90) and perhaps also to the cell exterior (space of Disse) (91). Lipid may accumulate in the cytosol as spheres with ER wrapped around them, almost completely [as after feeding rats ethyl-α-(p-chlorophenoxy) isobutyrate (CPIB) (92)] or less completely [as in untreated rats (92, 79)]. Under some conditions [after orotic acid (93) or in fetal rabbit liver (unpublished)] lipid spheres accumulate inside vesicles of ER. Jerome Nehemiah, in our laboratory, has recently described in hamster hepatocytes accumulation of lipid within dense body type lysosomes which retain their continuities with smooth ER. The term, lipolysosome, has been coined to describe the lipid-containing lysosomes. Lipofuscin may be considered as a member of the lipolysosome family. I have recently speculated (7, 79) that smooth ER may sweep up ferritin, glycogen and other substances from the cytosol into lipofuscin granules by microautophagy.

The observations on specialized areas of the ER, and the concepts and speculations to which they have led, were made possible by the cytochemical approaches and developments I have briefly described.

Acknowledgments

This work was supported by U. S. Public Health Service Research Grant R01-CA06576. The author is the recipient of U. S. Public Health Service Research Career Award 5K06-CA14923 from the National Cancer Institute.

I wish to acknowledge the participation in the recent work reported here of Phyllis M. Novikoff, Cleveland Davis, Nelson Quintana and Jerome Nehemiah. My thanks go also to Dr. Joseph C. Ehrlich for critical review of the manuscript, to Jack Godrich for preparation of the final photographs and to Fay Grad for superb secretarial work and typing of the manuscript.

References

1. Wachstein, M. and Meisel, E., 1957, Am. J. Clin. Path. 27, 13.

2. Novikoff, A.B., 1963, J. Cell Biol. 19, 88A.

3. Novikoff, A.B., 1963, in: Proceedings First International Congress of Histochemistry and Cytochemistry, Paris, 1960, (Pergamon Press) p. 465.

4. Novikoff, A.B., 1959, in: Subcellular Particles, ed. T. Hayashi (Ronald Press) p. 1.

5. de Duve, C., 1969, in: Lysosomes in Biology and Pathology, eds. J.T. Dingle and H.B. Fell (North-Holland, Amsterdam) p. 3.

6. Novikoff, A.B., 1963, in: Ciba Foundation Symposium on Lysosomes, eds. A.V.S. de Reuck and M.P. Cameron (Little, Brown & Co., Boston) p. 36.

7. Novikoff, A.B., 1973, in: Lysosomes and Storage Diseases, eds. G. Hers and F. Van Hoof (Acad. Press, New York) p. 1.

8. Novikoff, P.M. and Novikoff, A.B., 1972, J. Cell Biol. 53, 532.

9. Farquhar, M.G., Bergeron, J.J.M. and Palade, G.E., 1972, J. Cell Biol. 55, 72a.

10. Novikoff, A.B. and Masek, B., 1958, J. Histochem. Cytochem. 6, 217

11. Novikoff, A.B., Shin, W.-Y. and Drucker, J., 1961, J. Biophys. Biochem. Cytol. 9, 47.

12. Pearse, A.G.E., 1960, Histochemistry Theoretical and Applied, 2nd ed. (Little, Brown and Co., Boston).

13. Pearse, A.G.E., 1972, Histochemistry Theoretical and Applied, 3rd ed. (Williams & Wilkins, Baltimore).

14. Novikoff, A.B. and Essner, E., 1962, Fed. Proc. 21, 1130.

15. Novikoff, A.B., Essner, E., Goldfischer, S. and Heus, M., 1962, in: The Interpretation of Ultrastructure, ed. R.J.C. Harris (Acad. Press, New York) p. 149.

16. Novikoff, A.B., Novikoff, P.M., Davis, C. and Quintana, N., 1972, J. Histochem. Cytochem. 20, 1006.

17. Ernst, S.A., 1972, J. Histochem. Cytochem. 20, 13; 23.

18. Novikoff, A.B. and Goldfischer, S., 1961, Proc. Natl. Acad. Sci. 47, 802.

19. Novikoff, A.B. and Heus, M., 1963, J. Biol. Chem. 238, 710.

20. Yamazaki, M. and Hayaishi, O., 1968, J. Biol. Chem. 243, 2934.

21. Goldfischer, S., Essner, E. and Schiller, B., 1971, J. Histochem. Cytochem. 19, 349.

22. Bitensky, L., 1963, in: Lysosomes, eds. A.V.S. de Reuck and M.P. Cameron (Little, Brown & Co., Boston) p. 362.

23. Novikoff, A.B., Goldfischer, S., 1969, J. Histochem. Cytochem. 17, 675.

24. Hirai, K.-I., 1968, Acta Histochem. Cytochem. 1, 43.

25. Fahimi, H.D., 1969, J. Cell Biol. 43, 275.

26. Novikoff, A.B., Novikoff, P.M., Davis, C. and Quintana, N., 1972, J. Histochem. Cytochem. 20, 1006.

27. Herzog, V. and Fahimi, H.D., 1972, J. Cell Biol. 55, 113a.

28. Roels, F. and Wisse, E., 1973, C.R. Acad. Sci. Paris, 276, 391.

29. Novikoff, A.B., 1970, 7th Intl. Cong. of E.M., Grenoble, 1, 565.

30. Novikoff, A.B., Beard, M.E., Albala, A., Sheid, B., Quintana, N. and Biempica, L., 1971, J. Microsc. 12, 381.

31. Seligman, A.M., Shannon, W.A., Jr., Hoshino, Y. and Plapinger, R.E., 1973, J. Histochem. Cytochem. 21, 756.

32. Seligman, A.M., Karnovsky, M.J., Wasserkrug, H.L. and Hanker, J.S., 1968, J. Cell Biol. 38, 1.

33. Pearse, A.G.E., 1958, J. Clin. Path. 11, 520.

34. Novikoff, A.B., Albala, A. and Biempica, L., 1968, J. Histochem. Cytochem. 16, 299.

35. Graham, R.C. and Karnovsky, M.J., 1966, J. Histochem. Cytochem. 14, 291.

36. Clementi, F. and Palade, G.E., 1969, J. Cell Biol. 41, 33.

37. Feder, N., 1971, J. Cell Biol. 51, 339.

38. Venkatachalam, M.A. and Karnovsky, M.J., 1972, Lab. Invest. 27, 435.

39. Steinman, R.M. and Cohn, Z.A., 1972, J. Cell Biol. 55, 616.

40. Dvorak, A.M., Dvorak, H.F. and Karnovsky, M.J., 1972, Lab. Invest. 26, 27.

41. Simionescu, N., Simionescu, M. and Palade, G.E., 1973, J. Cell Biol. 57, 424.

42. Holtzman, E., Teichberg, S., Abrahams, S., Citkowitz, E., Crain, S.M., Kawai, N. and Peterson, E.R., 1973, J. Histochem. Cytochem. 21, 349.

43. Heuser, J.E. and Reese, T.S., 1973, J. Cell Biol. 57, 315.

44. Novikoff, P.M. and Novikoff, A.B., 1972, J. Cell Biol. 53, 532.

45. Novikoff, A.B., Novikoff, P.M., Quintana, N. and Davis, C., 1972, J. Histochem. Cytochem. 20, 745.

46. Fahimi, H.D., 1973, Fed. Proc. 32, 865A.

47. Archer, G.T. and Hirsch, J.G., 1963, J. Exptl. Med. 118, 277.

48. Leduc, E.H., Avrameas, S. and Bouteille, M., 1968, J. Exptl. Med. 127, 109.

49. Avrameas, S., 1970, Intl. Rev. Cytol. 27, 349.

50. Mason, T.E., Phifer, R.F., Spicer, S.S., Swallow, R.A. and Dreskin, R.B., 1969, J. Histochem. Cytochem. 17, 563.

50a. Sternberger, L.A., Hardy, P.H., Jr., Cuculis, J.J. and Meyer, H.G., 1970, J. Histochem. Cytochem. 18, 315.

51. Sternberger, L.A., 1974, Immunocytochemistry (Prentice-Hall, Englewood Cliffs, N.J.).

52. Bernhard, W. and Avrameas, S., 1971, Exptl. Cell Res. 64, 232.

53. Racker, E., 1970, in: Essays in Biochemistry, eds. P.N. Campbell and F. Dickens (Acad. Press, New York) p. 1.

53a. Roels, M.F., 1970, C.R. Acad. Sci. Paris, 270, 2322.

54. Beard, M.E. and Novikoff, A.B., 1969, J. Cell Biol. 42, 501.

55. Gomori, G., 1952, Microscopic Histochemistry, Principles and Practice (University of Chicago Press).

56. Reith, A. and Schüler, B., 1972, J. Histochem. Cytochem. 20, 583.

57. Cammer, W. and Moore, C.L., 1973, Biochem. 12, 2502.

58. Sabatini, D., Bensch, K. and Barrnett, R.J., 1963, J. Cell Biol. 17, 19.

59. Burstone, M.S., 1961, J. Histochem. Cytochem. 9, 59.

60. Nir, I. and Seligman, A., 1970, J. Cell Biol. 46, 617.

61. Nir, I. and Pease, D.C., 1973, J. Ultrastr. Res. 42, 534.

62. Arntzen, C.J., Dilley, R.A. and Crane, F.L., 1969, J. Cell Biol. 43, 16.

63. Vigil, E.L., Swift, H.H. and Arntzen, C., 1971, Abst. Am. Soc. Cell Biol., 314.

64. Essner, E., 1971, J. Histochem. Cytochem. 19, 216.

65. Herzog, V. and Miller, F., 1972, J. Cell Biol. 53, 662.

66. Nadler, N.J., 1965, Symp. Intl. Soc. Cell Biol., vol. 4, 303.

67. Tixier-Vidal, A., Picart, R., Rappaport, L. and Nunez, J., 1969, J. Ultrastr. Res. 28, 78.

68. Shin, W.-Y., Ma, M., Quintana, N. and Novikoff, A.B., 1970, 7th Intl. Cong. of E.M., Grenoble, 3, 79.

69. Novikoff, A.B. and Novikoff, P.M., 1973, Abst. Cytopharmacology of Secretion, Venice, 26.

70. Venkatachalam, M.A., Soltani, M.H. and Fahimi, H.D., 1970, J. Cell Biol. 46, 168.

71. Davis, C. and Novikoff, A.B., 1973, J. Histochem. Cytochem. 21, 412.

72. Wachstein, M., 1955, J. Histochem. Cytochem. 3, 246.

73. de Duve, C, 1973, J. Histochem. Cytochem., November (in press)

74. Novikoff, A.B., Novikoff, P.M., Davis, C. and Quintana, N., 1973, J. Histochem. Cytochem. 21, August (in press)

75. Hruban, Z., Vigil, E.L., Slesers, A. and Hopkins, E., 1972, Lab. Invest. 27, 184.

76. Novikoff, A.B. and Novikoff, P.M., 1973, J. Histochem. Cytochem., November (in press)

77. Lazarow, P.B. and de Duve, C., 1971, Biochem. Biophys. Res. Comm. 45, 1198.

78. Novikoff, P.M., Novikoff, A.B., Quintana, N. and Davis, C., 1973, J. Histochem. Cytochem. 21, 540.

79. Novikoff, A.B., Novikoff, P.M., Quintana, N. and Davis, C., 1973, J. Histochem. Cytochem., September (in press)

80. Leuenberger, P.M. and Novikoff, A.B. (in preparation)

81. Goldfischer, S., Roheim, P.S. and Edelstein, D., 1971, Science 173, 65.

82. Black, V.H. and Bogart, B.I., 1973, J. Cell Biol. 57, 345.

83. Reddy, J.K., 1973, J. Histochem. Cytochem., November (in press)

84. Goldfischer, S., Johnson, A.B., Essner, E., Moore, C. and Ritch, R.H., 1973, J. Histochem. Cytochem., November (in press)

85. Roels, F. and Goldfischer, S., 1971, J. Histochem. Cytochem. 19, 713.

86. Novikoff, A.B., Biempica, L., Beard, M.E. and Dominitz, R., 1971, J. Microsc. 12, 381.

87. Goldfischer, S., Villaverde, H. and Forschirm, R., 1966, J. Histochem. Cytochem. 14, 641.

88. Essner, E. and Novikoff, A.B., 1960, J. Ultrastr. Res. 3, 374.

89. Novikoff, P.M., Novikoff, A.B., Quintana, N. and Hauw, J.-J., 1971, J. Cell Biol. 50, 859.

90. Claude, A., 1970, J. Cell Biol. 47, 745.

91. Jones, A.L., Ruderman, N.B. and Guillermo Herrera, M., 1967, J. Lipid Res. 8, 429.

92. Novikoff, P.M., Roheim, P. and Novikoff, A.B., 1973, Fed. Proc. 32, 837 Abst.

93. Novikoff, A.B., Roheim, P. and Quintana, N., 1966, Lab. Invest. 15, 27.

94. Goldfischer, S., Novikoff, A.B., Albala, A. and Biempica, L., 1970, J. Cell Biol. 44, 3.

95. Smith, R.E. and Farquhar, M.G., 1965, RCA Sci. Instr. News 10, 13.

96. Beard, M.E. and Novikoff, A.B., 1969, J. Cell Biol. 43, 12a.

Electron microscopy and Cytochemistry, eds. E. Wisse, W.Th. Daems, I. Molenaar and P. van Duijn.
© 1973, North-Holland Publishing Company - Amsterdam, The Netherlands.

COLORIMETRIC AND CYTOCHEMICAL STUDIES FOR DETERMINATION OF OPTIMAL CONDITIONS FOR DEMONSTRATION OF CATALASE

V. Herzog and H.D. Fahimi

Harvard Pathology Unit, Mallory Institute of Pathology, Boston City Hospital, and Department of Pathology, Harvard Medical School, Boston, Massachusetts.

The influence of various parameters of fixation and incubation on visualization of the peroxidatic activity of catalase was reassessed using a recently developed colorimetric method for quantitation of the oxidation of 3,3'-diaminobenzidine (DAB) (2). This method is more sensitive than the o-dianisidine and the guaiacol procedures (3), and therefore more suitable for study of enzymes with low peroxidatic activity such as catalase.

1. Biochemical studies with Crystalline Beef Liver Catalase.

Crystalline beef liver catalase (0.5mg/ml) was treated for 30 minutes with various concentrations of glutaraldehyde and formaldehyde and was dialyzed against buffered saline at pH 7.0 for 48 hours at 4°C. The catalatic activity was determined according to the method of Lück (4). The pH optimum for the oxidation of DAB by aldehyde treated catalase was at pH 10.5. The peroxidatic activity was determined spectrophotometrically at this pH and at 450 nm, with 5×10^{-4}M DAB and 2×10^{-2}M H_2O_2. Under these conditions, the initial rate of reaction was linearly proportional to the amount of catalase protein.

Increasing concentrations of glutaraldehyde caused a gradual inhibition of the catalatic activity of catalase (Fig.1). Simultaneously, there was an increase of the peroxidatic activity, which reached its maximum after treatment with 6% glutaraldehyde. The treatment with formaldehyde caused slight inhibition of the catalatic activity and resulted, in comparison with glutaraldehyde, in a weak peroxidatic activity (Fig.1).

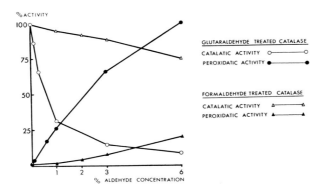

Figure 1:   Effect of various concentrations of glutaraldehyde and formaldehyde on the peroxidatic and catalatic activity of beef liver catalase.

2. <u>Cytochemical Studies with Isolated Rat Liver Peroxisomes</u>.

The significance of the described observations with glutaraldehyde for cytochemical demonstration of the peroxidatic activity of catalase was studied in isolated rat liver peroxisomes. Livers of male albino rats were homogenized in 0.25M sucrose and a crude peroxisomal fraction was obtained by differential centrifugation. Incubation of freshly prepared unfixed pellets for demonstration of peroxidatic activity of catalase (1) revealed no reaction product in peroxisomes (Fig.2). After the fixation in glutaraldehyde, however, a strong reaction was noted over the matrix of peroxisomes (Fig.3). The pH optimum for the oxidation of DAB by glutaraldehyde-fixed peroxisomal fractions, determined spectrophotometrically was at pH 10.5.

3. <u>Improved Conditions for Visualization of Catalase Used as Tracer</u>.

Since the original procedure for visualization of beef liver catalase, used as a tracer, is performed at pH 8.5 (5), and since our biochemical studies indicated that optimal pH for the oxidation of DAB by glutaraldehyde-treated catalase is at pH 10.5, we studied the importance of this observation in cytochemical preparations. Thirty minutes after intravenous injection of beef liver catalase, kidneys were fixed in glutaraldehyde and incubated at $22^{\circ}C$ in 1 mg/ml DAB with 0.02% H_2O_2 at pH 8.5,9.5,10.5 and 12 for up to 60 minutes. The intensity of reaction in glomerular capillaries was weakest at pH 8.5 (Fig.4), and increased slightly at pH 9.5 and was strongest at pH 10.5 (Fig.5). No reaction was observed after incubation at pH 12.

These findings contribute to the understanding of the mechanism of staining of catalase with DAB and demonstrate the usefulness of quantitative studies with DAB for assessment of optimal conditions of staining for enzymes with peroxidatic activity.

References

1. Fahimi, H.D., J. Cell Biol. 43,275, 1969.
2. Fahimi, H.D. and Herzog, V., J. Hisochem. Cytochem. 21,499, 1973.
3. Herzog, V. and Fahimi, H.D., Analyt. Biochem. (1973, In press).
4. Lück, H., in: Methods in Enzyme Analysis (H.U. Bergmyer, Ed.,) Verlag Chemie.
5. Venkatachalam,M.A. and Fahimi H.D., J. Cell Biol. 42,480, 1969.

Supported by N.I.H. grant NS 08533. Dr. Herzog is supported by Deutsche Forschungsgemeinschaft. Dr. Fahimi is the recipient of a Research Career Development Award from N.I.H., Bethesda, Maryland.

Figure Legends

<u>Figures 2 and 3</u> demonstrate the <u>effect of aldehyde fixation</u> on staining of catalase in peroxisomes. Crude peroxisomal fraction incubated in the DAB medium before (Fig. 2) and after (Fig.3) the fixation with glutaraldehyde.

<u>Figures 4 and 5</u> illustrate the <u>effect of pH</u> on staining of beef liver catalase used as tracer in mouse kidney. Note the difference in the intensity of reaction after incubation at pH 8.5 (Fig.4) and pH 10.5 (Fig.5).

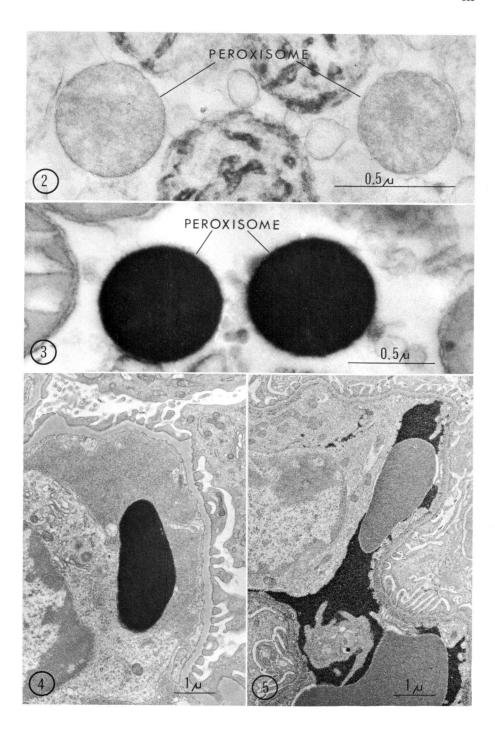

PEROXISOME

0.5 μ

②

PEROXISOME

0.5 μ

③

④ 1 μ

⑤ 1 μ

Electron microscopy and Cytochemistry, eds. E. Wisse, W.Th. Daems, I. Molenaar and P. van Duijn.
© 1973, North-Holland Publishing Company - Amsterdam, The Netherlands.

CYTOCHEMICAL DISCRIMINATION BETWEEN PEROXIDASES AND
CATALASES USING DIAMINOBENZIDINE

F. Roels, E. Wisse, B. De Prest, J. van der Meulen

Laboratorium voor Menselijke Anatomie, Rijksuniversiteit
Gent, Belgium ; and Laboratorium voor Electronenmicroscopie,
Rijksuniversiteit Leiden, The Netherlands

Summary. - When cells are incubated before fixation at pH 7,3 and
23°C with 0,003 % H_2O_2, peroxidases give a visible cytochemical stain,
and catalases do not. By prior fixation, and incubation at pH 9 or
9,7 and 37° with 0,03 to 0,07 % H_2O_2, catalase and peroxisome staining is strongly promoted. By comparing the results of incubations
carried out in either conditions, the enzymes can be discriminated.

Peroxidation of diaminobenzidine (DAB) can be produced by microbodies (peroxisomes) (16, 22) and crystalline catalase (13, 22), as
well as by peroxidases (10, 13, 17). As catalase is the marker enzyme for peroxisomes, while peroxidases have been located in several
other organelles, any study of either subject asks for a discrimination between the two enzymes. Thus far this was possible with the
available biochemical techniques, but not yet with DAB, the cytochemical reagent now widely in use. 3-Amino-1,2,4-triazole is of little
help in this respect, because it may inhibit both enzymes (6, 7, 11,
13, 17, 22) ; so does 2,6-dichlorophenolindophenol (18). We have
investigated the influence of prior aldehyde fixation, pH, H_2O_2 concentration and temperature on the DAB reaction in a number of systems
where the presence either of peroxidases, or catalases, had been as-
certained by biochemical methods. Partial results have been published (19).

Materials and methods. - The following materials have been examined :
crystalline catalase from bovine liver (Boehringer) and horseradish
peroxidase II (Sigma) adsorbed on filter paper ; fixed and unfixed
small blocks cut with a razor blade, cryostat sections and organprints
on coverslips of rat liver and kidney the microbodies of which are
known to be rich in catalase (2,4,5) ; also 50 µ Vibratomesections of
liver, and perfusion incubated liver (23) ; mouse liver and kidney,
the large granule fraction of which also contains sedimentable catalase (1, 15) ; small blocks and cryostat sections of rat thyroid, parotid and submaxillary glands where peroxidases but no catalase could
be demonstrated (17) ; air dried smears of heparinized human and rat
blood, and eosinophils in small blocks of rat stomach wall, the presence of peroxidases being known in large granules of neutrophils
(21) and eosinophils (3).

The fixatives used were : 3 % distilled glutaraldehyde (1,5 % for
perfusion fixation and blood smears) buffered with 0,1 M cacodylate
pH 7,4 containing 0,1 % $CaCl_2$; 4% depolymerized paraformaldehyde buffered with cacodylate + 1 % $CaCl_2$; and glutaraldehyde-formaldehyde
mixtures. For electron microscopy tissues were postfixed in 1 %
OsO_4, washed carefully, block stained with 0,5 % UAc at pH 5,0 for
30 min, and embedded in Epon. Ultrastructural preservation of and
localization of reaction product in tissues incubated prior to fixation are satisfactory.

DAB media (20 mg/10 ml) were freshly prepared in 0,05 M propanediol or Trisbuffer containing 7,5 % sucrose, or in Ringer solution
buffered with 0,025 M ; the pH was checked and H_2O_2 added immediately
before use. The concentration of the H_2O_2 stock solution was determined by the $KMnO_4$ method.

Results and discussion. - Prior glutaraldehyde fixation markedly stimulates staining of crystalline catalase. Stimulation of microbody staining too is observed when unfixed and unfrozen blocks, sections or prints, incubated at roomt° in DAB pH 9 + 0,035 % H_2O_2, are compared to controls preincubated in buffer for the same time, then fixed and incubated in DAB. With cryostat sections, or at 37°, the results are negative or equivocal. This may be partly due to diffusion of enzyme : when unfixed liver and kidney is kept in propanediol- or Trisbuffer + sucrose at 37°, or after freeze-thawing twice, catalase appears in the medium, as evidenced by the decomposition of H_2O_2 added to the supernatant ; at the same time microbody staining after subsequent fixation markedly decreases. This confirms earlier observations on mouse liver (1). Hence, absence of microbody staining in unfixed material, as reported (12, 14, 20), can be due to diffusion of enzyme. To eliminate this factor, controls as described above are necessary.

Prolonged incubation at roomt° and pH 9 + 0,035 % H_2O_2, or incubation at 37°, of blocks, sections and prints sometimes results in microbody staining without prior fixation.

Staining of crystalline catalase and microbodies in the 4 tissues is further promoted by increasing pH (tested up to 9,7) by increasing H_2O_2 concentration (tested up to 0,07 %) and by incubation at 37°C. No reaction is visible within microbodies after 90 min incubation of unfixed material at pH 7,3 and 23°C with 0,003 % H_2O_2 (Fig. 1 and 2).

In contrast, all peroxidases stain readily at roomt° without prior fixation ; reaction product is localized as described by previous authors, in the E.R., perinuclear cisterna, secretion granules, A-granules in thyroid, and azurophil and eosinophil granules in leucocytes. At pH 7,3 there is more reaction product that at 9, except for eosinophil granules where no difference is apparent. 0,035 % H_2O_2 is never stimulative compared to 0,003 %, and it strongly inhibits thyroid P.O. So, optimal peroxidase localization can be obtained in those conditions that exclude catalase and peroxisome staining (Fig. 3 and 4).

The influence of fixatives on peroxidases is variable. Salivary gland P.O. is particularly sensitive to glutaraldehyde, while formaldehyde is less inactivating in all three glands. When peroxidase containing tissues are fixed in glutaraldehyde and incubated at pH 9,7 and 37°C with 0,07 % H_2O_2, no uniform pattern is observed : with thyroid and salivary gland peroxidases the reaction is negative ; with HRPO and myeloperoxidase (neutrophils) it is decreased in comparison to optimal peroxidase conditions, and in eosinophil granules it is increased.

One should bear in mind that DAB staining can also be elicited by lipofuscins (9), myoglobin and hemoglobin (8, 9) (thermostable reactions), and by mitochondria.

Fig. 1 - 4. Small blocks incubated before fixation in DAB pH 7,3 + 0,003 % H_2O_2 at roomt°. Fig. 1 : Rat liver, 90 min DAB. Fig. 2 : Rat kidney, 1 h DAB. Microbodies (with visible core) show no reaction product. Counterstained with U and Pb. Fig. 3 : Acinar cells of rat submaxillary gland, 1 h DAB. Peroxidase localization in ER, perinuclear cisterna and secretion granules. Section not counterstained. Fig. 4 : Peroxidase localization in the secondary granules of rat eosinophils ; the central core remains negative. 30 min DAB, no counterstain. Mitochondrial membranes are reactive in all 4 preparations. Bar = 1 μ.

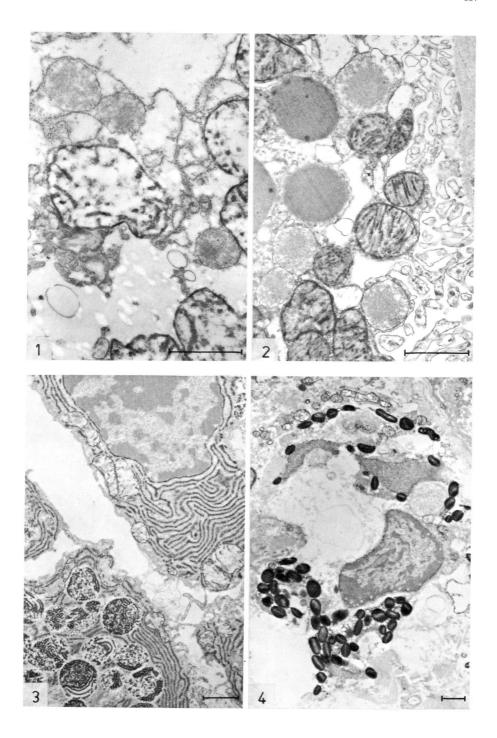

118

References

1. Adams, D.H. and Burgess, E.A., 1960, Biochem. J. 77, 247.
2. Allen, J.M., Beard, M.E. and Kleinbergs, S., 1965, J. Exp. Zool. 160, 329
3. Archer, G.T. and Hirsch, J.G., 1963, J. Exp. Med. 118, 277.
4. Baudhuin, P., Beaufay, H. and de Duve, C., 1965, J. Cell Biol. 26, 219.
5. Baudhuin, P., Müller, M., Poole, B. and de Duve, C., 1965, Biochem. Biophys. Res. Commun. 20, 53.
6. Castelfranco, P., 1960, Biochim. Biophys. Acta 41, 485.
7. Evans, W.H. and Rechcigl, M. Jr., 1967, Biochim. Biophys. Acta 148, 243.
8. Goldfischer, S., 1967, J. Cell Biol. 34, 398.
9. Goldfischer, S., Villaverde, H. and Forschirm, R., 1966, J. Histochem. Cytochem. 14, 641.
10. Graham, R.C. and Karnovsky, M.J., 1966, J. Histochem. Cytochem. 14, 291.
11. Heim, W.G., Appleman, D. and Pyfrom, H.T., 1956, Amer. J. Physiol. 186, 19.
12. Herzog, V. and Fahimi, H.D., 1972, J. Cell Biol. 55, 113a.
13. Herzog, V. and Miller, F., 1972, Histochemie 30, 235.
14. Hirai, K.I., 1968, Acta Histochem. Cytochem. 1, 43.
15. Holmes, R.S., 1972, Febs Letters 24, 161.
16. Novikoff, A.B. and Goldfischer, S., 1968, J. Histochem. Cytochem. 16, 507.
17. Novikoff, A., Beard, M.E., Albala, A., Sheid, B., Quintana, N. and Biempica, L., 1971, J. Microscopie 12, 381.
18. Roels, F., 1973, C.R. Acad. Sc. Paris, D, (in press).
19. Roels, F. and Wisse, E., 1973, C.R. Acad. Sc. Paris, D, 276, 391.
20. Sies, H., Herzog, V. and Miller, F., 1972, Proc. 5th European Congress Electron Microscopy, Manchester, p.274.
21. Schultz, J. and Kaminker, K., 1962, Arch. Biochem. Biophys. 96, 465.
22. Venkatachalam, M.A. and Fahimi, H.D., 1969, J. Cell Biol. 42, 480.
23. Wisse, E., Giphart, M., van der Meulen, J., Roels, F., Emeis, J.J. and Daems, W.Th., 1972, Proc. 5th European Congress Electron Microscopy, Manchester, p. 272.

Supported by the Belgian "Nationaal Fonds voor Wetenschappelijk Onderzoek".

First author's address : Ledeganckstraat 35, B-9000 Gent, Belgium.

Electron microscopy and Cytochemistry, eds. E. Wisse, W.Th. Daems, I. Molenaar and P. van Duijn.
© 1973, North-Holland Publishing Company - Amsterdam, The Netherlands.

PEROXIDATIC REACTION OF KUPFFER AND PARENCHYMAL CELLS
COMPARED IN RAT LIVER

E.Wisse, F.Roels[*], B.De Prest[*], J.v.d.Meulen, J.J.Emeis and W.Th.Daems

Laboratory for Electron Microscopy, Rijnsburgerweg 10, Leiden, The Netherlands,
and [*]Laboratory of Human Anatomy, Ledeganckstraat 35, Gent, Belgium.

Summary. Kupffer and parenchymal cells react differently in DAB-
containing media. The time required for the appearance of a visible
reaction product was studied as a measure of enzyme activity. Vibratome sections of 40 sec perfusion fixed rat livers were subjected to
time-series incubation. The composition of a "catalase" medium containing an elevated concentration of H_2O_2 under high pH and temperature conditions was determined on the basis of various modifications
of the "peroxidase" medium. In the peroxidase medium the Kupffer
cells in both fixed and unfixed tissue react positively after 20
min; in parenchymal cells the reaction product appears after 100
min. In the catalase medium the Kupffer cells do not react; the
peroxisomes of the parenchymal cells are positive in fixed tissue
after only 20 min incubation, the reaction being absent in unfixed
tissue. It is concluded that the peroxidatic reaction characteristics of Kupffer and parenchymal cells differ greatly, and that a
peroxidase is present in Kupffer cells.

Introduction. A diaminobenzidine (DAB) medium for the demonstration of peroxidatic activity, introduced by Graham and Karnovsky[2]
and modified by Novikoff and Goldfischer[3], was applied to liver tissue fixed by immersion in 3% glutaraldehyde for 3 hr. Frozen sections of this material were incubated at pH 9.0 and 37° C in a medium
containing 0.2% DAB and 0.02% H_2O_2. Microbodies in the parenchymal
cells, which are known to contain catalase, reacted positively.
Roels and Wisse[4,5] showed that prior fixation is needed for a positive reaction of catalase in DAB media, the peroxidase being already
reactive in the unfixed state. After perfusion incubation[7] of livers
with the DAB medium, reaction product was found in Kupffer cells;
the parenchymal cells were negative (Fig. 2). The experiments were
continued to investigate the characteristics of Kupffer and parenchymal cell peroxidatic activity by varying the fixation and incubation conditions.

Materials and methods. Rat livers were prepared according to the
following scheme:
Pre-perfusion: buffered physiological saline, 20 sec;
                        pH 7.4; mOsmol 300; room temp.; flow 10 ml/min.
Perfusion-fixation: 1.5% glutaraldehyde, 0.075M cacodylate, 40 sec;
                        pH 7.4; mOsmol 300; room temp.; flow 10 ml/min.
Perfusion-washing: 0.05M Tris-HCl, 7% sucrose, 60 sec;
                        pH 7.4; mOsmol 300; room temp.; flow 10 ml/min.
Tissue blocks (1 mm^3) were excised, mounted in the Vibratome, and
sectioned (10-20µ) while immersed in Tris-sucrose buffer at 4° C.
Thirty minutes after the start of fixation a large number of sections were incubated simultaneously in 25 ml of the peroxidase medium,
which contained: 0.05% DAB, 0.02% H_2O_2, 0.05M Tris-HCl, 7% sucrose
(pH 7.4; mOsmol 300; 25° C; shaking waterbath). At 20 min intervals
one section was washed in Tris-sucrose buffer, mounted in Hydramount
and viewed in the light microscope. For electron microscopy the sections were post-fixed in 1% phosphate buffered OsO_4. The moment of
appearance of reaction product was first recorded light-microscopic-

ally for both Kupffer and parenchymal cells and then checked in the electron microscope. The following variations were studied: unfixed liver tissue, liver tissue fixed by perfusion for 40 sec and 30 min, several concentrations of DAB (0.025-0.2%) and H_2O_2 (0.001-1.0%), different pH values (6.0-9.7), and two temperatures (25° and 37° C). Above pH 9.0 propanediol buffer was used.

Results. At the shortest incubation time (20 min) the Kupffer cells showed a positive reaction in the nuclear envelope, rough endo plasmic reticulum, and annulate lamellae (see also Fig. 2). In the light microscope, the reaction product appeared as a diffuse light-brown staining of the cytoplasm and gave a characteristic delineation of the nucleus. Some erythrocytes remaining in the sinusoids also reacted positively. The parenchymal cell reaction commenced at 100 min. Tissue fine structure proved to be good after only 40 sec of fixation. Visible reaction product appeared in Kupffer cells (Kc) and parenchymal cells (Pc) after the following intervals (in min):

	Kc	Pc			Kc	Pc
unfixed	20	neg	0.001%	H_2O_2	80	140
control*)	20	100	0.1%	H_2O_2	80	120
fixed for 30 min	60	60	1.0%	H_2O_2	neg	100
0.025% DAB	20	125	pH 6.0		20	160
0.10% DAB	20	100	pH 8.0		20	80
0.20% DAB	20	80	pH 9.0		60	60
37° C	20	40	pH 9.7		neg	40

*) fixed for 40 sec; 0.05% DAB; 0.02% H_2O_2; pH 7.4; 25° C.

It is evident that the Kupffer cell reaction was reduced by longer fixation, higher H_2O_2 concentration, and high pH values, and that the peroxidatic reaction of the peroxisomes in the parenchymal cells was enhanced by the same factors. After the incubation of unfixed tissue the peroxisomes showed no reaction.

On the basis of these observations we composed a medium combining most of the factors found to have a stimulatory effect on the peroxisomal reaction; this medium contained 0.05% DAB, 0.1% H_2O_2, 0.05M propanediol buffer (pH 9.7), and 5% sucrose, applied at a temperature of 37° C. Time-series incubations in this "catalase" medium of Vibratome sections of briefly fixed livers showed negative Kupffer cells (at 120 min) and positive peroxisomes (at 20 min)(Fig.3); the

Fig.1 Light micrograph (transmitted dark field illumination) of a 20 μ Vibratome section incubated for 20 min in the peroxidase medium. The reaction product, present in Kupffer cells (Kc) and absent in parenchymal cells (Pc), possesses light-scattering properties permitting easy recognition of the Kupffer cells by their brightness on a dark ground. In this material the shape and distribution of the Kupffer cells is clearly distinguishable in the light microscope due to their selective staining by the peroxidase medium. s = sinusoids.

Fig.2 Perfusion incubation[7] of liver tissue for the demonstration of peroxidatic activity gives reaction product in the RER cisternae (arrows) and nuclear envelope (asterisk) of the Kupffer cell (Kc). Peroxisomes (mb) of the parenchymal cell (Pc) are negative. Vibratome sections incubated in the peroxidase medium showed the same reaction pattern after incubation times of up to 80 min. SD = Space of Disse.

Fig.3 Vibratome section incubated for 20 min in the catalase medium, showing positive peroxisomes (mb) in a parenchymal cell (Pc). In this material the Kupffer cells did not show reaction product. bc = bile capillary.

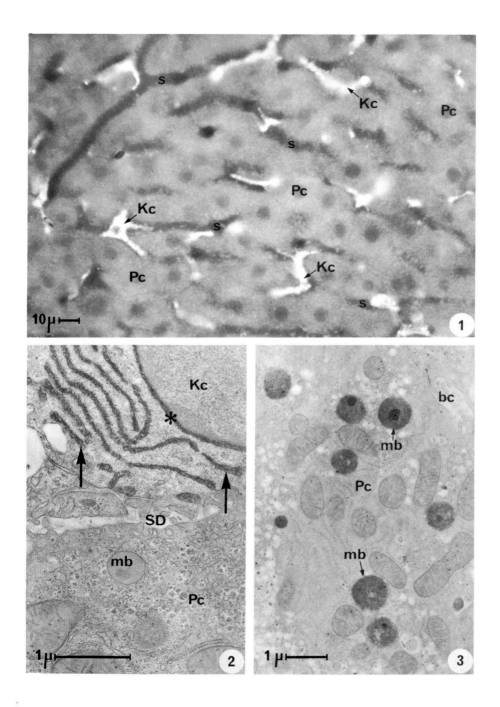

erythrocytes were also positive at 20 min. Vibratome sections of unfixed livers showed only positive erythrocytes, the Kupffer and parenchymal cells both being negative.

Discussion. Time-series incubations gave more clearly differenti-
ated results than the use of a single standard incubation time, the
latter providing only an all-or-none effect or a basis for comparison of amounts of reaction product. The appearance time of visible
reaction product proved to be a good parameter for the comparison of
the peroxidatic activity of Kupffer and parenchymal cells. An attempt
was made to minimize several sources of artifacts. The sections were
prepared such that the enzymes were kept as much as possible under
physiological conditions. Thin Vibratome sections (10-20µ) gave fewer
diffusion difficulties of medium components. Since both types of cell
are present in one section, the incubation circumstances are identical. The short fixation time was chosen to minimize enzyme inactivation by the fixative and to approach the natural unfixed state as
closely as possible. Inactivation seems to reduce the reaction in
Kupffer cells after 30 min of fixation.

There is a distinct reverse effect of increased fixation time,
H_2O_2 concentration, and pH level on the peroxidatic activity, the
parenchymal cells being stimulated and the Kupffer cells inhibited.
Unfortunately, stimulation of the Kupffer cell reaction could not be
recorded, because the cells of the control were already positive at
the shortest incubation time. The reverse effects, together with the
findings in unfixed tissue, permit histochemical distinction between
the peroxidatic activity of peroxidase and catalase[4,5], the Kupffer
cells possessing peroxidase, the parenchymal cells catalase.

As a practical advantage of the method, the peroxidase medium can
be used as a specific Kupffer cell staining, permitting recognition
of these cells even at the light microscopical level, since the endo-
thelial, fat-storing, and parenchymal cells do not react (Fig.1).

With 40 sec of fixation the parenchymal cell peroxisomes react
positively in the catalase medium, whereas unfixed ones are negative.
Even this short fixation time must therefore be considered responsible for the inactivation of the catalatic activity of the catalase
molecule in favor of the peroxidatic activity on the same molecule[1].

References.
1. Feinstein,R.N., Savol,R. and Howard,J.B., Enzymologia 41, 345
   (1971).
2. Graham,R.C. and Karnovsky,M.J., J.Histochem.Cytochem. 14, 291
   (1966).
3. Novikoff,A.B. and Goldfischer,S., J.Histochem.Cytochem. 17, 675
   (1969).
4. Roels,F. and Wisse,E., Compt.Rend.Acad.Sci.Paris 276,391(1973).
5. Roels,F., Wisse,E., De Prest,B. and v.d. Meulen,J., this proceedings.
6. Wisse,E., J.Ultrastruct.Res. to be published (1974).
7. Wisse,E., Giphart,M., v.d. Meulen,J., Roels,F., Emeis,J.J. and
   Daems,W.Th., Proc.5th.Eur.Regional Conf.Electron Microscopy,
   Manchester, p.272 (1972).

Electron microscopy and Cytochemistry, eds. E. Wisse, W.Th. Daems, I. Molenaar and P. van Duijn.
© 1973, North-Holland Publishing Company - Amsterdam, The Netherlands.

QUANTITATIVE ASPECTS OF CYTOCHEMICAL PEROXIDASE REACTIONS WITH
3,3'-DIAMINOBENZIDINE AND 5,6-DIHYDROXY INDOLE AS SUBSTRATES

M.van der Ploeg, J.G.Streefkerk, W.Th.Daems, and P.Brederoo

Department of Histochemistry, University of Leiden;
Department of Histology, Free University of Amsterdam;
Laboratory for Electron Microscopy, University of Leiden

Summary.Cytochemical procedures for the demonstration of peroxidase
activity were studied in a model system consisting of polyacrylamide films into which enzyme was incorporated. The quantities of dye
formed with 3,3'-diaminobenzidine (DAB) or 5,6-dihydroxy indole
(DHI) as the chromogenic substrates proved, under given conditions,
to be a measure for the activity of the enzyme. The relative turnover number of DHI is calculated to be about 60 times higher than
that of DAB. At the ultrastructural level, too, DHI shows certain
advantages over DAB; in addition, DHI appears to discriminate between the endogenous peroxidase of guinea pig resident peritoneal
macrophages and exudate monocytes.

Introduction.Reliable and sensitive cytochemical peroxidase methods
are prerequisites for investigations on the distribution and function of endogenous peroxidases in different types of cells. The
same holds for immuno-enzyme cytochemistry using peroxidase as the
marker enzyme and for studies of protein transport with peroxidase
or methemoglobin as tracer. Such investigations would gain in value
if the cytochemical procedures determined enzymatic activity not
only qualitatively but also quantitatively.

The study of a cytochemical staining reaction in biological objects is difficult, because the chemical composition of biological
material is often complicated and insufficiently known. But test
tube experiments are not completely relevant either, due to the
differences between (heterogeneous) cytochemical and (homogeneous)
biochemical systems. However, a model system consisting of polyacrylamide films containing the enzyme (9), permits evaluation of
the reliability of cytochemical staining procedures under conditions
approximating those prevailing when sections or smears are stained.
Since 3,3'-diaminobenzidine (DAB) (4) and 5,6-dihydroxy indole (DHI)
(2,3,8) are chromogenic substrates with suitable properties convenient for light and electron microscopical peroxidase cytochemistry, model system studies were performed to investigate the applicability of these compounds for quantitative purposes. The results were
compared with those obtained in electron microscopical enzyme cytochemistry.

Material and methods.Polyacrylamide films (9) incorporating purified horseradish peroxidase (HRP) or a guinea pig neutrophilic
granulocyte homogenate containing myeloperoxidase were used. The
homogenate was prepared from exudate harvested from the peritoneal
cavities of guinea pigs 16 hours after intraperitoneal injection of
20 ml 0.6M NaCl solution. HRP was obtained from Sigma (types II and
VI) and Boehringer (nr.II). Reactions with DAB were carried out at
room temperature, according to Graham and Karnovsky (4). Incubation
with DHI was performed at room temperature in 0.05M tris-HCl buffer
(pH 7.6), the final H_2O_2 concentration being 0.01% (w/v) (7). DHI
solutions were prepared from DL 3,4-dihydroxy phenylalanine according to the method described elsewhere (8). The solution can be
stored in small portions under N_2 at $-20^{\circ}C$ for several years.

For EM cytochemical studies, guinea pig peritoneal cells were harvested 0, 16, and 30 hr after intraperitoneal injection of 10 ml 0.15M NaCl solution. The cells were fixed in 1.5% cacodylate-buffered glutaraldehyde solution (pH 7.4; 330 mOsm; 4°C) for 10 min, washed, and incubated in media containing either DAB (see 1) or DHI, at pH values of 7.4 (tris-HCl buffer 0.05M), 9.0 or 10.5 (ammediol buffer 0.05M). Post-fixation was performed for 10 min in 1% osmium tetroxide in 0.1M phosphate buffer (pH 7.2; 350 mOsm; 4°C).

<u>Results and discussion</u>. Model experiments showed that under specific conditions (e.g. enzyme concentrations not too high in relation to substrate concentration or incubation time) a linear relationship between peroxidase activity and the amount of reaction product can be obtained for DAB (Fig.1). Similar observations resulted from experiments with incubation media containing DHI (7). For films with

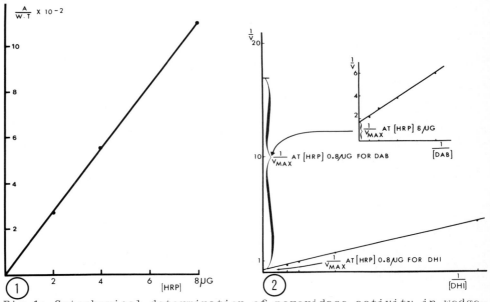

<u>Fig.1</u>: Cytochemical determination of peroxidase activity in wedge-shaped films containing various amounts of HRP stained for different periods with DAB. The curve shows the relationship between the measured activity ($\frac{A}{w \cdot t}$) and the HRP concentration.

<u>Fig.2</u>: Lineweaver-Burk plots for films containing equal concentrations of HRP stained with DAB or DHI, showing that V_{max} for DHI exceeds V_{max} for DAB.

<u>Fig.3</u>: A cell of the plasmacellular series in rat lymph node containing antibody against human IgG, demonstrated with a direct immunofluorescence technique using FITC-labeled human IgG.

<u>Fig.4</u>: The same cell, visualized immuno-enzymatically by the use of human IgG labeled with HRP, which was stained with DAB.

<u>Fig.5</u>: Detail of a guinea pig eosinophilic granulocyte after incubation in a DHI-containing medium (pH 7.4) for 60 minutes.

<u>Figs.6-9</u>: Details of guinea pig resident peritoneal macrophages after incubation for 60 minutes in:

    Fig.6 DAB, pH 7.4 (inset: monocyte granule); Fig.7 DAB, pH 10.5;
    Fig.8 DHI, pH 7.4 (inset: monocyte granule); Fig.9 DHI, pH 10.5.

Note the presence of a microperoxisome-like body in Fig.7 (arrow).

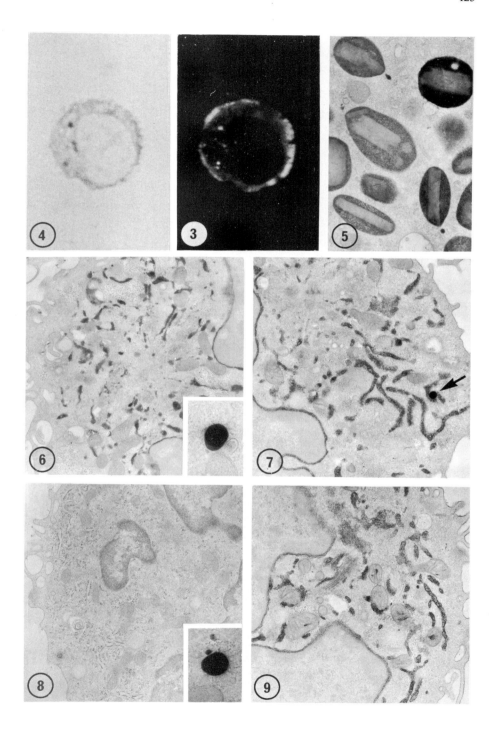

126

the same enzyme concentration, the stain intensity was much higher
after DHI incubation than could be obtained using DAB. For both
chromogenic compounds at different substrate concentrations, the
reactions followed the Michaelis-Menten kinetics, and so V_{max} could
be calculated. For incubation in DHI, the V_{max} in films with equal
enzyme activity was up to 60 times higher than with DAB-containing
media under optimal conditions (Fig.2). The cell populations used
for electron microscopy were derived from unstimulated peritoneal
cavities (containing resident macrophages and eosinophilic granulocytes) and from stimulated ones (which in addition contain exudate
monocytes and neutrophilic granulocytes); all of the cell types having peroxidatic activity (1).

In monocytes, peroxidatic activity was found in cytoplasmic granules in the Golgi region in both DAB- (inset Fig.6) and DHI-con-
taining media (inset Fig.8) at all of the pH values mentioned.

In neutrophilic and eosinophilic granulocytes (Fig.5), too, peroxidatic activity was located in granules after both DAB and DHI
incubations. The preservation of the fine structure of the granulocytes was better after DHI incubation than after DAB.

In resident peritoneal macrophages incubated in DAB-containing
media, the reaction product was as usual (1) present in the endoplasmic reticulum and the Golgi apparatus (Figs.6 and 7) at all of
the pH values. No reaction product was found with DHI at pH values
of 7.4 (Fig.8) or 9.0, whereas melanin was observed with a location
similar to that described above after incubation in a DHI-contain-
ing medium at pH 10.5 (Fig.9).

In general, the melanin precipitate formed during the DHI reaction is finely granular and shows a sharp localization. "Bleeding"
phenomena (5) were not observed, even in heavily stained preparations.

In conclusion it may be said that a linear relationship between
peroxidase activity and a measured amount of dye resulting from cytochemical peroxidase reactions with DAB or DHI as the chromogenic
substrate, can be obtained under special conditions. This opens
perspectives for the photometric cytochemical quantitation of peroxidases in situ. In addition, the results obtained so far indicate
that DHI is able to discriminate between the endogenous peroxidases
of resident peritoneal macrophages and exudate monocytes of the
guinea pig.

References
1. Daems,W.Th., Brederoo,P., and Van Lohuizen,E.J., 1973,
        Z.Zellforsch., in press.
2. Daems,W.Th. and Van der Ploeg,M., 1966, in:Electron microscopy
        1966. Proc.6th Int.Congress for Electron Microscopy, vol.2,
        ed. R.Uyeda (Maruzen Co., Tokyo) p.83.
3. Daems,W.Th., Van der Ploeg,M., Persijn,J.-P., and Van Duijn,P.,
        1964, Histochemie 3, 561.
4. Graham,R.C. and Karnovsky,M.J., 1966, J.Histochem.Cytochem.14,29.
5. Novikoff,A.B., Novikoff,P.M., Quintana,N., and Davis,C., 1972,
        J.Histochem.Cytochem.20, 745.
6. Streefkerk,J.G. and Van der Ploeg,M., 1973,
        J.Histochem.Cytochem.21, in press.
7. Streefkerk,J.G., Van der Ploeg,M., and Van Duijn,P., 1973,
        in preparation.
8. Van der Ploeg,M. and Van Duijn,P., 1964,
        J.Roy.Microscop.Soc.3, 405 and 415.
9. Van Duijn,P. and Van der Ploeg,M., 1970, in:Introduction to quan-
        titative cytochemistry, vol.2, eds. G.L.Wied and G.F.Bahr,
        (Acad.Press, New York and London) p.223.

IMMUNOCYTOCHEMISTRY

Electron microscopy and Cytochemistry, eds. E. Wisse, W.Th. Daems, I. Molenaar and P. van Duijn.
© 1973, North-Holland Publishing Company - Amsterdam, The Netherlands.

ULTRASTRUCTURAL LOCALIZATION OF TISSUE ANTIGENS WITH THE
PEROXIDASE-LABELED ANTIBODY METHOD

P. K. Nakane

Department of Pathology
University of Colorado Medical Center, Denver, USA

1. Introduction

The majority of cellular macromolecules lack unique morphological features or specific cytochemical reactions and are ultrastructurally indistinguishable from each other. However, since many of them are antigenic, several investigators have adapted the immunohistochemical method developed by Coons and his colleagues[1] for electron microscopy to localize these macromolecules by their very specific immunologic reactions.

As in immunofluorescent methods, success at the ultrastructural level requires specific and well-characterized antibodies against the macromolecules to be localized. The label or marker for this antibody must satisfy the following requirements to be useful at the ultrastructural level:

First, the label conjugated to antibody must have the ability to scatter electrons. Fluorescein-isocyanate lacks this ability and cannot be used for electron immunohistochemistry. Second, the label must be attached to antibody molecules by means of stable linkages without altering the nature of the antibody. Third, the label must be small enough that it will not interfere with the mobility of the labeled antibody in inter- and intracellular spaces. This is particularly important because electron immunocytochemistry often requires impregnation of tissue sections or slices with the labeled antibody. Fourth, the label must be stable under the adverse physical and chemical conditions presented by electron microscopy.

In search of a better label, it was postulated that if histochemically demonstrable enzymes could be conjugated to antibodies,
the antigens in tissues could be localized at both light and electron microscopic levels by the reaction product of the enzymes.
Ideally, the enzyme should not exist in the material being studied
or it should be in well-defined situations so that the sites of
antigen may not be confused with those of endogenous enzyme.

Since the introduction of cytochemical methods for the localization of peroxidase at the ultrastructural level by Graham and
Karnovsky[2] in 1966, horseradish peroxidase has been used as a model
protein in studies dealing with protein transport. This enzyme
appeared to be an ideal marker for immunoglobulin in immunohistochemistry. Since the histochemical reaction product of peroxidase
is visible by both light and electron microscopes, peroxidase-
labeled antibodies should be applicable for both light and electron
microscopic localization of tissue antigens. Since horseradish
peroxidase is a relatively small globular glycoprotein, peroxidase-
labeled immunoglobulin can be expected to penetrate through tissues
as easily as unlabeled immunoglobulins. In addition, the enzyme
is not consumed in the histochemical reaction and many molecules
of electron dense reaction product are deposited at antigenic
sites, making the method extremely sensitive.

2. Labeling of immunoglobulin with peroxidase

In the ideal conjugation reaction each immunoglobulin would be
labeled with one molecule of peroxidase without loss of the enzymatic and immunologic activities. This ideal conjugate penetrates
tissues easily - a process which is not interfered with by the
presence of larger conjugates (with multiple peroxidases or immunoglobulins). Loss of enzymatic activity in the conjugation reaction

may not be tolerated. If peroxidase activity is lost, immunoglobulin labeled with enzymatically inactive peroxidase will act
as unlabeled immunoglobulin and compete for antigenic sites. However, some loss of immunologic activity may be tolerated since the
immunoglobulin without immunological activity will not compete for
the antigenic sites.

Horseradish peroxidase is a heme-containing glycoprotein with
molecular weight of 40,000, an amino acid content of about 300
residues, and a carbohydrate content of about 18%[3]. Three different approaches have been used to attach it to immunoglobulin.

The most commonly employed method utilizes bifunctional reagents
such as difluorodinitrodiphenol sulfone[4], glutaraldehyde[5], carbodiimide[6] or diisocyanate[6] to chemically bridge peroxidase and
immunoglobulin. This method of coupling never has been an ideal
one for the following reasons: 1) The majority of α and ε amino
groups and hydroxyl groups of commercially prepared peroxidase are
not available for coupling since during isolation of peroxidase as
many as 99.5% of these groups are blocked by alloisothiocyanate[4].
2) Since the conjugates are usually formed as a result of random
coupling, only 2-3% are found to be peroxidase-labeled immunoglobulin while the majority are polymers of immunoglobulin[6]. More
recently this randomness in coupling has been reduced and the efficiency has been increased somewhat[6,7], but the lack of active
groups is still a limiting factor.

The second approach utilizes the ability of anti-peroxidase
immunoglobulin to couple peroxidase[8,9]. The resulting soluble
peroxidase-anti-peroxidase (PAP) complex has a molecular weight of
410,000-429,000 and an estimated diameter of 205 Å. The utilization of PAP complexes in immunohistochemical identification of
antigens requires the sequential application of specific rabbit
antiserum, specific antiserum against rabbit immunoglobulin G and

PAP. Because of its size, the usage of PAP is limited to the localization of antigens on the cell surface or on ultrathin sections. Furthermore, the experimental conditions required to leave one antigen-binding site of the antibody (against rabbit immunoglobulin G) free to couple with rabbit anti-peroxidase and the conditions required to prevent significant loss of the enzymatic activity of peroxidase in PAP have not yet been well established.

The third approach utilizes the carbohydrate chains located on the surface of peroxidase. They are not involved with enzymatic activity since peroxidase which contained little or no carbohydrate had enzymatic activity similar to that with carbohydrate. It was thus recognized that the carbohydrate portion of peroxidase might be modified to serve as a bridge for conjugation to protein[10]. Such modification of the carbohydrate moiety can be achieved by formation of aldehyde groups by periodate oxidation. Moreover, complete blocking of α and ε amino groups with fluorodinitrobenzene[11] prior to oxidation prevents self-coupling and the peroxidase aldehyde can couple only to added protein having available amino groups. With this approach, peroxidase may be coupled to an immunoglobulin with any desired efficiency between 0-100%. When peroxidase was reacted with immunoglobulin in the molar ratio of 3 to 1, 99% of the immunoglobulin was labeled with at least one peroxidase. Furthermore, 98% of peroxidase activity and 92% of immunologic activity were retained. This third approach essentially satisfies the criteria for the ideal conjugation reaction.

3. Fixation

Common to immunohistochemical techniques in general is the problem of stabilization of antigens _in situ_ without destroying their antigenicity while retaining relatively good tissue and cellular morphology. For this purpose the majority of investigators have adapted fixation methods which have been used for conventional light or electron microscopes with varying degrees of success.

For example, Leduc et al. observed in a study on the intracellular localization of immunoglobulin with enzyme-labeled antibodies that 2.5% glutaraldehyde and 4% formaldehyde inhibit any subsequent antigen-antibody reaction whereas fixation in 1% formaldehyde for 30 minutes to one hour conserves it[12]. In our experience, antigenicity of epithelial basement membrane antigen of the mouse can be retained by fixation in cold organic solutions (such as acetone) with inadequate tissue preservation, or in 2% paraformaldehyde with somewhat better preservation but the antigenicity is lost when the tissues are fixed in 1 or 2% glutaraldehyde or in mixtures of glutaraldehyde and paraformaldehyde[13]. On the other hand, antigenicity of vasopressin can be retained when posterior pituitary glands are fixed in a mixture of glutaraldehyde and paraformaldehyde[14]. Growth hormone and prolactin retain their antigenicities when anterior pituitary glands are initially fixed in picric acid-paraformaldehyde mixtures and postfixed in 1% OsO_4[15], but not when they are initially fixed in 2% glutaraldehyde.

As demonstrated by the examples, a general fixative which will preserve both antigenicity and cellular morphology is not yet available. The usual fixatives for electron microscopy are designed to fix proteins. Since the majority of antigens are either proteins or glycoproteins, they too are denatured. Hence a new type of fixative should be designed. It should avoid as much as possible the fixation of proteins and should be directed against lipid and carbohydrate components of the cell structure. To this end, we are currently studying the preservation of antigenicity and morphology in tissues fixed by the cross-linking of carbohydrates and lipids rather than of proteins. In order to fix the carbohydrate components, their hydroxyl groups are oxidized with periodate to form aldehydes. These aldehydes will react with diamino compounds such as lysine and thereby cross-link the carbohydrates. Since paraformaldehyde and lysine will not interact at near neutral pH, the fixative may be composed of all three ingredients. Our preliminary results indicate this periodate-lysine-paraformaldehyde solution is extremely useful in general electron microscopy[16], and some weak antigens such as epithelial basement membrane may be localized in relatively well-preserved tissues (Fig. 1).

4. a. Localization of antigens in tissue slices

Slices or small pieces of fixed tissue were obtained either by chopping manually or mechanically with a razor blade or by sectioning in a cryostat. Nakane and Pierce[13] localized basement membrane antigen intracellularly in endoplasmic reticulum of formaldehyde-fixed parietal yolk sac carcinoma of the mouse. They utilized mechanically chopped tissues with a thickness of 50 μ. Various hormones were localized in rat anterior pituitary glands which were fixed and sectioned frozen at 20-40 μ in thickness[17]. Leduc et al.[12] used manually chopped lymph nodes of the rabbit for the localization of immunoglobulin.

The rate of penetration of labeled as well as unlabeled antibodies to the intracellular antigenic sites varies considerably depending upon the tissues used and the manner in which the tissues are processed. With 50 μ thick parietal yolk sac carcinoma sections, two to three days exposure to the labeled antibodies were required to assure complete penetration of the antibodies into the

Fig. 1. Parietal Yolk Sac Epithelium of Mouse Embryo (8 days gestation). The cell was fixed in periodate-lysine-paraformaldehyde, frozen, and sectioned. The basement membrane antigen was localized with anti-mouse basement membrane labeled with peroxidase. The antigens are found in the endoplasmic reticulum as well as in extracellular basement membrane[21].

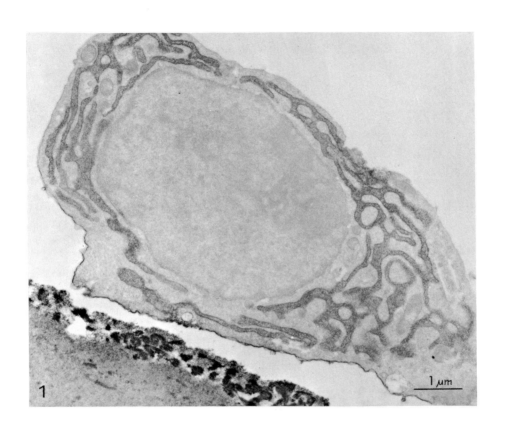

1   1 μm

tissues[13]. With 20 μ thick frozen sections of the anterior pituitary, a 24 hour exposure to the antisera was required[17]. When
manually chopped lymph node was used, no more than two or three
cell layers reacted with the antibodies after 48 hours exposure[12].

Because of the length of time required and difficulty involved
with sampling, the use of slices or small pieces of tissue has been
discouraged in our laboratory for the ultrastructural localization
of tissue antigens.

b. Localization of antigens on ultrathin sections

With the use of tissue slices, the conjugate appeared to penetrate into every component of the cell. However, there is always
the lingering suspicion that the reactants, antibodies and substrates, may not penetrate adequately. To completely eliminate any
problem of penetration, one can localize tissue antigens directly
on ultrathin sections. Furthermore, since serially sectioned
material may be used, one antigen may be localized on one section
and other antigens on succeeding sections, resulting in better
control. Only small amounts of tissue and antisera are required
to carry out the procedures, and the results may be compared
directly with the light and electron microscopes.

Kawarai and Nakane[18] localized growth hormone, prolactin and
luteinizing hormone on ultrathin sections of the anterior pituitary
gland of the rat which had been fixed and embedded in methacrylate
(Fig. 2). Growth hormone and prolactin were also localized on
ultrathin sections of the gland embedded in Epon[15], (Fig. 3), and

Fig. 2. Anterior Pituitary Gland of the Rat. The gland was fixed in
paraformaldehyde-picric acid and embedded in methacrylate. Prolactin was localized on ultrathin sections using 4-Cl-1-naphthol and
H_2O_2 as substrates[18]. The hormone was found in secretion granules.

Fig. 3. Anterior Pituitary Gland of the Rat. The gland was fixed
in paraformaldehyde-picric acid, postfixed in 1% OsO_4, and embedded
in Epon. Prolactin was localized on ultrathin sections using 4-Cl-
1-naphthol and H_2O_2 as substrates[15]. The hormone was found in
secretion granules.

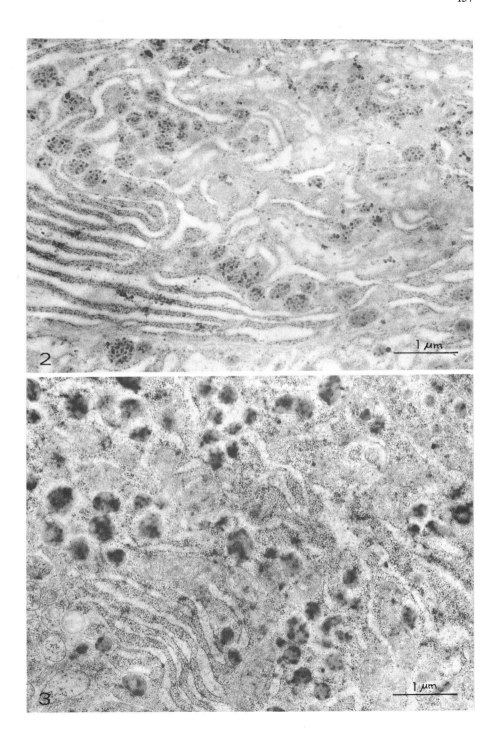

ACTH has been localized on ultrathin sections of the gland embedded in Araldite 6005[19]. Leduc et al.[12] have attempted to localize immunoglobulin on frozen ultrathin sections of lymph nodes, but concluded that peroxidase-labeled antibodies were not useful because of the resulting high background, although some success was reported with the use of alkaline phosphatase labeled conjugate.

In spite of these reports, there are some disadvantages to the localization of antigens on ultrathin sections. The size of reaction product deposited at the antigenic sites was considerably larger than that deposited in tissues when tissue slices were used. Furthermore, the reaction products had a tendency to deposit non-specifically elsewhere on the surfaces of the ultrathin section unless the reaction was carried out in a constant flow of substrate. The antigens localized so far have been known to be resistant to denaturation by fixation and polymerization of the embedding medium. However, this may not be the case for other antigens.

Provided that the antigens withstand the tissue treatment and the antibodies satisfy the immunologic criteria for specificity, this approach is a powerful tool for the ultrastructural localization of antigens.

    c. Localization of antigens in tissue sections mounted on
        histological glass slides

The localization of antigens in tissue slices or on ultrathin sections often presents difficulties in obtaining a large area of tissue for sampling as well as difficulties in correlating between light and electron microscope observations.

Fig. 4. Anterior Pituitary Gland of the Rat. The gland was fixed in paraformaldehyde-picric acid and embedded in polyethylene glycol. Luteinizing hormone was localized on sections, 4-6 μ thick. The sections were embedded in Epon and processed for electron microscopy[20]. The hormone was found in secretion granules of various sizes and textures.

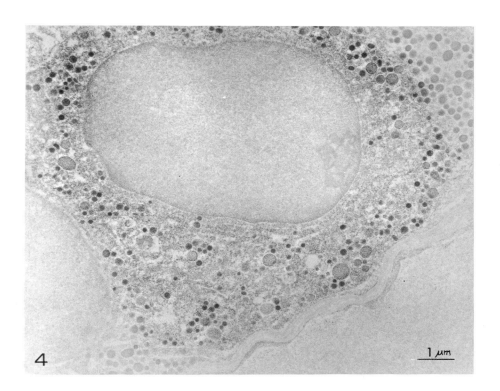

4   1 μm

One way to overcome these problems is to reuse light microscopic tissue preparations for electron microscopy. With this approach, large areas of tissue can be screened and desired areas can be examined with the electron microscope.

Tissue sections 4-8 μ in thickness are usually prepared either from fixed tissues embedded in polyethylene glycol[20] or from frozen prefixed tissues[21]. The sections are mounted on albuminized glass slides and antigens are localized on the sections with a similar method to the localization of antigens at the light microscopic level. The stained sections while on the slide are embedded in Epon by inverting an Epon-filled gelatin capsule over them. After polymerization, the capsule is removed from the slide and the desired area of tissue sectioned by an ultramicrotome for electron microscopic observation (Fig. 1, Fig. 4).

The use of polyethylene glycol as an embedding medium for tissues has many advantages. Polyethylene glycol is soluble in both water and alcohol and has a low melting point, yet is firm enough to be sectioned at room temperature. In addition to immunohistochemical staining, the sections may be used for standard histologic stains such as hematoxylin and eosin and Mallory's trichrome as well as for histochemical stains such as periodic acid-Schiff and demonstration of acid and alkaline phosphatase activities.

Currently in our laboratory, tissue antigens are routinely localized at the ultrastructural level by utilizing tissue sections mounted on glass slides. One such method is outlined:

1. Place tissue slices (1 mm or less in thickness) in a fixative which is known to preserve the antigen and morphology at 4°C under gentle agitation for an optimal time duration.

2. Wash the fixed tissues with several changes of phosphate buffered sucrose (0.01 M phosphate, pH 7.2, osmotic pressure adjusted to equal that of the fixative with sucrose) overnight at 4°C under gentle agitation.

3. Wash the tissue with the phosphate buffered sucrose solution plus 1∿10% dimethylsulfoxide (DMSO minimizes damage from ice crystals) at 4°C for 1∿2 hours.

4. Place the washed tissue in an aluminum boat with supporting medium (O.C.T. Lab-Tek) and partially submerge it in a mixture of ethanol and dry ice. Sections, 6-8 μ thick, are then cut in a cryostat set at -25 to -27°C.

5. Place the frozen section on an egg-albuminized glass slide at room temperature and air dry for several minutes.

6. Wash the slide in phosphate buffered saline (0.01 M phosphate, pH 7.2, 0.09% NaCl) (PBS) for 10 minutes with a minimum of three changes at 4°C.

7. React the section with specific antiserum for two hours in a moist chamber at room temperature. Then wash with three changes of PBS for a total of 30 minutes at 4°C.

8. React the section with peroxidase-labeled anti-immunoglobulin for two hours in a moist chamber at room temperature. Wash with three changes of PBS for a total of 30 minutes at 4°C.

9. Fix the section with 3% glutaraldehyde in PBS for 30-45 minutes at 4°C. Wash with three changes of PBS for a total of 30 minutes at 4°C.

10. Place the section in Karnovsky's solution[2] without hydrogen peroxide but with 1% DMSO for 30 minutes at room temperature. Then place the section in complete Karnovsky's solution with 1% DMSO for 1-5 minutes at room temperature. Wash the section with three changes of PBS for a total of 15 minutes at room temperature.

11. Place a drop of 2% OsO_4 in phosphate buffer (0.05 M, pH 7.2) on the stained tissue section for 30 minutes at room temperature. Wash the section with two changes of PBS for a total of 5 minutes at room temperature.

12. Dehydrate the section in graded ethanol series up to 100% ethanol.

13. While the section is still wet from 100% ethanol, invert a gelatin capsule filled with Epon over the tissue and polymerize the Epon at 60°C until desired hardness is obtained.

14. Heat the block in an oven set at 90°C for 30 minutes. The capsule should be removed from the slide easily by hand.

15. Under the dissecting microscope, trim around the desired tissue area and section for electron microscopy. Observe the section without counterstaining with the electron microscope.

142

5. Conclusion

Efficient conjugation of peroxidase to immunoglobulin can be achieved through the carbohydrate moiety of peroxidase without significant loss of either enzymatic or immunologic activities. Such conjugates are suitable for the localization of specific antigens at both the light and electron microscopic levels.

Development of a general fixative which preserves both antigenicity and morphology should make the method more readily applicable. To this end, preliminary results with the periodate fixative are promising.

Through our experience, the initial study at the light microscopic level has been rewarding. In this way one can quickly assess the resistance of antigenicity, the specificity of antisera and the adequacy of controls.

References

1. Coons, A., 1958, in: General cytochemical methods, ed. J.F. Danielli (Academic Press, New York) p. 399.

2. Graham, R.C. and Karnovsky, M.J., 1966, J. Histochem. Cytochem. 14, 291.

3. Shannon, L.M., Kay, E. and Lew, J.Y., 1966, J. Biol. Chem. 241, 2166.

4. Nakane, P.K. and Pierce, G.B., 1966, J. Histochem. Cytochem. 14, 929.

5. Avrameas, S., 1969, Immunochemistry 6, 43.

6. Clyne, D.H., Norris, S.H., Modesto, R.R., Pesce, A.J. and Pollak, V.E., 1973, J. Histochem. Cytochem. 21, 233.

7. Avrameas, S., 1972, Histochem. J. 4, 321.

8. Mason, T.E., Phifer, R.F., Spicer, S.S., Swallow, R.A. and Dreskin, R.B., 1969, J. Histochem. Cytochem. 17, 190.

9. Sternberger, L.A. and Cuculis, J.J., 1969, J. Histochem. Cytochem. 17, 190.

10. Kawaoi, A. and Nakane, P.K., 1973, Fed. Proc. 32, 840.

11. Gluysen, J.-M., Tipper, D.J. and Strominger, J.L., 1966, Methods in enzymology, vol. VIII (Academic Press, New York) p. 685.

12. Leduc, E.H., Scott, G.B. and Avrameas, S., 1969, J. Histochem. Cytochem. 17, 211.

13. Nakane, P.K. and Pierce, G.B., 1967, J. Cell Biol. 33, 307.

14. Rufener, C. and Nakane, P.K., 1973, Anat. Rec. 175, 432.

15. Nakane, P.K., 1971, Acta Endocrin. Suppl. 153, 190.

16. McLean, I. and Nakane, P.K., unpublished observation.

17. Nakane, P.K., 1970, J. Histochem. Cytochem. 18, 9.

18. Kawarai, Y. and Nakane, P.K., 1970, J. Histochem. Cytochem. 18, 161.

19. Moriarty, G.C. and Halmi, N.S., 1972, J. Histochem. Cytochem. 20, 590.

20. Mazurkiewicz, J. and Nakane, P.K., 1972, J. Histochem. Cytochem. 20, 969.

21. Martinez-Hernandez, A., Pierce, G.B. and Nakane, P.K., unpublished observation.

Acknowledgements

I wish to thank Drs. Kawarai, Kawaoi, Martinez-Hernandez, Mazurkiewicz, Pierce and Rufener and Ms. Wilson and Van Der Schouw for their encouragement and technical cooperation. This work was supported in part by Grants AI-09109 and AM-13112 from the U.S.P.H.S. and Grant E-105 from the American Cancer Society. P.K.N. is a Career Development Awardee of the U.S.N.I.H. Grant GM-46228.

Electron microscopy and Cytochemistry, eds. E. Wisse, W.Th. Daems, I. Molenaar and P. van Duijn.
© 1973, North-Holland Publishing Company - Amsterdam, The Netherlands.

PLASMA PROTEIN SYNTHESIS BY HUMAN LIVER CELLS: ITS
MORPHOLOGICAL DEMONSTRATION WITH ANTIBODIES LABELLED WITH
PEROXIDASE.

G. Feldmann

Unité de Recherches de Physiopathologie Hépatique (INSERM), Hôpital Beaujon,
92110 Clichy, France.

Although it is well established that the liver synthesizes albumin and
ceruloplasmin[10] the nature and the number of the cells which produce these
plasma proteins and their location within the hepatic lobule in the normal
human subject   are controversial[5, 7, 11]. The aim of this work, based upon
ultrastructural identification of albumin and ceruloplasmin by antibodies
labelled with peroxidase[1], was to ascertain which cells contain albumin or
ceruloplasmin, whether the cells which contain these proteins also synthesize
them and what proportion of cells synthesize these two plasma proteins.

Material

Fifteen surgical biopsies were used for the demonstration of albumin, and
5 other surgical biopsy specimens for the demonstration of ceruloplasmin.
These biopsies were obtained from adults with normal liver.

Methods

1) Immunological methods. Antibodies against albumin and ceruloplasmin
were obtained by rabbit immunization, using purified human albumin or
ceruloplasmin. The antibodies were purified by immunoadsorption, according
to the technique of Avrameas and Ternynck[2], using an immunoadsorbent
prepared with purified albumin for anti-albumin antibodies, and a human serum
completely deficient in ceruloplasmin (obtained from a patient with Wilson's
disease) for anti-ceruloplasmin antibodies. The antibodies were labelled with
horseradish peroxidase (R.Z.3.0, Sigma Chemical Co, USA)[5]. Normal
rabbit gammaglobulins were purified by chromatography[4] and labelled with
peroxidase as described above.

2) Histological and ultrastructural methods. The specimens were cut into
1 mm thick slices and at once fixed in paraformaldehyde[8]; fixation was optimal
when it was carried out at 4°C for between 8 and 12 h. After fixation, the liver
pieces were washed for 24 to 48 h at 4° C. Six micron thick sections, cut with
a cryostat were incubated at room temperature for 1 h in the solution of coupled

antibodies.   After incubation the sections were washed and the peroxidase was demonstrated by the technique of Graham and Karnovsky[6].   The sections were then placed on a thin plate of Epon, fixed in 1.5% osmium tetroxide embedding in Epon by the technique of Zeitoun and Lehy[12], and studied in the light microscope without additional staining.   Ultrathin sections, cut at the level of the cells in which a reaction has been noted in light microscopy, were examined unstained in electron microscope.

Control specimens were made 1) with normal rabbit gammaglobulins labelled with peroxidase and 2) with the solution employed by Graham and Karnovsky[6] in the absence of exogenous peroxidase.

Results

1) Albumin[4]

Light microscopy: the reaction indicating the presence of albumin (expressed as a dark brown deposit) was located in the cytoplasm - never in the nuclei - of a proportion of hepatocytes (fig.1).   Hepatocytes containing these deposits seemed to be randomly distributed in the hepatic lobule.   Their numbers varied from one lobule to another and from one subject to another. The mean percentage of hepatocytes containing albumin in the 15 normal livers was about 36%.   No reaction was observed in the Kupffer cells and in the biliary cells.   No reaction was detected in the control sections 1 and 2.

Electron microscopy: the electron dense deposits which constitute the reaction for albumin were visible in the cytoplasm of certain hepatocytes. These deposits were seen mainly on the rough endoplasmic reticulum (RER) (fig.2), not only on the ribosomes, but also on the membranes of this organelle. The membranes of the smooth endoplasmic reticulum (SER) were also sometimes positive.   The lumina of the RER and the SER were negative.   No deposit was detected in the cell nuclei and in the other cytoplasmic organelles. No deposit was seen in the Kupffer cells and in the biliary cells; in particular, in the Kupffer cells, the RER was always negative.   The control reactions 1 and 2 were always negative.

Fig.1.   Light microscope appearance of normal human liver, showing albumin confined to cytoplasm of hepatocytes.

Fig.2.   Electron microscope appearance of normal human liver.   The ribosomes of the rough endoplasmic reticulum (RER) contain albumin but the lumina are empty.

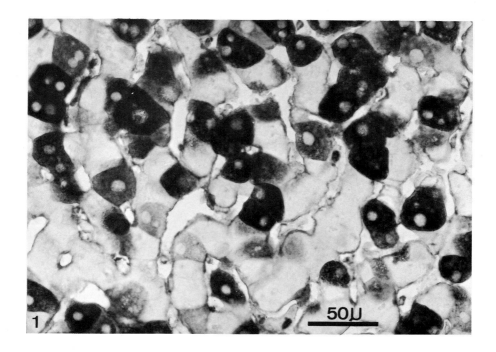

1   50µ

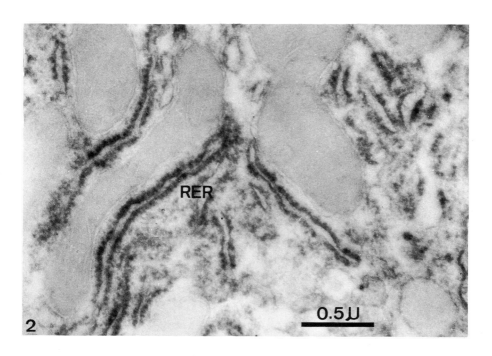

2   RER   0.5µ

2) Ceruloplasmin

Light microscopy: similar results were obtained for ceruloplasmin. The reaction indicating the presence of ceruloplasmin was located only in the cytoplasm of the hepatocytes. The number of positive cells was variable from one hepatic lobule to another and the mean percentage of hepatocytes containing ceruloplasmin in five normal liver was about 10%. No preferential location of positive cells was observed in the hepatic lobule. The others hepatic cells (Kupffer cells, biliary cells) are negative. The control reactions 1 and 2 were negative.

Electron microscopy: in the ceruloplasmin containing hepatocytes, the electron dense deposits indicating the presence of ceruloplasmin were mainly visible on the ribosomes and membranes of the RER. Some parts of the SER were also positive. Other cytoplasmic organelles were negative. No deposit was detected in the Kupffer cells and in the biliary cells. The control reactions 1 and 2 were negative.

Discussion

These data indicate that in the liver, albumin and ceruloplasmin are located in the hepatocytes only, not in the Kupffer cells or in the biliary cells. Most workers who have studied the problem of albumin or ceruloplasmin location with immunofluorescence have likewise demonstrated albumin or ceruloplasmin in the hepatocytes[5, 7, 11], but some have also found them in the Kupffer cells[5, 7]. The divergences may be due to several technical reasons, exposed in detail elsewhere[4].

Our findings of albumin on the ribosomes and the membranes of the RER of human hepatocytes is consistent with its presence noted by other workers in the microsomal fraction of rat livers[3, 9]. Although this fact is not so firmly established for ceruloplasmin, it is well know that this protein is synthezised by the liver[10] and is probably exported through the endoplasmic reticulum, as other proteins exported from the liver cells into the plasma. The location of albumin and ceruloplasmin on the ribosome, an organelle known to be the intracellular site of protein synthesis, suggests that the hepatocytes which contain albumin or ceruloplasmin, are in fact those which manufacture them.

Several hypotheses could be proposed to interpret the fact that a portion of the hepatocytes are apparently engaged in albumin and/or ceruloplasmin synthesis at a given time: a) is protein synthesis intermittent ? ; b) are some clones of hepatocytes specialized in the synthesis of proteins ?; c) does the immunological procedure detect the proteins only when the polypeptide chain is long enough ? ; d) does the immunological procedure detect the proteins only if the presence time of the proteins on the ribosomes is long enough ?

This static and purely morphological study cannot decide between these hypotheses.

Summary

Albumin and ceruloplasmin - two proteins synthesized by the liver and exported into the plasma - were demonstrated in normal human hepatocytes using specific antibodies labelled with peroxidase. These proteins were detected only in a part of the hepatocytes. These proteins were demonstrated mainly on the ribosomes of the rough endoplasmic reticulum, a location which suggests that the cells which contain these plasma proteins have also synthesized them.

References

1. Avrameas, S., 1969, Immunochemistry 6, 43.

2. Avrameas, S. and Ternynck, T., 1969, Immunochemistry 6, 53.

3. Campbell, P.N., Greengard, O. and Kernot, B.A., 1960, Biochem.J. 74, 107.

4. Feldmann, G., Penaud-Laurencin, J., Crassous, J. and Benhamou, J.-P., 1972, Gastroenterology 63, 1036.

5. Gitlin, D., Landing, B.H. and Whipple, A., 1953, J. Exp. Med. 97, 163.

6. Graham, R.C. Jr and Karnovsky, M.J., 1966, J. Histochem.Cytochem. 14, 291.

7. Hamashima, Y., Harter, J.G. and Coons, A.H., 1964, J.Cell.Biol. 20, 271.

8. Karnovsky, M.J., 1965, J.Cell.Biol. 27, 137 A.

9. Peters, T. Jr, 1962, J.Biol.Chem. 237, 1181.

10. Schultze, H.E. and Heremans, J.F., 1966, Molecular biology of human proteins with special reference to plasma proteins, vol.1 (Elsevier publishing Company, Amsterdam).

11. Shaposhnikov, A.M., Zubzitski, Yu.N. and Shulman, V.S., 1969, Experientia 25, 424.

12. Zeitoun, P. and Lehy, T., 1970, Lab.Invest. 23, 52.

Electron microscopy and Cytochemistry, eds. E. Wisse, W.Th. Daems, I. Molenaar and P. van Duijn.
© 1973, North-Holland Publishing Company - Amsterdam, The Netherlands.

ULTRASTRUCTURALLY LOCALIZATION OF HUMAN CHORIONIC SOMATO-MAMMOTRO-
PHIC HORMONE (HCS) IN THE HUMAN PLACENTA BY THE PEROXIDASE IMMUNO -
CYTOENZYMOLOGICAL METHOD.

Liliana K. de IKONICOFF

Laboratoire de Chimie Hormonale, Maternité de Port-Royal,
123, Bld de Port-Royal, Paris 14e (France).

1 Summary

HCS was detected in the syncytium of the human placenta at the
ultrastructural level by a direct immunocytological method using
peroxidase. The immunoreaction was localized in the ribosomes of the
dilatedergastoplasmie cisternae, and in the plasma membrane of the
microvilli. The Golgi apparatus and the mitochondria are always negative as are the cytotrophoblastic Langhans cells. The results
suggest that the syncytium is responsible for the production of HCS

Introduction

Detection of placental proteic hormones : human chorionic somato
mammotrophic (HCS) and gonadotrophic (HCG) hormones by immuno-his-
tochemical technics for light microscopy was reported previously (1)
(2) (3). In the present study we have adapted the peroxidase immuno
cytological method to thin and ultrathin sections for electron microscopy.

Materials and Methods

Fifteen full term placentas were obtained as soon as possible
after delivery, the villi were dissected, minced in I mm slices and
fixed for 1 h to 24 hrs at + 4°C in different fixatives with constant agitation : a) buffered formaldehyde (4 %), sucrose solution ;
b) glutaraldehyde (2,5 %) ; c) glutaraldehyde (1,25 %), formaldehyde (1 %). Frozen sections cut in cryostat at -20°C do not give satisfactory results.

After washing, the villi were exposed to sequential immunological incubations at + 4°C to evidence the tissue antigen HCS by direct method : 1) diluted specific rabbit antisera against HCS for
30 mm to 48 hrs ; 2) diluted sheep anti-rabbit 1 g G serum coupled
to HRP by glutaraldehyde following Avrameas method ; 3) the visualization of antigen localization was carried out by the reaction of
Graham and Karnovsky using DAB. Many washings between incubations re-

moved the excess of antibodies fixed in tissues.

The villi were then post-fixed with O_4O_S (2 %) for 1 h, dehydrated in decreasing alcohol dilutions and embedded in Epon.

The specificity of antisera was verified by immuno-electrophoresis and Ouchterlony's double immunodifusion. Section controls where made at each step of the reaction with systematic elimination of each serum. Incubation with DAB without previous treatment gives also negative results.

The placental villi are surrounded by a multinuclear syncytium of fused cells. At the apical surface numerous microvilli project into maternal space.

Rare Langhans' cells are found between syncytium and basement membrane. The syncytial cytoplasm contains many secretory organelles

Results

Specimens processed for demonstration of HCS disclosed a dense layer of peroxidase reaction product, indicative of immuno-staining, covering the maternal surface of the syncytial microvilli (arrow). The irregular cisternae of the rough endoplasmic reticulum (re) limite by a ribosome studded membrane contained dense, punctate or less dense, floculent material, the electron opacity of which also reflected immunoreactivity. See the electron micrographs of placenta

The dilated cisternae are especially abundant in the middle and upper thirds of the syncytium. Occasionally pseudopodes (syncytial microvilli) projected in the maternal space enclosing also cisternae of ergastoplasm. The Golgi elements and mitochondria lacked evidence of specific reaction.

Conclusion

The ultrastructural observations confirm our earlier studies which localized HCS in the syncytiotrophoblast and not in the cytotrophoblast cells. The results moroever attribute the presence of HCS in the syncytium to its production in this site and not only to its transport by the syncytiotrophoblast.

References

1. IKONICOFF L.K. de, HUBERT Ch. CEDARD L. C.R.AC.Sc Paris,272,2938 (1971)
2. IKONICOFF L.K. de, HUBERT Ch, CEDARD L. C.R.Ac.Sc Paris,274,3431 (1972).
3. IKONICOFF L. K.de, CEDARD L. Am. J. Obst. Gynec (in press) 1973

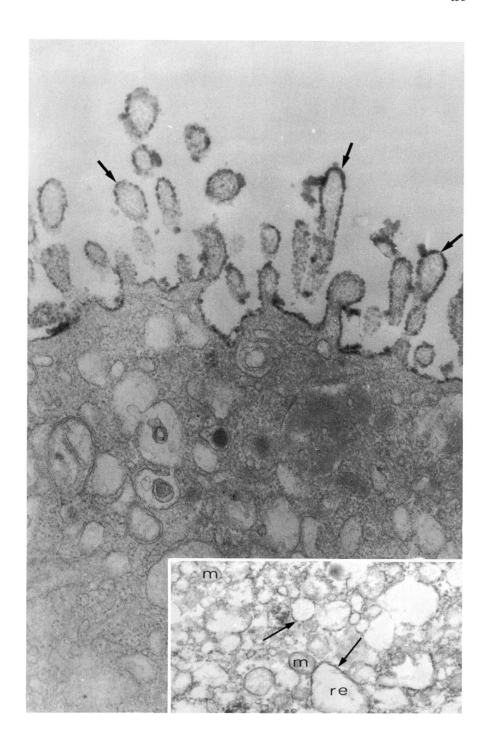

Electron microscopy and Cytochemistry, eds. E. Wisse, W.Th. Daems, I. Molenaar and P. van Duijn.
© 1973, North-Holland Publishing Company - Amsterdam, The Netherlands.

ULTRASTRUCTURAL LOCALIZATION OF ANTIGENS BY PEROXIDASE-LABELLED ANTIBODIES

W.D. Kuhlmann

Medizinische Universitätsklinik Heidelberg
and
Institut für Nuklearmedizin, DKFZ Heidelberg (Germany)

Summary

Lymph nodes of rats and rabbits were fixed with formaldehyde and/or glutaraldehyde and processed for immunocytochemical staining of Ig_G globulins. Specific sheep antibodies and their Fab fragments were labelled with peroxidase in a two-step procedure using glutaraldehyde for intermolecular cross-linking. Best intracellular antigen staining was observed with thick frozen sections prepared from well-fixed specimens.

1. Introduction

Immunocytochemical methods were studied to explore procedures with which intracellular antigens can be detected at the ultrastructural level. For this purpose lymph nodes of rats and rabbits were used in order to stain intracellular Ig_G globulins within immunocompetent cells using specific sheep antibodies coupled with horseradish peroxidase as a marker (Kuhlmann, W.D. et al. ms. in prep.).

2. Materials and Methods

Rats and rabbits were immunized by injection of bovine serum albumin into both hind foot pads. 15-20 days after primary immunization, or 3 days after a booster injection the animals were bled and popliteal lymph nodes dissected out. Prefixation of lymph nodes and cell suspensions was performed at 4°C using (a) 4 % formaldehyde/24 hours ; (b) 1-1.5 % glutaraldehyde/1-1.5 hours ; (c) 1 % formaldehyde + 1.5 % glutaraldehyde/1-1.5 hours.

Specific sheep antirat and antirabbit Ig_G antibodies were obtained by immunization with rat Ig_G and rabbit Ig_G respectively followed by passage of the corresponding immunsera on homologous water-insoluble immunoadsorbents[2]. Pure antibodies and their Fab fragments were conjugated with horseradish peroxidase

(HRP, RZ 3) by a two-step procedure using glutaraldehyde as the coupling reagent[3]. Peroxidase-labelled antibody and Fab were separated from uncoupled HRP and antibody by gelfiltration on Sephadex G 200. In the same manner, normal sheep Ig_G and its Fab fragments were conjugated with HRP, and then purified.

Prior to incubation in conjugates, the tissue blocks were reduced to small fragments with a razor blade or a Tissue Chopper ; frozen sections (5-40 µ) were cut in a cryostat[1]. The specificity of the immunocytochemical reactions was examined by incubation with HRP alone and with normal Ig_G globulins or their Fab fragments labelled with HRP. Only the known and already described endogenous peroxidase activity to certain cells, and which do not synthesize Ig_G, was seen[4].

3. Results and Discussion

The intracellular localization of Ig_G globulins in immunocompetent cells corresponded to what has been described in the perinuclear space (PNS), the cisternae of the rough-surfaced endoplasmatic reticulum (RER) and the Golgi apparatus using either antienzyme antibody techniques[4,5] or specific antibodies coupled with HRP[6]. In addition, Ig_G globulins were also observed in the extracellular space and on cell surface membrane.

With the employed fixation procedures, ultrastructural detail for immunocytochemical work was well preserved and antigenic determinants of Ig_G globulins could be stained as was confirmed by combined light and electron microscopic observations of thick frozen sections. Experimentation with the different fixation procedures and tissue sampling techniques demonstrated that staining of intracellular Ig_G with specific anti-Ig_G conjugates was mainly governed by the penetration of the immunocytochemical reagents. In the case of hand-cut small tissue fragments only border-lining cells came into contact with the incubation media, however, these failed to penetrate reproducibly into the cells, and staining was obtained preferentially in the extracellular space of superficial cell layers. When using cryostat sections, we observed that antibody-conjugates diffused completely 40 µ thick specimens, but in the depth of such sections most often intracellular antigen tagging was not seen ; cells lining the section surface exhibited intracellular staining. Thus, it was deduced that cell membranes act as barriers which must be broken, and consistent results were obtained with 10-15 µ

Figs. 1-2. Specific staining of Ig_G within ergastoplasmic cisternae (◄—) and perinuclear space (◄). Note staining reaction in the extracellular space and on cell surface membranes. Lymph node fixed in 4 % formaldehyde ; 40 µ frozen section incubated in Fab-HRP conjugate. No counterstain.

Fig. 1. inset : semithin section. Note positive plasma cells (◄—). Countstained with toluidine-blue. Original x 250.

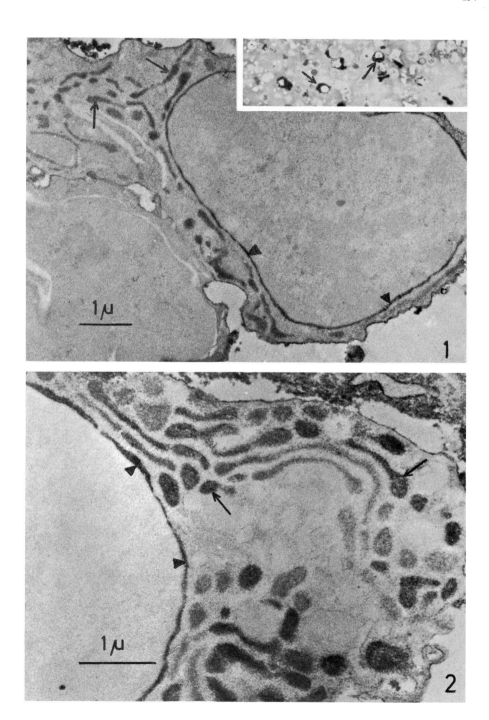

thick frozen sections.

Simultaneously to tissue blocks and frozen sections, the stainability of intracellular Ig_G globulins was examined on single cells of prefixed lymph node suspensions. Apart from positive reactions on plasma membranes of a certain number of cells, their intracellular compartments, i.e. PNS, RER and Golgi apparatus, were most often completely negative. These observations underline the necessity to alter the limiting plasma membranes when intracellular antigen staining is required. In our hands, best results were obtained by using frozen sections prepared from well-fixed specimens (Figs. 1, 2). Weak aldehyde fixation as recommended for immunocytological work may enhance penetration of conjugates, but ultrastructural detail is poorly preserved. Furthermore, proteins will leak out from unsufficiently fixed tissue, thus increasing the possibility of staining artifacts.

Acknowledgement

Much of this work was done in collaboration with Dr. S. Avrameas when the author was at the Institut de Recherches Scientifiques sur le Cancer in Villejuif in Dr. W. Bernhard's Department of Electron Microscopy.

References

1. Kuhlmann, W.D. and Miller, H.R.P., 1971, J. Ultrastr. Res. 35, 370.
2. Avrameas, S. and Ternynck, T., 1969, Immunochemistry 6, 53.
3. Avrameas, S., 1972, Histochem. J. 4, 321.
4. Leduc, E.H., Avrameas, S. and Bouteille, M., 1968, J. Exptl. Med., 127, 109.
5. Kuhlmann, W.D. and Avrameas, S., 1972, Cell. Immunol., 4, 425.
6. Leduc, E.H., Scott, G.B. and Avrameas, S., 1969, J. Histochem. Cytochem., 17, 211.

Electron microscopy and Cytochemistry, eds. E. Wisse, W.Th. Daems, I. Molenaar and P. van Duijn.
© 1973, North-Holland Publishing Company - Amsterdam, The Netherlands.

A STUDY OF THE INTRACELLULAR AND EXTRACELLULAR
LOCALIZATIONS OF THE LYSOSOMAL PROTEASE CATHEPSIN D
USING IMMUNOHISTOCHEMICAL METHODS

A. R. Poole

Strangeways Research Laboratory,
Wort's Causeway, Cambridge, U.K.

1. Summary

Intracellular cathepsin D was localized, with immunofluorescence, within
lysosomes in many cell types. Extracellular cathepsin D was detected with a new
'antibody capture' method. It was used to study the secretion of this enzyme
from osteoblasts, osteocytes, osteoclasts, fibroblasts and chondrocytes in
cultured embryonic and adult tissues. Many cells in synovia of knee joints of
rabbits with an experimental arthritis and from patients with rheumatoid
arthritis actively secrete this enzyme, unlike chondrocytes which appear
relatively inactive.

2. Introduction

Immunohistochemical studies of the lysosomal protease cathepsin D are being
made with a view to gaining a further insight into the involvement of this enzyme
in the intracellular and extracellular digestion of proteins, particularly
proteoglycan. Previously, there were no available methods for the histochemical
detection of this enzyme.

Monospecific antisera have been raised in sheep against chicken, rabbit and
human cathepsin D using enzyme purified from liver[1,2]. These antisera are species
specific but not organ or tissue specific.

3. Intracellular localization

Intracellular cathepsin D was localized with Fab' antibody fragments in
tissues fixed in 4% formaldehyde in PBS for up to 30 min. using immunofluorescence
microscopy. Use of Fab' ensured excellent penetration of fixed membranes, unlike
the very poor penetration observed for native IgG. It also avoided the
considerable non-specific attachment of native IgG via the Fc piece, seen with
frozen sections.

We have studied the localization of cathepsin D in a variety of cell types[3,4]
including synovial cells, chondrocytes, macrophages, fibroblasts and cardiac
muscle. Staining for cathepsin D revealed that it was predominantly detectable
in secondary lysosomes where it is involved in intracellular protein digestion[3].
Diffuse cytoplasmic staining was usually also seen in healthy cells to a much
lesser extent. Only in pathological situations, such as in cardiac muscle of
starved rabbits[5], was this diffuse staining intense: here it probably
represented either newly synthesised enzyme bound to organelles, which were too

small to be resolved, or soluble enzyme free in the cytoplasm. The appearance of this intense diffuse staining in myofibres was associated with degenerative ultrastructural changes in the organization of myofilaments.

4. Extracellular localization

The secretion of cathepsin D and its extracellular localization have been studied in living tissues by using a recently developed 'antibody capture' technique.

This method has been used to investigate the secretion of cathepsin D from osteoblasts, osteocytes and osteoclasts in ossifying tissues[6,7].

The secretion of cathepsin D from chondrocytes has been demonstrated in both normal cartilage and in cartilage cultured with an excess of retinol, which induces an excessive release of this enzyme[8]. Studies of experimental arthritis in the rabbit have revealed that by introducing the antibody into the joint cavity one could capture extracellular enzyme in vivo and demonstrate that cathepsin D was being secreted by many synovial cells. Alternatively, by culturing small pieces of synovia freshly explanted from these diseased joints and culturing for the same period with antisera to cathepsin D, we obtained similar results[9].

The extracellular involvement of this enzyme in the erosion of joint cartilage in rheumatoid arthritis in man has been studied with short-term culture of tissues isolated during surgery. In diseased joints, unlike normal joints, often many cells in the hyperplastic synovia and in pannus tissue invading the cartilage are actively releasing cathepsin D. There is little evidence for a release of cathepsin D from chondrocytes[10]. This evidence supports the hypothesis that the secretion and extracellular activity of proteolytic enzymes of this kind contributes to the erosion of articular and patellar cartilage observed in this disease. We are at present studying the nature of the macrophage-like cells releasing this enzyme.

5. Conclusions

Cathepsin D is normally mainly localized within lysosomes. It is secreted

---

Fig. 1. Method for the demonstration of extracellular cathepsin D. Tissues were cultured with either immune sera to cathepsin D or non-immune sera for periods from 30 min to 24 hr. During this time antibody (◖) can react with secreted extracellular enzyme (●) at or remote from cell surfaces with the formation of extracellular immunoprecipitates. The latter can be subsequently demonstrated by localization (✗) of immunoglobulin G present within the precipitate.

Fig. 2. Capture by antibody of cathepsin D secreted from cells in invasive synovium isolated from a knee joint of a patient with rheumatoid arthritis exhibiting gross erosion of articular cartilage. Arrows indicate immunoprecipitates (green) formed at the surfaces of cells (counterstained red) releasing the enzyme which has been demonstrated with a fluorescein-labelled antibody.

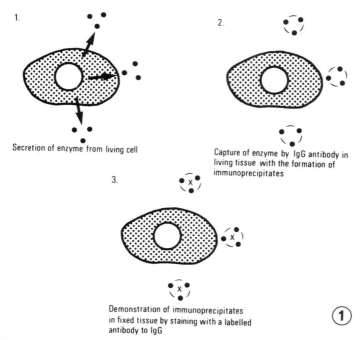

1.

Secretion of enzyme from living cell

2.

Capture of enzyme by IgG antibody in
living tissue with the formation of
immunoprecipitates

3.

Demonstration of immunoprecipitates
in fixed tissue by staining with a labelled
antibody to IgG

(1)

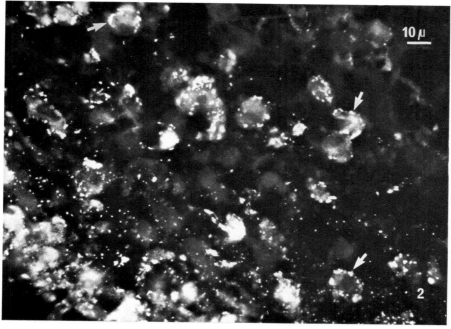

10 μ

2

by a variety of cell types to extracellular sites where it is thought to be involved in the degradation of cartilage proteoglycan in normal and diseased states.

References

1. Dingle, J.T., Barrett, A.J. and Weston, P.D., 1971, Biochem. J. 123, 1.

2. Weston, P.D., 1969, Immunology 17, 421.

3. Dingle, J.T., Poole, A.R., Lazarus, G.L., and Barrett, A.J., 1973, J. exp. Med. 137, 1124.

4. Weston, P.D. and Poole, A.R., 1973, Lysosomes in Biology and Pathology, vol. 3, ed. J.T. Dingle (Elsevier/North Holland; Amsterdam) p.426

5. Wildenthal, K., Poole, A.R., Glauert, A.M. and Dingle, J.T., 1973, Manuscript in preparation.

6. Poole, A.R., Hembry, R.M. and Dingle, J.T., 1973, Calcified Tissue Res. In press.

7. Poole, A.R. and Reynolds, J.J., 1973, Unpublished results.

8. Poole, A.R., Hembry, R.M. and Dingle, J.T., 1973, J. Cell Science. In press.

9. Poole, A.R., Hembry, R.M. and Dingle, J.T., 1973, Manuscript in preparation.

10. Poole, A.R., Hembry, R.M., Pinder, I., Dingle, J.T., Ring, F.R.J. and Cosh, J., 1973, Manuscript in preparation.

Electron microscopy and Cytochemistry, eds. E. Wisse, W.Th. Daems, I. Molenaar and P. van Duijn.
© 1973, North-Holland Publishing Company - Amsterdam, The Netherlands.

IN SITU IMMUNOCHEMICAL STAINING OF GONADOTROPIC CELLS IN PRIMARY
CULTURES OF RAT ANTERIOR PITUITARY CELLS WITH THE PEROXIDASE
LABELLED ANTIBODY TECHNIQUE. A LIGHT AND ELECTRON MICROSCOPE STUDY

Tougard, C.[*], Picart, R.[*], Tixier-Vidal, A.[*], Kerdelhué, B.[**] and
Jutisz, M.[**]

[*] Laboratoire de Biologie Moléculaire, Collège de France, Paris 5e
[**] Laboratoire des Hormones Polypeptidiques, CNRS, Gif-sur-Yvette

1. Introduction

Previous study of the kinetic of LH and FSH secretion in primary
cultures of dispersed anterior pituitary cells showed a sharp
decrease of LH and FSH media contents during the first two weeks,
followed by a low sustained level (1). Whether this decrease is due
to numerical or to functional modification of the gonadotropic cells
in culture is not yet known. To answer this question a method was
developed for the in situ detection of cells immunoreactive with
guinea-pig antisera to ovine FSH (A-oFSH), ovine LH (A-oLH) and its
two subunits (A-oLHβ, A-oLHα) using the peroxidase labelled antibody
technique (2).

2. Material and Methods

Primary cultures of anterior pituitary cells dispersed by a
trypsin collagenase mixture were initiated in small plastic Petri
dishes (1 inch 0,5 x 10^5 cells/ dish) in serum-enriched F 10 medium
(1). After 5, 14, 21, 27 and 35 days of culture the monolayers
were fixed in situ with paraformol-picric acid for 4 hrs, and rinsed
overnight in 0.1M Sorensen's phosphate buffer + 10% sucrose (pH 7.3).

The immunochemical staining was performed in the culture dishes
according to the following schedule : 1) rinse with saline phosphate
buffer 0.01M (pH 7.45) (PBS) ; 2) incubation with specific antisera
1.1/2 hrs. ; 3) rinse with PBS ; 4) incubation with peroxidase bound
anti-guinea pig gamma globulin 1.1/2 hrs. ; 5) rinse with PBS ;
6) fixation with 1% glutaraldehyde in 0.2M cacodylate buffer (pH 7.4)
+ 1% sucrose, 2 hrs. 4° C ; 7) rinse overnight in cacodylate buffer
+ 0,15M sucrose ; 8) detection of peroxidase according to Graham and
Karnovsky (3) ; 9) rinse in tris-maleate buffer (0.05M, pH 7.6).
The percentage of immunoreactive cells was then calculated from a
population of 1000 glandular cells/ dish. Other cultures underwent

the following control treatment : 1) DAB reaction without incubation with antisera ; 2) normal guinea-pig serum instead of specific anti-serum ; 3) effect of hormonal absorption of the specific antisera. Nonspecific staining was only found over necrotic cells and some nuclei.

Cultures were then embedded _in situ_ in Epon according to Brinkley et al. (4). The cell colonies containing immunoreactive cells were located in the light microscope, cut out, and mounted on araldite blocks. Ultrathin sections were examined in the electron microscope with no further staining.

3. Results

Quantitative study showed that the percentage of immunoreactive gonadotropic cells generally varied between 2% and 5% for both A-oLH and A-oFSH. It was higher for A-oLHβ and for A-oLHα. It did not dramatically decrease between 5 and 26 days in culture and even in one series it increased. This contrasts with the sharp decrease

Fig. 1 - Five-day monolayer of anterior pituitary cells. A gonado-tropic cell immunochemically stained with anti-oLH. At this stage of culture these cells retain their in vivo typical features (see text). The positive reaction is seen on secretory granules and ground cytoplasm and is restricted to small areas of the peripheral cytoplasm (arrows).

Fig. 2 - 30-day monolayer of anterior pituitary cells. A gonado-tropic cell immunochemically stained with anti-oLHβ. One notices the ultrastructural modifications as compared to figure 1 (see text). The positive reaction is seen on the same intracellular sites than in fig. 1, but is concentrated in a large area of the cytoplasm.

Fig. 3 - 12-day monolayer of anterior pituitary cells : details of dense bodies. a) Immunochemical staining with A-oLH : the large dense bodies are negative.   b) Acid-phosphatase method : the large dense bodies are positive.

Fig. 4 - 30 day-monolayer of anterior pituitary cells. Details of dense bodies. a) Immunochemical staining with A-oLHβ, same cell than fig. 2 : the dense bodies are negative and in close contact with some positive secretory granules x 40000. b) Acid phosphatase method. The reaction product is seen around the dense bodies and on some secretory granules in the way of degradation.

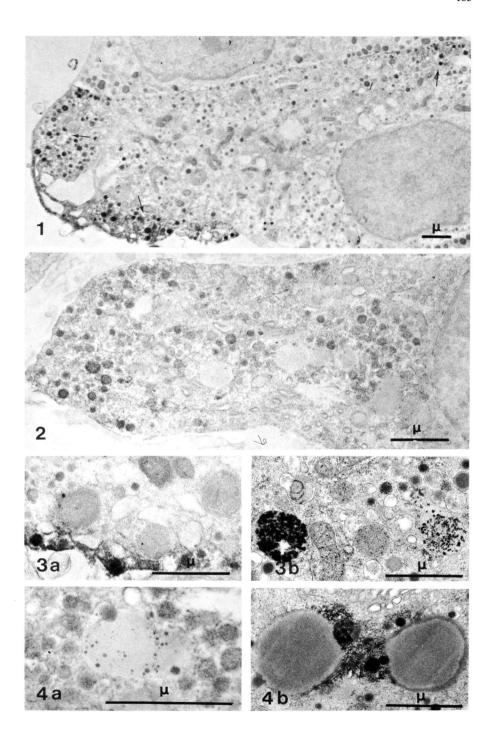

of the LH and FSH medium content. The decreasing ability of the
cultures to secrete LH and FSH results therefore more from a
modification of the secretory activity than from a numerical
regression of the gonadotropic cells. Electron microscopic study of
the immunoreactive cells revealed ultrastructural modifications of
the gonadotropic cells with time in culture. In 5-day monolayers
the two types of gonadotropic cells already described in vivo by
electron microscopic immunocytoenzymology (5) could be recognized.
As in vivo the main intracellular antigenic sites were, for the
four antisera, the secretory granules and ground cytoplasm. The
positive reaction was however often restricted to small areas of
the peripheral cytoplasm (fig. 1). Besides, an important proportion
of the gonadotropic cells were negative at this stage of culture.
In 12, 18 and 27-day monolayers gonadotropic cells with two classes
of secretory granules and dilated cisternae progressively disappeared. Finally a single gonadotropic cell remained. It was
characterized by one class of small secretory granules (125 to
200 mμ diam.), linear ergastoplasmic cisternae and large round or
oval dense bodies (600 mμ diam.). With the four antisera (oLH, oLHβ,
oLHα and oFSH), the small secretory granules and the ground
cytoplasm were strongly positive, but the large dense bodies were
always negative (fig. 2, 3a, 4a). In such cells the positive
reaction was often strikingly localized to a large area of the cell
cytoplasm (fig. 2) and the positive small secretory granules were
very numerous.

In parallel study of the monolayers with acid phosphatase method
the large dense bodies were strongly positive and some pictures
suggested an intracellular degradation of secretory granules (fig.
3 b, 4 b).

References

1. Tixier-Vidal, A., Kerdelhué, B. and Jutisz, M. 1973, Life Sci.
   12, part I, 499.
2. Nakane, P. 1970, J. Histochem. Cytochem, 18, 9.
3. Graham, R.C. and Karnovsky, M.J. 1966, J. Histochem. Cytochem.
   14, 291.
4. Brinkley, B.R., Murphy, P. and Richardson, C. 1967, J. Cell Biol.
   35, 279.
5. Tougard, C., Tixier-Vidal, A., Kerdelhué, B. and Jutisz, M.
   J. Cell Biol. in press (septembre 1973).

Electron microscopy and Cytochemistry, eds. E. Wisse, W.Th. Daems, I. Molenaar and P. van Duijn.
© 1973, North-Holland Publishing Company - Amsterdam, The Netherlands.

ULTRASTRUCTURAL VISUALIZATION OF KLH ANTIGEN ON
SELECTED MOUSE SPLEEN LYMPHOCYTES

Joseph R. Goodman, Michael H. Julius, and Leonard A. Herzenberg: Veteran Administration Hospital, San Francisco and Department of Genetics, Stanford University, School of Medicine, Stanford, Calif.

Summary

An ultrastructural study of antigen-binding spleen lymphocytes from primed mice, concentrated by a fluorescence-activated electronic cell separator, (FACS) visualized the antigen (KLH) on localized portions of the cell surface. The distribution of the KLH molecular aggregates is consistent with the "patching" and "capping" phenomenon seen by fluorescence microscopy.

The KLH aggregates were arranged in variable orientation to the cell surface. Some lay flat on the cell surface while others were perpendicular to the cell.

Introduction

De Petris and Raff (1) have recently shown the ultrastructural distribution of surface bound immunoglobin (Ig) on mouse lymphocytes by its binding to anti-mouse Ig antibody which had been conjugated to ferritin. Their studies have shown the location of the ferritin tag under differing experimental conditions of incubation temperature and metabolic inhibition, and confirm the capping phenomenon of fluorescent microscopists'. These ultrastructural studies have used poly-specific anti-mouse Ig antisera, a reagent that produces surface labeling on 30 to 50% of mouse spleen lymphocytes (the "B" cells).

We were interested in the surface distribution and behavior of antigen receptors which specifically bind a single antigen, Keyhole limpet hemocyanin (KLH). Fluorescent microscopic study of primed mouse spleen cells in this system showed 0.2 to 2% of cells which bind KLH. However, an improved version of the fluorescence-activated cell sorter (FACS) (2) developed at Stanford Medical School provides a means of purifying the fluorescent positive cells up to 100%. Fluorescein conjugated KLH (^{F}KLH) labelled cells selected with this machine were used to examine the ultrastructure of cells binding this antigen.

Methods

Spleen cell suspensions were prepared from adult BALB/cN mice primed 2-3 months earlier with KLH as described by Julius et al (3).

A flow diagram of the cell sorter is shown in Fig. 1.

Fluorescent cells, no fluorescent cells and an aliquot of the original cell suspension which had been treated with ^{F}KLH were processed for electron microscopy. KLH preparations were examined by negative staining.

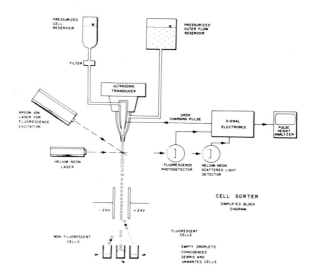

Fig. 1. Block diagram of fluorescence-activated cell sorter (FACS). The live cell suspension is introduced into the center of the fluid from the outer reservoir. The stream of cells passes the laser beams in single file and when a fluorescent cell, which conforms to preset limits intercepts the laser illumination, it will be detected. The stream is charged at the point of droplet formation. Droplets are formed by mild ultrasonic shaking at a rate of 40,000 p/sec. Charged droplets are deflected electrostatically. Reproduced from Clin.Chem. [2]

Results

The initial cell suspension had 0.7% fluorescent cells.

A selected cell population was re-run through the machine and produced a cell suspension (84% positive fluorescent) which was processed for ultrastructural studies.

The ^{F}KLH used in one experiment showed uniform molecular aggregates when negatively stained. Fig. 2a.

These aggregated molecules, 320 Å in cross section and 180-500 Å long, were produced consistently when either KLH or ^{F}KLH were treated with glutaraldehyde in a procedure by Thompson and Wofsy (4). KLH is seen on the surface of the cells in Fig. 2b and 2c.

The predominant fluorescent distribution in this preparation was a capping

Fig. 2. A - negative stained ^{F}KLH treated with glutaraldehyde. scale 0.1 micron. B and C - KLH aggregates (arrows) on a lymphocyte surface in a capped area (c) scale 1 micron.

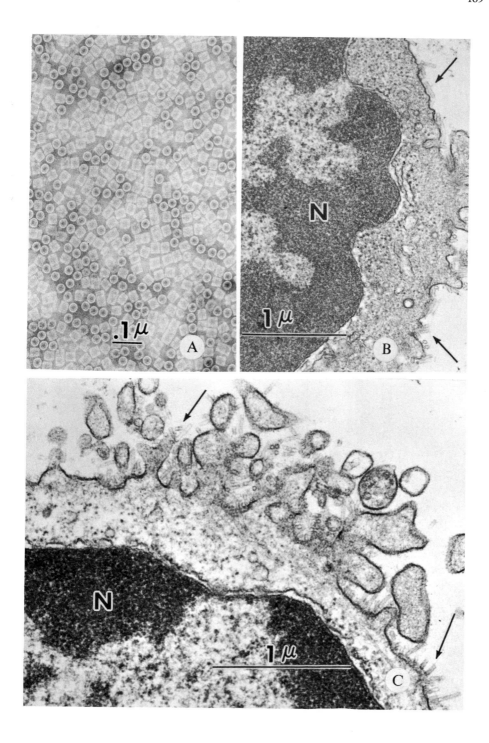

pattern and these cuts are apparently thru a capped area.

No KLH molecular aggregates were seen on the surface of the cells in the non fluorescent fraction.

A morphological examination, by E.M., of the fluorescent and non fluorescent groups showed no marked difference.

Discussion

These observations demonstrate clearly the position of antigen bound to lymphocytes presumably on an antibody molecule produced by the lymphocyte. The distribution of these KLH molecules is consistent with patching and capping processes seen by fluorescent microscopy in these experiments, and also in fluorescent and ultrastructural studies of poly-specific surface Ig distribution. The varied orientation of the KLH aggregates to the surface of the lymphocytes is also consistent with the concept of the multivalent character of the KLH molecular aggregates.

References

1. De Petris, S. and Raff, M.C., 1972 Eur. J. Immunol. 2, 523.

2. Hulett, H.R., Bonner, W.A., Sweet, R.G. and Herzenberg, L.A., 1973, in press Clin. Chemist. Aug.

3. Julius, M.H., Masuda, T. and Herzenberg, L.A., 1972, Proc. Nat. Acad. Sci. (Wash.) 69: 1934.

4. Thompson, K. and Wofsy, L., Personal communications.

Electron microscopy and Cytochemistry, eds. E. Wisse, W.Th. Daems, I. Molenaar and P. van Duijn.
© 1973, North-Holland Publishing Company - Amsterdam, The Netherlands.

FERRITIN-ANTIBODY STAINING OF ULTRATHIN FROZEN SECTIONS

S. J. Singer, R. G. Painter, and K. T. Tokuyasu

Department of Biology, University of California at San Diego
La Jolla, California

1. Earlier Studies

Ferritin-antibody conjugates have been used extensively as specific stains in electron microscopy, particularly where localization of macromolecules with high spatial resolution is required. In order that these stains be used at their fullest potential, it is desirable that simple and reproducible methods be developed to stain antigens that are inside cells, cell organelles, viruses, and other structures ordinarily impermeable to the conjugates. It was early recognized[16] that for this purpose the staining reaction would generally best be carried out directly on ultrathin sections of fixed tissues or cells, so that antigens would be exposed and accessible to the specific ferritin-antibody conjugates. We called this procedure "post-embedding staining"[15]. It was also early recognized that this procedure had to overcome several obstacles if it were to be successful.

(a) Fixation. The macromolecular components of cells and tissues must be fixed with the maximum possible retention of their capacity to bind to specific antibody. If the antigenic determinants of interest happen to be oligosaccharide, there is little problem, since the commonly used fixatives do not modify carbohydrate structures. If, however, the antigen in question is a protein whose determinants (as they usually do) reflect the native three-dimensional conformation of the molecule, fixation may alter the conformation, and destroy the binding capacity of the antigen. It has been found, for example, that subjecting several soluble proteins to the conditions used in osmium and glutaraldehyde fixation markedly changes their native conformations[8], with the latter fixative causing less alteration than the former. Thus, even the use of 0.5% glutaraldehyde for 16 hr at room temperature produces a 40% loss of the capacity of bovine trypsinogen to bind to its antibodies[7]. On the other hand, 2% glutaraldehyde treatment for 1 hr at 0°C has little effect on the antibody binding of bovine ribonuclease[14].

Several devices may be useful in those cases where even mild glutaraldehyde fixation largely inactivates an antigen. Bifunctional imidates, such as diethylmalonimidate[4] (DEM), or dimethylsuberimidate[3], have been used as cross-linking reagents in protein chemistry, because it is known that even extensive reaction with imidates often has only minimal effect on protein activities[6,18]. After extensive studies of the reactions of DEM with proteins and intact erythrocytes (A. Dutton and S. J. Singer, to be published), we have used DEM as a fixative for electron microscopy in the case of erythrocytes[11] and several other cells and tissues (J. D. McLean and S. J. Singer, unpublished experiments). With erythro-cytes, the antigenic properties of hemoglobin appeared to be largely unaltered[11], corresponding to the finding that its oxygen binding capacity and the spectra of the oxy- and deoxy-forms of the molecule were also unaltered (A. Dutton and S. J. Singer, to be published). DEM is less rapidly reactive than glutaraldehyde, however, and in our preliminary observations, it does not completely fix certain tissues. In specific cases, nonetheless, DEM or related compounds may prove use-ful as fixatives in ferritin-antibody staining procedures.

It is also possible to pre-treat a solubilized protein with glutaraldehyde, and immunize with this modified, partly denatured protein. Some of the antibodies so produced might be directed to specific linear sequences on the protein which are not exposed in its native conformation. Such antibodies might then bind spec-ifically to the glutaraldehyde-fixed antigen within the cells or tissue being examined.

(b) Embedding. For the purposes of post-embedding staining with ferritin-antibody conjugates, the embedding procedure: a) must not only satisfactorily preserve and delineate the ultrastructure involved, but b) must not itself alter the antigenic properties of the macromolecule to be stained; and c) must not pro-mote nonspecific staining nor inhibit specific staining by the conjugates. We quickly learned that with all conventionally embedded specimens, the sections were heavily nonspecifically stained with ferritin-antibody conjugates. Over the last decade or so, therefore, our efforts have been directed to developing a pro-cedure satisfying these conditions.

Our first partial success was with the polyampholyte embedding procedure[10, 15], utilizing the two water-soluble ionic vinyl monomers dimethylaminoethylmethacryl-ate (positively charged) and methacrylic acid (negatively charged), along with a bifunctional monomer as a cross-linker. This produces a hydrophilic polymer which exhibits little nonspecific ferritin staining, but which is difficult to section because of wetting problems. Some years later, the cross-linked protein procedure was introduced for embedding[5, 11]. In this method, a highly water-soluble protein such as bovine serum albumin (BSA) serves as the embedding matrix. The fixed cells or tissue are suspended in 30-50% BSA. The water is then removed from this suspension by dialysis against a hygroscopic high polymeric substance until a block of the appropriate consistency is produced; after which, glutaraldehyde is diffused into the block to cross-link the BSA molecules. The blocks of cross-linked BSA can be reproducibly sectioned, and in several systems the sections allowed specific ferritin-antibody staining to occur[11]. With some modifications, this method has also been used in studies by Kraehenbuhl and Jamieson[7], and their work with the method is extensively described in this volume. We will therefore not discuss this procedure further, but will proceed to a more recent development from our laboratory, the preparation of ultrathin sections by ultracryotomy. Although the cross-linked BSA method should be useful in a variety of electron microscopic investigations[12], we have encountered two problems with it which have not yet been entirely solved. One is that the ultrastructure, while preserved, is often not sharply delineated. If one treats cells or tissues with the usual osmium fixation conditions, washes, and suspends the specimen in the 30-50% BSA for the embedding process described above, no blackening of the specimen occurs such as is observed in the course of alcohol dehydration for polymer embedding procedures. Post-staining of the cross-linked BSA sections has not been entirely effective either. A second problem is that since the BSA does not penetrate glutaraldehyde-fixed cells, the removal (by dialysis) of the water from the cells or tissue suspended in the 30-50% BSA solution appears to cause an intracellular compaction. This not only contributes to the difficulty in delineating ultra-structural details, but also may make it difficult for ferritin-antibody conjugates to permeate into the interior of the ultrathin sections, which may result in only superficial ferritin-antibody staining of the sections.

2. Ultracryotomy and Ferritin-Antibody Staining

In attempting to improve still further upon specimen preparation techniques for ferritin-antibody staining, Tokuyasu[17] has devised a simple and reproducible procedure for the ultracryotomy of fixed cells and tissues without any prior embedding. As this procedure has recently been described in detail[17, 14], it will be only briefly summarized here. The glutaraldehyde-fixed cells or tissue are infused with buffered sucrose solutions, with the sucrose concentration between 0.6 to 1.6 M, depending on the tissue (see below). Tissue pieces or droplets of the cell suspension on the flat ends of short copper rods are rapidly frozen by immersion in liquid N_2. The rods are attached to the precooled head of the microtome arm maintained at -50°C to -90°C with a cryokit attachment[1] to the Sorvall MT-2 microtome. After ultrathin sectioning, the frozen sections are picked up from the knife-edge on a droplet of saturated sucrose on an eyelash probe. The droplet remains liquid long enough to effect the transfer. Upon removal from the cryokit bowl, the section melts and spreads on the droplet. The section may then be transferred to a coated grid by touching the droplet lightly to the grid surface. Without allowing the section to dry, it is then washed with a droplet of buffer and then floated on a solution of 5% BSA to condition the surface against nonspecific ferritin staining[11]. Following this, a large droplet of ferritin-antibody conjugate (for direct staining) is applied to the grid for 5-10 minutes, the grids are washed, and then finally floated on 2% glutaraldehyde to fix the antibody stain to the section. Finally, the grids are stained with 0.2% phosphotungstic acid, washed, and dried. The electron micrographs shown in this paper were obtained with a Philips EM 300 electron microscope at an accelerating voltage of 60 kv.

Both frozen cell suspensions and tissue blocks can be studied by this technique, as is demonstrated below. The concentration of sucrose used to infuse the tissue blocks before freezing, and the temperature of sectioning of the frozen block, are adjusted to the properties of the specimen. In general, for less densely packed tissue, higher sucrose concentrations and lower sectioning temperatures yield better results. Specimens of a wide variety of properly fixed cells and tissues prepared by this technique show excellent structural preservation and detail[17]. This method of ultracryotomy promises to be useful in many different applications in electron microscopy, but in this paper its use in connection with ferritin-antibody staining is emphasized.

Two different macromolecular species have so far been stained in situ with specific ferritin-antibody conjugates using this technique. One system involved the enzyme ribonuclease (RNase) in cells of bovine pancreas[14]. Fragments of pancreas were fixed in buffered 2% glutaraldehyde for 1 hr at 0°C, and then infused with 1.1 M sucrose-buffer, frozen, and sectioned at -75°C. The sections were then stained directly with a ferritin-antibody conjugate directed against bovine pancreatic RNase. The rabbit anti-RNase antibodies were first purified by affinity chromatography on columns of RNase coupled to Sepharose 4B[2]. In Fig. 1 is shown a section prepared by ultracryotomy of bovine pancreas stained with ferritin-anti-RNase. (The ultrastructure is not particularly well preserved, presumably because of post-mortem autolysis which occurred before fixation of the tissue.) The ferritin particles are found over all the zymogen granules and the cisternae of the endoplasmic reticulum, but not over the cell nucleus, cytoplasm, or mitochondria. As a control (Fig. 2) sections of rat pancreas prepared and treated in identical fashion (and well preserved by rapid fixation) showed no significant staining with the ferritin-anti-RNase conjugate, corresponding to the fact that rat and bovine pancreatic RNase do not cross-react immunologically[14]. Another control for the specificity of the ferritin-antibody staining in Fig. 1 was the finding that pretreatment of the section of bovine pancreas with unconjugated anti-RNase antibodies prevented any significant staining upon subsequent treatment with the ferritin-antibody conjugate[14].

Fig. 1 Ultrathin frozen section of a bovine pancreatic acinar cell stained with ferritin-conjugated antibodies to RNase, as described in text. The zymogen granules (Z) and the cisternae (C_t) of the endoplasmic reticulum are stained, but not the cytoplasm (Cy).

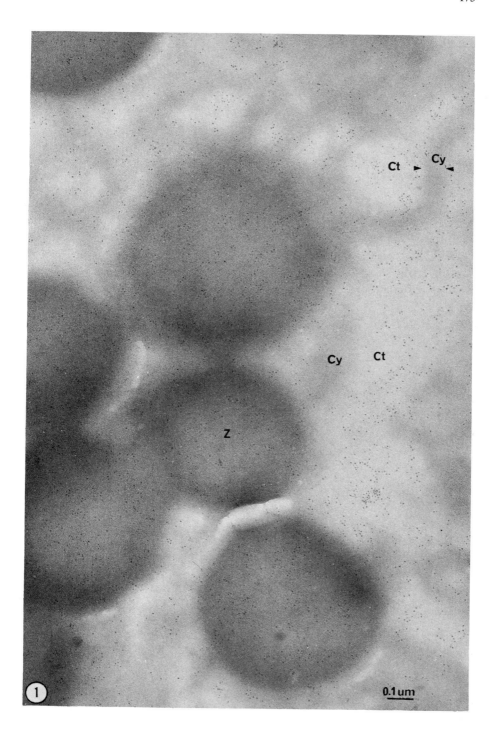

These results are similar to the findings of Kraehenbuhl and Jamieson[7] (see also this volume), using the cross-linked BSA embedding method and an indirect ferritin-antibody staining procedure for the zymogen trypsinogen, and confirm their conclusion that the different secretory proteins of the pancreatic acinar cells are each found in every zymogen granule.

The second system we have studied involves the protein spectrin[9], which is peripherally attached to the inner, cytoplasmic surface of erythrocyte membranes [13]. Human erythrocytes were fixed with isotonic buffered 0.5% glutaraldehyde for 1 hr at 0°C. (The binding of spectrin to its antibodies is 50% retained in this treatment, but is completely inactivated by 2% glutaraldehyde, as determined by separate complement fixation experiments). After being washed and suspended in 1.1 M sucrose-buffer, the cells were centrifuged into a pellet, and a droplet of the pellet was frozen and sectioned at -60°C as described above. In transferring the sections to the coated grids, individual cell sections become firmly attached, and were not removed by subsequent treatments. The sections were conditioned with 5% BSA, and then treated for 5 minutes at room temperature with a droplet of unconjugated rabbit anti-spectrin antibodies that had been purified by affinity chromatography on columns of spectrin-Sepharose 4B. After thorough washing, the sections were treated with a droplet of ferritin-conjugated horse anti-rabbit IgG antibodies (indirect stain). (The horse anti-rabbit IgG antibodies had also been purified by affinity chromatography prior to conjugation with the ferritin). Following this staining step, the remainder of the procedure was the same as described above, except that the negative staining with 0.2% PTA was omitted.

An example of the results is shown in Fig. 3A. The ferritin-antibody staining is largely confined to the region near the cytoplasmic surface of the membrane of the intact erythrocyte, as expected from our earlier studies on erythrocyte ghosts[13]. As a control on the specificity of the staining, a parallel experiment was carried out, but with normal rabbit IgG used instead of the rabbit anti-spectrin antibodies in the direct staining step. This was then followed by the indirect staining with the ferritin-conjugated horse anti-rabbit IgG antibodies. As shown in Fig. 3B, essentially no ferritin was present over the erythrocytes in this control.

In these two cases, one involving tissue and the other isolated cells, specific ferritin-antibody staining of an intracellular antigen has been achieved by the

---

Fig. 2 Ultrathin frozen section of a rat pancreatic acinar cell treated in the same manner as the specimen in Fig. 1. No significant ferritin staining is observed. The ultrastructure is well preserved compared to Fig. 1 because of rapid post-mortem fixation.

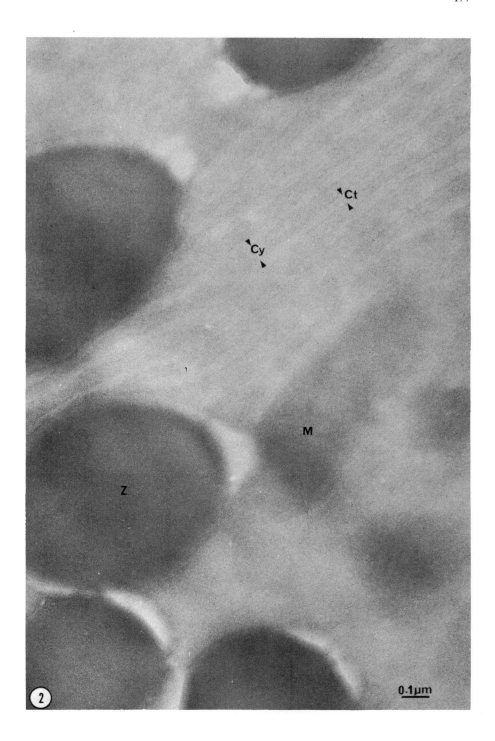

2 0.1µm

178

method described. It appears likely, therefore, that the method will be generally useful in the localization of intracellular macromolecules at high resolution. The method has a number of advantages. Ultracryotomy, with the appropriate refinements[17], is simple and reproducible. Serial sections can routinely be obtained. It is chemically and physically a very gentle method of specimen preparation, and is likely to permit retention of most of the molecular and ultrastructural properties of the specimen. The intracellular domains within the section remain relatively porous compared to the sections prepared by the cross-linked BSA procedure, because no dehydration step is involved. The method in its present stage of development, however, still has a few problems. In some cases, the intracellular matrix in the ultrathin section is not well supported after the sucrose is washed out of the section, and some ultrastructural distortions may result. Procedures are being investigated to introduce an appropriate intracellular matrix that will not be washed out of the section in order to circumvent this problem where it arises. Certainly, other refinements are bound to be developed as well. We are satisfied, however, that we now have available a basically sound and generally useful procedure for the ferritin-antibody localization of intracellular components.

References

1. Christensen, A. K., 1971, J. Cell Biol. 51, 772.

2. Cuatrecasas, P., 1970, J. Biol. Chem. 245, 3059.

3. Davies, G. E. and Stark, G. R., 1970, Proc. Nat. Acad. Sci. U.S.A. 66, 651.

4. Dutton, A., Adams, M., and Singer, S. J., 1966, Biochem. Biophys. Res. Communs. 23, 730.

5. Farrant, J. L. and McLean, J. D., 1969, Abstracts 27th Meeting of the Electron Microscopy Society of America, 422.

6. Hunter, M. J. and Ludwig, M. L., 1962, J. Am. Chem. Soc. 84, 3491.

7. Kraehenbuhl, J. P. and Jamieson, J. D., 1972, Proc. Nat. Acad. Sci. U.S.A. 69, 1771.

8. Lenard, J. and Singer, S. J., 1968, J. Cell Biol. 37, 117.

9. Marchesi, S. L., Steers, E., Marchesi, V. T., and Tillack, T. W., 1970, Biochemistry 9, 50.

Fig. 3 Ultrathin frozen sections of intact human erythrocytes treated with A) rabbit anti-spectrin antibodies; and B) normal rabbit IgG; in both cases followed by ferritin-conjugated horse antibodies to rabbit IgG.

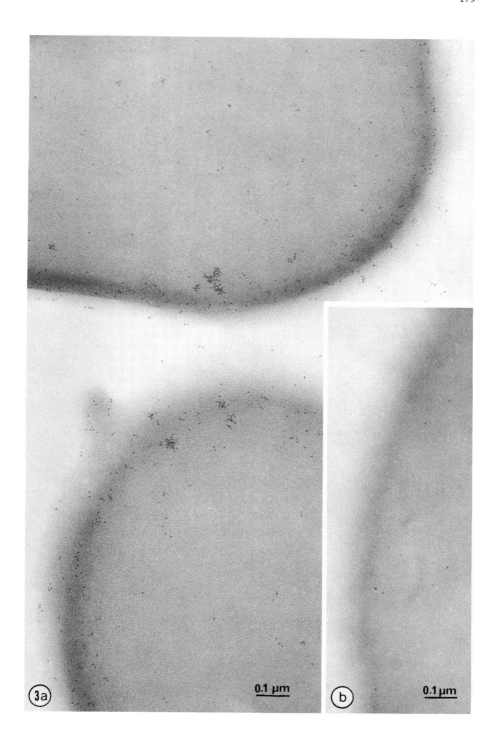

3a   0.1 µm

b   0.1 µm

180

10. McLean, J. D. and Singer, S. J., 1964, J. Cell Biol. 20, 518.

11. McLean, J. D. and Singer, S. J., 1970, Proc. Nat. Acad. Sci. U.S.A. 65, 122.

12. Nicolson, G. L., 1971, J. Cell Biol. 50, 258.

13. Nicolson, G. L., Marchesi, V. T., and Singer, S. J., 1971, J. Cell Biol. 51, 265.

14. Painter, R. G., Tokuyasu, K. T., and Singer, S. J., 1973, Proc. Nat. Acad. Sci. U.S.A., in press.

15. Singer, S. J. and McLean, J. D., 1963, Lab. Invest. 12, 1002.

16. Singer, S. J. and Schick, A. F., 1961, J. Biophys. Biochem. Cytol. 9, 519.

17. Tokuyasu, K. T., 1973, J. Cell Biol. 57, 551.

18. Wofsy, L. and Singer, S. J., 1963, Biochemistry 2, 104.

Electron microscopy and Cytochemistry, eds. E. Wisse, W.Th. Daems, I. Molenaar and P. van Duijn.
© 1973, North-Holland Publishing Company - Amsterdam, The Netherlands.

LOCALIZATION OF INTRACELLULAR ANTIGENS USING

IMMUNOELECTRON MICROSCOPY

J.P. Kraehenbuhl and J.D. Jamieson

Yale University, School of Medicine, New Haven, Conn. 06510

I.. Introduction

Immunocytochemical techniques for the localization of intracellular antigens at
the electron microscope can be classified as surface localization procedures (post-
embedding staining) or diffusion localization procedures (pre-embedding staining).
In the diffusion technique, labeled antibodies or antibody fragments are allowed
to penetrate fixed tissues and interact with intracellular antigenic sites prior
to embedding and thin sectioning. Antigenic sites are subsequently recognized by
virtue of the inherent density of the immunologic tracer or by electron opaque
reaction products previously formed in the tissue in the case of enzyme-tagged
antibodies. In the surface localization procedure, thin sections of fixed and embedded tissue are prepared prior to the detection of antigenic sites by application
of the immunologic tracers. These may consist either of tracer molecules (electron
dense particles such as ferritin or small viruses, or enzymes) covalently linked to
antibodies, or tracers non covalently linked to antibodies (unlabeled antibody
techniques).

For the diffusion technique, the main practical limitation has been the problem
of obtaining tracers of sufficiently small size that they will readily penetrate
all barriers present in fixed cells (plasma membrane, membranes bounding intracellular compartments, etc.). These diffusion barriers generally preclude the use of
particle-tagged complete antibodies, except for some specific cell types (e.g. IgG
in plasmacytes). The advantage of the diffusion technique is that high sensitivity
can be expected when enzyme-tagged antibodies or antibody fragments are used, the
catalytic activity of which results in an amplifying effect due to the local
accumulation of reaction product.

For the surface localization procedure, no diffusion restrictions are to be
expected because the immunological reagents are applied directly to the surface of
the thin sections. This procedure should be particularly useful for the detection
of antigens enclosed in membrane bound compartments (e.g. secretion granules,
lysosomes, etc.) since during sectioning these compartments are opened, diffusion
barriers are removed and antigenic sites are exposed to tagged antibodies. When
the tracers employed are particle-tagged antibodies, such as ferritin-labeled

antibodies, the resolution of the technique is expected to be high in that only a small number of tracer molecules can react with antigenic sites on the section surface, but by the same token the sensitivity of the method will be low.

Based on these considerations it would be desirable to use both procedures together for a given problem since the results obtained will be complementary and will enable one to assess the problem of false positive and false negative results.

In both procedures (surface and diffusion localization techniques), each step has been systematically examined and will be dealt sequentially: a) the effect of fixation on antigenicity; b) the nature and quality of the first step antibodies employed (an indirect localization sequence is used); c) the preparation of tagged antibodies; d) examples of the use of both procedures.

2. Fixation and preservation of antigenicity

To obtain adequate immunocytochemical localization, the fixatives employed should stabilize cellular structures and prevent extraction, displacement, and secondary relocation of antigens. At the same time they should not seriously inter-fere with the ability of antigens to interact with antibodies. These two sets of requirements lead to an apparent paradox in that currently used fixation procedures for electron microscopy compromise antigenicity. To circumvent this problem, it will be important to devise fixation conditions which adequately preserve antige-nicity and to see if these are commensurate with preservation of cell fine struc-ture, including retention of antigens in situ. Among the fixatives commonly used for electron microscopy, the aldehydes fulfill the conditions mentioned above.

While the importance of secondary and tertiary structure of proteins in antigen-antibody reaction is well established, the effect of aldehydes on these interactions has been less extensively examined. Habeeb (3) in 1969, showed that formaldehyde treatment of bovine serum albumin caused little inhibition of its ability to precipitate the corresponding antibody. The results with glutaraldehyde showed much more severe effects.

Recently we have devised a simple procedure for testing the effect of glutaral-dehyde fixation on preservation of antigenicity of a number of proteins and pep-tides. For this purpose, antigens are insolubilized on agarose beads and subsequen-tly exposed to varying concentration of glutaraldehyde for times ranging from 5 min to 16 hrs. After removal of unreacted fixative by washing and reducing free aldehyde groups with sodium borohydride, the adsorbents are tested for their abi-lity to extract specific antibodies from antisera. For a given concentration of glutaraldehyde (0.5 %), the loss of antigenicity is progressive for exposure times up to 1 hr, but thereafter little further loss is obtained (Fig. 1). At a fixed time (16 hrs) of glutaraldehyde exposure, antigenicity generally decreases with

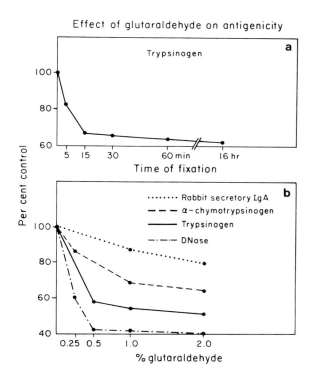

Fig. I. a) Time course of effect of 0.5 % glutaraldehyde on antigenicity of bovine trypsinogen.

b) Effect of various concentrations of glutaraldehyde on antigenicity of several insolubilized antigens. The adsorbents were exposed 16 hrs to glutaraldehyde.

increasing concentration of the fixative although the response of individual antigens varies quantitatively. In general inactivation proceeds progressively and markedly up to 0.5 % glutaraldehyde, but from 0.5 to 2 % antigenicity is not further severely depressed. While these data serve as a general guide for expected retention of antigenicity, it should be borne in mind that they do not mimic the situation in the cell.

Based on these biochemical data it would appear that the use of low concentrations of glutaraldehyde is desirable for primary fixation. Since, however, such low concentrations are not routinely used in electron microscopy fixation, it was important to determine whether cellular fine structure is adequately preserved under these conditions. Two cell types, the hepatic parenchymal cell and the pancreatic exocrine cell were studied in this regard since they have been both the subject of extensive morphological investigations. The osmolarity of the fixative was adjusted with sucrose to a level equivalent to that present in fixatives containing 2 % glutaraldehyde. In the case of the liver, glutaraldehyde concentrations as low as 0.1 % give adequate fine structural preservation, which is comparable to that of tissue fixed with 2 % glutaraldehyde. In contrast for the pancreas, important extraction particularly of zymogen granule content occurs when concentrations lower than 0.5 % are used. Using formaldehyde as a primary fixative, extraction of zymogen granules in pancreas occurs even with concentration as high as 4 %, whereas in the liver, the fine structure is preserved well at 2 % with no obvious extraction.

Based on these considerations, we have routinely used 0.5 % glutaraldehyde as a primary fixative, but have of necessity used it exclusively for the surface localization procedure. Due probably to its superior cross-linking properties, glutaraldehyde fixation is not usable for the diffusion localization procedure, since impenetrable barriers result. As a consequence, 4 % formaldehyde, a less efficient cross-linker, has been employed.

3. Nature and quality of the first step reagent

In our studies, we have chosen to use an indirect localization sequence in which a common tagged second step antibody is used to reveal antigenic sites which have interacted with specific first step antibodies. This type of localization is economical in that only small amounts of specific first step antibodies are required and these need not be individually conjugated to the tracers. In addition the results are more readily comparable since a common second tagged antibody is used. In this indirect procedure the quality and nature of the first step antibodies determine whether or not adequate localization is obtained with minimal background. Since in our procedure both monovalent (Fab) and divalent antibodies are used in the first step, it is important to emphasize that the affinity of monovalent versus bivalent antibodies is generally much less for antigens and that the affinity, particularly of the monovalent antibody, varies according to the time of immunization (10). Consequently in our laboratory we have routinely used antisera from hyperimmunized animals.

For the diffusion technique, the efficiency of localization depends on the nature of the first step antibody (6). The efficiency of labeling with monovalent antibody fragments has been shown to be independent of their concentration after an optimum level is reached whereas for bivalent antibodies, the efficiency is titer dependent, decreasing after an optimum is surpassed. In our hands, the best results have been obtained using Fab fragments purified by immunoadsorption, although papain digested antisera or even whole antisera at optimum titer can also be used successfully.

For the surface localization technique, the nature and the technique of preparation of the first step antibodies are particularly important. From a practical point of view, the use of whole serum and of ammonium sulfate fractions of gamma-globulins results in high backgrounds, which is probably due to soluble aggregates in these preparations. DEAE-cellulose purified IgG or antibodies prepared by immunoadsorption appear to have fewer of these soluble aggregates and hence produce lower non specific backgrounds. As some of the background may, in addition, be due to non specific interactions of the Fc fragment with the section surface, the use of F(ab')$_2$ provides an additional advantage. The use of affinity purified Fab in the first step is ineffective over a large concentration range, even in the absence of protein conditioning of the grid, possibly because the antigenic sites of the Fab molecule after specific interaction of the section surface are masked and unavailable for recognition by the bulky tracer. The concentration of antibodies in the reagent is also important. For the antitrypsinogen system applied to thin sections of bovine pancreas we have found that in the range of 1 to 10 mg/ml of specific antibodies the intensity of specific ferritin staining in the second step remains constant. With higher concentrations of antibodies, however, soluble aggregates form which result in high backgrounds. For each system the minimal concentrations of antibodies commensurate with adequate localization should be determined.

4. Preparation of tagged antibodies

Two different approaches have been used for immunocytochemical visualization of antigens. In one, antibodies are covalently linked to tracer molecules which include directly visualizable particles such as ferritin or small viruses, or enzymes, the reaction product of which can be subsequently seen in the electron microscope. The second class of procedures involves the use of hybrid antibodies (4) or of bridge techniques (12), where the visualizing agents are the same as for the previous techniques. In the latter method no covalent linkage of tracers antibodies is required. Although the hybrid technique is elegant and particularly useful for topographic mapping, for simplicity we decided to choose classic conjugation methods.

In the past, most conjugating agents have been used in a one step batch reaction which inevitably leads to random polymerization of tracers and antibodies. Fewer polymers can be expected to result if a two step reaction is used for conjugation. In a two step procedure, one of the reaction partners (usually the tracer molecule) is first exposed to the bifunctional reagent and the activated intermediate, largely freed of the remaining reagent is subsequently allowed to conjugate with the antibody in a second step. The first bifunctional reagent to be used in a two step reaction was toluene-2,4-diisocyanate which was introduced by Singer in 1959 (11).

For most of the batch procedures described in the literature, the antigenic binding sites of the antibody molecule are potentially able to participate in the conjugation reaction thus destroying or masking the antibody active sites. This has clearly been shown by Borek and Silverstein (1) and Vogt and Kopp (13) who measured the loss of antibody precipitating activities following conjugation of antibody to ferritin. They suggested that this loss was due to the conversion of divalent antibodies into univalent antibodies during conjugation.

In an attempt to overcome the loss of antibody activity which accompanies conjugation, we have devised a procedure in which the antibody active site is involved in a specific interaction with its antigen during conjugation and conse-quently should be protected (7). In addition, the antigen is insolubilized on a solid support which facilitates the mechanics of the conjugation procedure. So far we have applied this solid phase procedure for the preparation of Fab-ferritin conjugates although in principle it should be useful for any antibody-tracer sys-tem. The procedure consists of the following steps. 1. Antigens are insolubilized onto agarose beads using CNBr coupling (7). 2. The adsorbent is used to extract specific Fab fragments from whole antisera digested with papain. 3. After washing to remove non specific proteins the charged adsorbent is exposed to ferritin pre-viously activated with toluene-2,4-diisocyanate. 4. After conjugation is completed excess reagents are washed out, the conjugate is eluted from the adsorbent with 3.0 M KSCN, and final purification takes place by gel filtration. The conjugate consists of one ferritin molecule carrying 3 - 4 Fab fragments. As described later it has been used for our surface localization procedure.

For the diffusion localization procedure, the main practical limitation as mentioned previously, has been to obtain immunological tracers of sufficiently small size that they will penetrate fixed tissues. To this end, we have conjugated a heme-octapeptide possessing peroxidatic activity to Fab fragments. The conjugate has a molecular weight of approximately 50,000 daltons and does not differ signi-ficantly in size from the Fab fragment which is the smallest immunological unit able to react specifically with an antigen.

The heme-octapeptide contains 8 amino acids and carries a heme bound by thio-

---

Fig. 2. Localization of trypsinogen on thin section of bovine pancreas embedded in BSA. First step reagent: 2.5 mg/ml of rabbit antitrypsinogen antibodies; second step reagent: sheep Fab antirabbit Fab conjugated to ferritin. Heavy ferritin staining of zymogen granules (Z) in two adjacent cells and staining (arrows) of elements of the Golgi complex (G). Sections stained after localization with uranyl acetate and lead citrate. N: nucleus; M: mitochondrion.

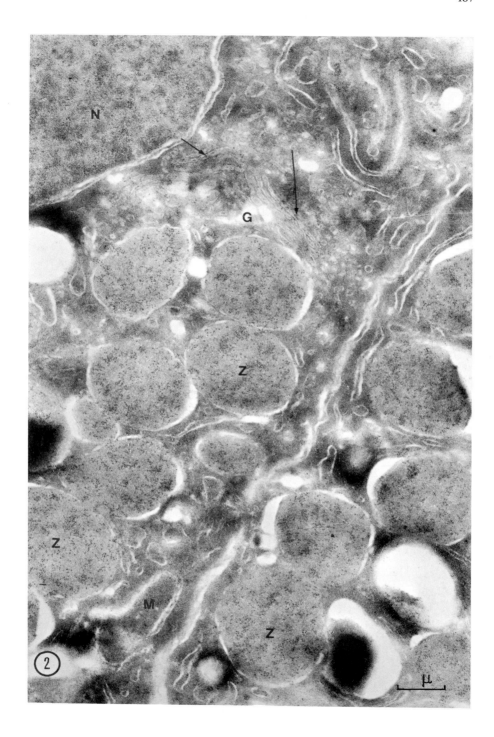

2

ether linkages to the cysteines of the peptide. It is prepared from cytochrome c by sequential pepsin and trypsin digestion (5). In our procedure (8), the pepsin digestion mix is chromatographed on Biogel P6 and the main peak containing a heme-undecapeptide is collected and digested with trypsin which removes 3 more residues from the amino terminus of the molecule. The trypsin digestion mix is then purified subsequently by gel filtration on Biogel P6 and countercurrent distribution. The heme-octapeptide obtained is pure and homogeneous, and is characterized by a molecular weight of 1550 daltons and a peroxidatic activity with a V_{max} of 4 mmoles H_2O_2 decomposed per min and per mg of heme-peptide, a K_m of 0.2 M and a pH optimum of 7.0 (8).

In order to couple the heme-octapeptide to Fab fragments we developed a two-step procedure which takes advantage of the presence of the unique amino group on the N-terminal cysteine of the peptide. First an active ester is formed between N-hydroxy-succinimide and para-formylbenzoic acid using carbodiimide under anhydrous conditions. The active ester is in turn coupled to the amino group of the heme-peptide and the derivatized heme-octapeptide is purified by gel filtration on Sephadex G 25. It is a stable intermediate which can be lyophilized and stored at $- 20^{\circ}$C. In a second step the formyl group introduced on the heme-peptide is allowed to form a Schiff base with the amino groups of the Fab fragment at pH 9.5. The Schiff bases are subsequently reduced using 0.1 M sodium borohydride.

In order to obtain a conjugate free of uncoupled molecules, the conjugation mix is purified both by gel filtration to remove the free heme-octapeptide and by ion exchange chromatography to separate free from conjugated Fab.

The conjugate consists of 2 heme-octapeptides on average conjugated to an Fab fragment. Its peroxidatic activity is characterized by a V_{max} of 0.4 mmoles H_2O_2 decomposed per min and per mg of attached heme-octapeptide, a K_m of 0.4 M and a pH optimum of 7.0. 70 % of the antigen binding capacity is retained (8).

5. Examples

Surface localization procedure. The example consists of the localization of secretory proteins in the bovine exocrine pancreatic cell. We have chosen this system to validate the surface localization procedure on a complex cell type,

---

Fig. 3-4. Thin section of rabbit mammary gland fixed 4 hrs with 4 % formaldehyde, 0.1 % picric acid in 0.1 M cacodylate pH 7.4. After fixation, the tissue was exposed sequentially to goat antibodies antirabbit secretory IgA, followed by rabbit Fab directed against goat IgG and coupled to the heme-octapeptide. The peroxidatic activity was obtained using diaminobenzidine and H_2O_2. N: nucleus; M: mitochondrion.
Fig. 3. Plasmacytes in the interstitial space beneath the epithelium (E). Note the staining of the internal spaces of the rough endoplasmic reticulum (arrows).

Fig. 4. Epithelial cell with its luminal surface (L). Reaction product is observed in vesicles (arrows) and in the Golgi complex (G).

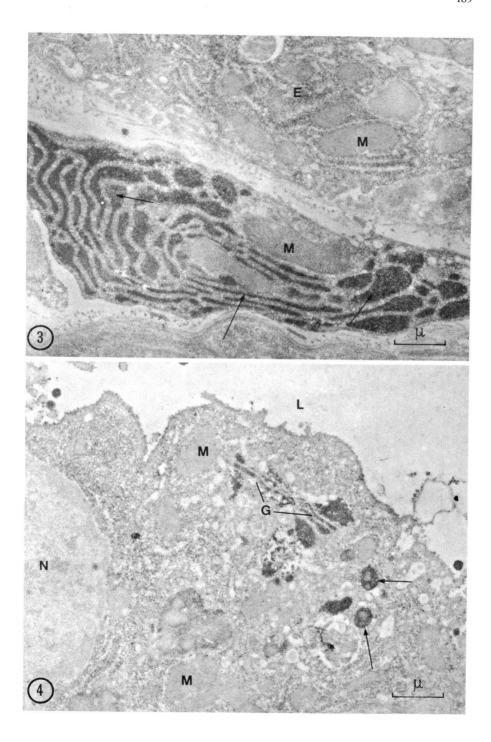

because it has been well established in earlier studies that the secretory proteins are present in high concentration in isolated zymogen granules, and because these proteins are commercially available in highly purified form for use as antigens. The specific antigen that we have chosen is trypsinogen A since this protein represents on a weight basis, about 23 % of the granule content.

In our studies, the tissue is fixed 4 hrs at 20°C in 0.5 % glutaraldehyde, buffered with 0.1 M cacodylate containing 5 % sucrose and embedded in bovine serum albumin (BSA) according to a modification of the technique of Farrant and Mc Lean (2) and Mc Lean and Singer (9). We should mention that the first step antiserum which we have raised in rabbits is monospecific. This is particularly important because of the sequence homologies of trypsinogen and chymotrypsinogen. No cross reactivity was observed, most likely because of the affinity purification involved in the preparation of the antitrypsinogen antibodies. The first step reagents and the conjugate (ferritin - sheep Fab antirabbit IgG) are applied directly onto sections mounted on grids (7).

Fig. 2 illustrates the presence of trypsinogen as marked by ferritin staining in all the granules in this field. The antigen is also located in the acinar lumen, in the Golgi complex and in condensing vacuoles. The labeling of the RER cisternae, however, is weak. A slight degree of background is observed over mitochondria, cytosol and the nucleus.

A similar localization has been found in the granules and the elements of the Golgi complex for ribonuclease A, DNase, chymotrypsinogen and carboxypeptidase A. These results suggest that all granules in all exocrine cells contain at least these secretory proteins, which tend to rule out specialization of product packaging at least at the level of the zymogen granules. These results, however, do not rule out the possibility that regional specialization of the synthesis of specific proteins exists on the attached polyribosomes. However, as is evident from results that we have to date, one of the limitation of the surface localization technique appears to be low sensitivity, particularly in regard to the ER which contains a low concentration of antigens.

Diffusion localization procedure. This example consists of the localization of rabbit secretory immunoglobulin A in the mammary gland of lactating rabbits. The procedure we use is an indirect sequence in which the first step reagent consists of goat serum directed against rabbit secretory IgA prepared from milk and colostrum. After adsorption with rabbit IgG, the antiserum reacts with antigenic determinants on the alpha chain of IgA and with the secretory component. The second step reagent consists of rabbit Fab directed against goat immunoglobulins and conjugated to the heme-octapeptide as described previously. The tissue is fixed 4 hrs in 4 % formaldehyde and non-frozen sections, 50 µ thick, are obtained and incubated 8 hrs with each immunological reagent separated by an overnight wash in 0.1 M phosphate buffer, pH 7.5. The peroxidatic activity is revealed with 3,3' diamino-benzidine, 0.05 % H_2O_2 in a 0.1 M Tris buffer, pH 7.0. The sections are finally postfixed in osmium and flat embedded in Epon.

Rabbit secretory IgA is localized in two different cell types: first in the plasmacytes situated in the interstitium of the mammary gland, but also in some subcellular compartments of the epithelial cells. Fig. 3 illustrates localization in plasmacytes. The reaction product is confined to the cisternae of the rough endoplasmic reticulum, to the perinuclear space and to the elements of the Golgi complex. The cytoplasm and the nucleus are free of reaction product. These results indicate that the local plasmacytes are involved in the synthesis of the molecule, and according to other works are responsible of the production of the IgA moiety. Fig. 4 illustrates the reaction product in the epithelial cells. On this micro-graph one can recognize a part of the alveolar lumen, the nucleus, mitochondria, rough endoplasmic reticulum, the Golgi complex and some apical vesicles. The reaction product is observed in the vesicles and saccules of the Golgi complex, in small vesicles in the apical cytoplasm, and also in lysosome-like bodies located basally in the cell. The controls were negative and no endogeneous peroxidatic activity was detected in the compartments mentioned. These data suggest that both cell types are involved in the processing of secretory of IgA but definitive con-clusions concerning the biological significance of this process will have to be correlated with the use of antisera reacting with each of the subunits of the secretory IgA molecule.

Our results, however, clearly indicate that the immunological tracer that we have developed is useful for the detection of intracellular antigens by the dif-fusion technique. In the two cell types mentioned, the tracer apparently is able to penetrate several different membrane-bounded compartments.

References

1. Borek, F., and Silverstein, A.M. (1961). J. Immunol. 87: 555.
2. Farrant, J.L., and Mc Lean, J.D. (1969). Abstracts 27th Meeting of the E.M. Soc.Am. p. 422.
3. Habeeb, A.F.S.A. (1969). J. Immunol. 102: 457.
4. Haemmerling, U., Aoki, T., de Harven, E., Boyse, E.A., and Old, L.J. (1968). J. Exp. Med. 128: 1461.
5. Harbury, H.A., and Loach, P.A. (1960). J. Biol. Chem. 235: 364.
6. Kraehenbuhl, J.P., de Grandi, P., and Campiche, M.A. (1971). J. Cell Biol. 50: 432.
7. Kraehenbuhl, J.P., and Jamieson, J.D. (1972). Proc. Nat. Acad. Sci.(USA) 69: 1171.
8. Kraehenbuhl, J.P., Galardy, R.E., and Jamieson, J.D. (1973). In press.

9. Mc Lean, J.D., and Singer, S.J. (1970). Proc. Nat. Acad. Sci. (USA) 65: 122.

10.Rosenstein, R.W., Nisonoff, A., and Uhr, J.W. (1971). J. Exp. Med. 134: 1431.

11.Singer, S.J. (1959). Nature (London) 183: 1523.

12. Sternberger, L.A., and Cuculis, J.J. (1969). J. Histochem. Cytochem. 17: 190.

13.Vogt, A., and Kopp,R. (1964). Nature (London) 2o2: 1350.

Supported by U.S.P.H.S. Grant AM 1o928.

Electron microscopy and Cytochemistry, eds. E. Wisse, W.Th. Daems, I. Molenaar and P. van Duijn.
© 1973, North-Holland Publishing Company - Amsterdam, The Netherlands.

IMMUNO-ELECTRON MICROSCOPE STUDY ON TWO TYPES OF
STREPTOCOCCAL CARBOHYDRATE ANTIGENS.
A COMPARISON OF TWO DIFFERENT INCUBATION TECHNIQUES

W.H. Linssen, J.H.J. Huis in 't Veld, C. Poort, J.W. Slot and J.J. Geuze

Centre for Electronmicroscopy, Faculty of Medicine,
State University, Beetsstraat 22, Utrecht, The Netherlands.

Summary.

Ferritin labeled antibodies were used to localize a group antigen Z_3 and a type antigen III in two strains of streptococci. The results obtained by incubating whole or disrupted cells as compared to those obtained by incubating ultrathin sections of glycolmethacrylate embedded bacteria are discussed.
With the aid of the section technique it could be proved that the type antigen is present throughout the entire cell wall whereas the group antigen is located in a thin layer just outside the cellmembrane. The section technique disclosed an intracellular deposit of type III antigen in distinct parts of the cytoplasm.

1. Introduction.

Two strains of streptococci classified as Z_3III and Z_3 have a carbohydrate in common which acts as a group specific antigen. In addition Z_3III has a type antigen III, the mutant strain Z_3 is deprived of it. Serologically there are strong indications that in Z_3III the outer part of the cell wall contains the type antigen III and that Z_3 carries the group antigen Z_3 in its outer layer[1]. No morphological data were available on the localization of the group antigen in the strain Z_3III, although serological experiments suggest that this antigen is not available in the surface layer of the bacteria.

2. Materials and methods.

The organisms were grown for 30 hrs at 37^0 C in Todd-Hewitt medium[2]. The bacteria were collected and washed by centrifugation. Anti-group (anti Z_3) and anti-type (anti III) sera were obtained by immunization of rabbits[2].

The gamma globulin fraction was obtained by ammonium sulphate precipitation and DEAE-cellulose column chromatography[3].

The ferritin coupling was carried out using the method described by Sri Ram et al.[4], modified by Linssen[3].

<u>Incubation of the bacteria with the labeled antibodies in "bulk"</u>

A suspension of bacteria was incubated in a solution of labeled antibodies (0,5%) at pH 7.55 for 20 hours at 4^0 C either before, or after fixation. The organisms were fixed in a solution of 1-3% formaldehyde in 0,07 M sodium phosphate pH 7.4.

This was followed after 3 min by the addition of 1% glutaraldehyde. The post-fixation was carried out with 1% osmium tetroxide buffered with veronal. The de-hydration was in aethanol, the embedding in Epon 812.

Incubation of the bacteria with the labeled antibodies "on the section".

Bacteria were fixed in 1-3% formaldehyde followed by glutaraldehyde as described above. As osmium tetroxide interferes with the polymerization of the embedding medium, no postfixation was allowed. The dehydration occurred in increasing con-centrations of glycolmethacrylate.
The specimen were embedded in a mixture of glycolmethacrylate 92%, water 3% and divinyl benzene 5%[3].
Ultrathin sections were incubated with 10 μl of ferritin-labeled antibodies (o,5%) in a special trough[3]. The pH was adjusted to 7.5-7.6. After 2 hrs the sections were exhaustively washed at pH 7.5-7.6. The sections were pre-incubated in some cases in a solution containing 1% apoferritin (pH 7.5-7.6) to avoid non specific binding of the ferritin conjugate.
For blocking, inhibition and other controls, unlabeled antibodies and non speci-fic labeled rabbit serum globulins were used.
Sections were stained with uranyl acetate.

3. Results

The bulk labeling of Z_3III bacteria in anti III-ferritin-labeled globulin re-sulted in deposits over the outer half of the cell wall (fig. 1). On the other hand, incubation of sections with anti-III labeled globulin yielded a labeling of the cell wall throughout its whole thickness as well as on some distinct sites in the cytoplasm (fig. 2). The intracellular distribution as depicted, was obtained after preincubation of the sections with apoferritin. Without this measure there was a more diffuse and apparently aspecific labeling of the cytoplasm.

After bulk labeling of the Z_3 mutant strain with ferritin labeled anti Z_3 an-tibody a relatively dense labeling was found over the cell wall. In this strain

Fig. 1. Streptococcus Z_3III incubated with ferritin labeled anti-type III serum globulin. Reaction in bulk.
Fig. 2. Streptococcus Z_3III incubated with ferritin labeled anti-type III serum globulin. Reaction on the section. Apoferritin preincubation.
Fig. 3. Streptococcus Z_3 incubated with ferritin labeled anti-group Z_3 serum globulin. Reaction on the section, without apoferritin preincubation.
Fig. 4. Streptococcus Z_3III incubated with ferritin labeled antigroup Z_3 serum globulin. Reaction on the section, without apoferritin preinucbation.

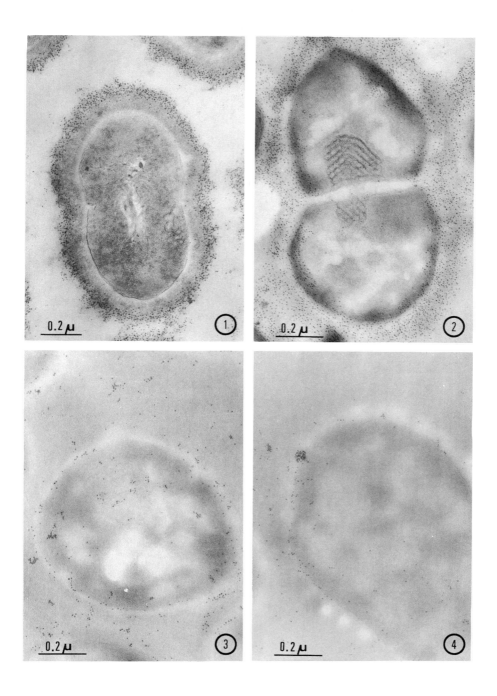

the cell wall is much thinner than in the Z_3III strain and seems to correspond to its innermost layer. In sections of Z_3 bacteria the much weaker labeling of the cell wall with Z_3 antibody follows closely the cell membrane (fig. 3).

The label found throughout the cell seems to be unspecific to some extent as a similar pattern is found after the incubation with labeled non specific globulins.

While Z_3III bacteria were not labeled upon incubation in bulk with ferritin labeled anti Z_3 antibody the incubation of sections resulted in a labeling similar to that of sectioned Z_3 bacteria (fig. 4).

4. Discussion.

Comparison of the incubation techniques of whole bacteria vs. section clearly demonstrates the advantage of the latter in accurately localizing less accessible antigens.

However, the incubation of sections often results in much weaker reactions. This is partly because the fixation and embedding interferes with the specific binding capacities of the antigen. Partly a restriction of the reaction intensity is given by the balance in chosing incubation conditions that must be made between the conflicting demands of an optimal antigen-antibody reaction and avoidance of unspecific binding of the conjugate to the section.

Preincubation of the section with apoferritin has been found to be a useful measure in reducing aspecific labeling.

References.

1. Willers, J.M.N., Deddish, R.A. and Slade, H.D., 1968, J. Bacteriol. 96, 1225.

2. Huis in 't Veld, J.H.J., 1973, in: Studies on the structure of the streptococcal cell envelope. Thesis, University of Utrecht, p. 37.

3. Linssen, W.H., 1971, in: Over de toepasbaarhid van de ferritinetechniek bij de submicroscopische lokalisatie van antigenen. Thesis, University of Utrecht.

4. Sri Ram, J., Tawde, S.S., Pierce, G.B. and Midgley, A.S., 1963, J. Cell Biol. 17, 673.

5. Huis in 't Veld, J.H.J. and Linssen, W.H., 1973, J. Gen. Microbiol. 74, 315.

Electron microscopy and Cytochemistry, eds. E. Wisse, W.Th. Daems, I. Molenaar and P. van Duijn.
© 1973, North-Holland Publishing Company - Amsterdam, The Netherlands.

LOCATION OF BACTERIAL ANTIGENS

P.D. Walker, J.H. Short and R.O. Thomson
Department of Bacteriology
Wellcome Research Laboratories, Beckenham, Kent.

1. Introduction

Unlike corresponding studies with tissues, a study of the immuno-chemical localisation of antigens of bacteria presents formidable problems. Pieces of tissue, even when cut as thin frozen sections, remain a unity and, therefore, can be stained by various cytochemi-cal, histochemical and immunochemical procedures prior to embedding and sectioning. On the other hand, a bacterial suspension is a collection of balls or rod-like particles without a supporting matrix and any attempt to cut sections for post-embedding staining techniques without supplying an artificial matrix holding the particles together is impossible. The necessity, therefore, to develop post-embedding staining techniques with labelled antibodies for staining bacteria is absolutely vital if internal antigens are to be located.

2. Results of pre-embedding staining

Our initial studies first published in 1966 used as a model spore-forming bacteria. Using antisera prepared against the vegetative cell and spore labelled with ferritin we were able to demonstrate the difference in the surface antigens of the spore and vegetative cell, and to locate vegetative cell antigens within partially disintegrated spores.

More recently we have labelled antisera from animals vaccinated with various vaccines with ferritin and used such antisera to locate antigens associated with immunogenicity. Young cells of Fusiformis nodosus, an organism capable of causing foot-rot in sheep, are heavily piliated (Fig.1) and ferritin labelled antisera from vaccina-ted animals show specific location of ferritin particles along the pili (Fig.2). The presence of high titres of antipili antibody in sheep is correlated with protection and the results of labelled anti-body studies confirm the role of pili and the agglutination reaction. Similarly, experimental vaccines prepared from Neisseria gonorrhoea in the appropriate phase of growth produce antibody in animals specifically directed against the pili.

These observations have significance in the area of development of vaccines and demonstrate the use of pre-embedding staining techniques to confirm the type of antibody being produced in vaccinated animals.

3. Results with post-embedding staining techniques

Attempts to avoid the non-specific attraction of ferritin to commonly employed embedding media have been directed along two main lines. In the first instance, we have investigated the non-ionic embedding media described by Singer, polymerised bovine serum albumin, and glycol methacrylate. These studies have been combined with different staining procedures paying particular attention to keeping the specimen wet during staining. Although successful results have been obtained on occasions these have, in general, been disappointing.

More success has been achieved using enzyme labelled antibodies. Sporulating cells of Bacillus cereus fixed in formaldehyde and embedded in glycol methacrylate were post-stained with acid phosphatase antibody to the vegetative cell. Specific location of antibody was seen on the cell wall and developing cortex of the spore (Fig.3). These results confirmed earlier studies with ferritin labelled antibody. Several factors appear to affect the consistency of staining from day to day even with the same solutions. Absolute cleanliness during staining is necessary and in order to achieve this sections have been cut on to acetone and no wax has been used in sealing boats to the knife.

References

1. Thomson, R.O., Walker, P.D. and Hardy, R.D. 1966. Nature 210, 760.
2. Walker, P.D., Baillie, A., Thomson, R.O. and Batty, I. 1966. J.Appl.Bact. 29, 512.
3. Walker, P.D., Short, J., Thomson, R.O. and Roberts, D.S. 1973. J.Gen.Microbiol. 77 (In press).
4. Walker, P.D., Thomson, R.O. and Short, J. 1971. In A.N.Barker, G.W. Gould, and J. Wolf (ed.) Spore Research 1971. Academic Press Inc., London.

Fig.1 Negatively stained preparation of young culture of Fusiformis nodosus.

Fig.2. Negatively stained preparation of young culture of F.nodosus stained with ferritin labelled antibody from vaccinated animals

Fig.3. Ultrathin section of sporulating cells of Bacillus cereus fixed in formaldehyde and embedded in glycol methacrylate, post-stained with acid phosphatase labelled vegetative cell antibody.

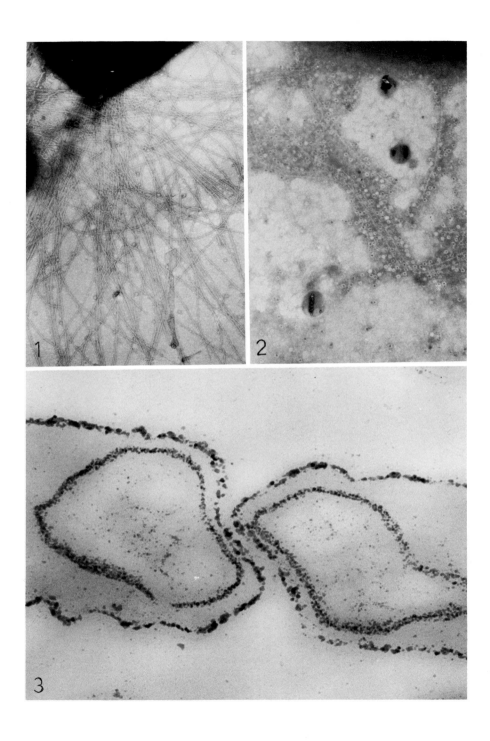

X-RAY MICROANALYSIS

Electron microscopy and Cytochemistry, eds. E. Wisse, W.Th. Daems, I. Molenaar and P. van Duijn.
© 1973, North-Holland Publishing Company - Amsterdam, The Netherlands.

THE USE OF WAVELENGTH DISPERSIVE X-RAY
MICROANALYSIS IN CYTOCHEMISTRY

John A. Chandler

Tenovus Institute for Cancer Research,
Welsh National School of Medicine,
Heath, Cardiff CF4 4XX

Summary

The principles of X-ray microanalysis are outlined and various instrumental arrangements are described using wavelength spectrometers and energy dispersive analysers. X-ray microanalysis has an important function in cytochemistry since it enables quantitive determinations of cellular components to be determined. The technique depends on the detection of one or more elements in the tissue thereby obviating the exclusive need for electron dense stains and opening up the possibilities of using well established optical stains.
Specimen preparation is discussed in relation to microanalysis and a number of applications illustrates the usefulness of the technique in cytochemical work.

1. Introduction

"The nature of X-ray microanalysis".

X-ray microanalysis is a technique which allows a chemical analysis to be performed on biological tissue (and other material) within very small and well defined regions of the specimen. It makes use of the fact that atoms, when struck by electrons from an external source, yield X-rays which are characteristic of those atoms and are used to identify and quantify the elements present.

The idea of generating X-rays from atoms with electrons for identification was first proposed in 1943 but it was not until the late 1940's that it became a practical reality when the two techniques of X-ray spectrometry and electron optics were combined. Since then, and especially in the last 15 years, many hundreds of instruments employing X-ray microanalysis have been produced and used in such diverse scientific areas as metallurgy, physics, electronics, mineralogy, environmental pollution, geology, and lately in pathology, zoology, biochemistry and other biological fields. However, the number of instruments being used in biological research is a tiny proportion of the whole, due in part to the somewhat forbidding nature of the technology, and also to the peculiar physical and chemical properties of biological material.

The purpose of this paper is to enlighten the cytochemist who wishes to gain an understanding of and greater confidence in the principles of X-ray microanalysis and to help him determine the potential such a technique might have in his own field of research.

A range of applications is included in an attempt to encourage the biological reader to believe that sophisticated technology need not be divorced from the real business of understanding his tissue.

"The nature of the problem and the need in cytochemistry."
Modern electron optical instruments are able to provide

information from biological material at resolutions of a few
angstrom units in bulk(thick),thin(1-2 μm) and ultrathin(<1000 AU)
specimens. Ultrastructural detail can now be provided such that the
limitations lie not so much in instrumental factors but in methods
of specimen preparation and interpretation of the information provided.

There also frequently arises the need to complement the available
morphological information with a chemical analysis from the same
areas of interest in order to determine the relationship between
ultrastructural changes during a dynamic phenomenon and variations
in chemical composition of the tissue. Without local chemical
analysis the electron microscopist must interpret as best he can
from the ultrastructure what biochemical events are occurring. For
many years the microscopist has been happy to do this,indeed to such
an extent that ultrastructure has become as important as the biochemistry and in many cases more so.

To the investigator concerned with a complete understanding of
the nature of tissue and cell processes, the combination of ultrastructural detail and subcellular chemical analysis within the same
specimen is a very important subject.

Several classes of problem can occur as listed briefly below.
(i) Natural elemental content of tissues.
The distribution of elements throughout subcellular regions
can be demonstrated by analysis of normal physiological levels. As
will be seen this requires a great deal of attention to be paid to
methods of specimen preparation.
(ii) Accidentally introduced foreign material.
The distribution of particles,toxic chemicals and strange
elements can be located and identified within tissue.
(iii)Deliberately introduced elements.
The results of administration of drugs and experimental substances can be traced throughout the tissue and related to morphological changes which occur therein.
(iv)Histochemistry and immunochemistry.
This field is perhaps the most promising one for it allows
the analysis of biochemical events in situ.

Although X-ray microanalysis, as defined above, allows the identification of individual elements,we are not always concerned with
elements as such. Molecular groups and protein structures can be
identified by specific labelling methods,e.g.by precipitating or
substituting the molecular group in situ within the tissue,or by
reacting protein in the tissue against specific antibodies to which
enzymes and other labels have been coupled.

Other proteins and some smaller molecules such as steroids may
also be labelled by coupling to heavy metals as tracers in the
tissue suitable for analysis by X-ray methods.

2. X-ray Production
"Characteristic radiation"
Each element in the periodic table has a very well defined
distribution of electrons within the atom. X-ray microanalysis is
based upon the excitation of these electrons to produce an emitted
X-ray spectrum which is characteristic of the element concerned.

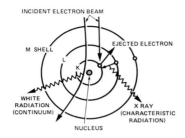

Fig 1.
Simple schematic representation of the atom.

In fig.1 the nucleus, comprising neutrons and protons, is
surrounded by orbital electrons distributed quantum mechanically in
various energy levels. There may be many energy levels depending on
the size and state of the atom and these are grouped into major units
called shells. Elements are thus characterized by their nuclear
charge, and number and energy distribution of orbital electrons. The
element can be defined or identified by the electron energy levels
or the energy difference between these levels. Large atoms (from
heavy elements) contain greater numbers of electron orbits and shell
units. If one of the orbital electrons is removed from its energy
level by an incident electron then the atom is said to be in an
excited state, or ionised. When this occurs an electron from a
higher energy state will fall down into the gap to stabilise the
atom. Because of the difference in potential energy levels the excess
energy is emitted during this electron transition as an X-ray photon.
This photon represents the exact difference in energy levels
between the two electron shells. Thus if an electron in the K shell
is removed, a second electron from the L shell may instantaneously
replace it, thus giving off its excess energy as a photon of energy
$E_L - E_K$, generally called K_α radiation. Now the filling of the vacancy in the K shell by an electron from the L shell will also produce a vacancy in the L shell. In turn this may be filled with one
from the M shell and so on, each with the production of X-ray photons
of energies determined by the orbital energies. Thus a single ion-
isation can give rise to a whole spectrum of characteristic X-rays
and this energy spectrum identifies the atom.

"X-ray Continuum"
Referring again to fig 1. a primary electron beam, instead of
interfering with the orbital electrons to produce characteristic
X-rays from the atom, may interact with the nucleus. As the incoming
electron beam is decelerated by the field of nuclear charge it
radiates energy. This loss of energy can be anything from the
maximum originally carried by the electrons to a small fraction of it.

Fig.2. shows the energy spectrum produced by this effect with the
maximum energy being that of the primary electron beam E_O. This
general spectrum is called "X-ray continuum" or "continuous radiation", "white radiation" or "bremstrahllung". It forms the

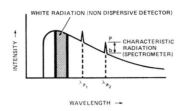

Fig 2. Energy spectrum produced by electron interaction with
matter. Characteristic lines are superimposed on the
white radiation.

background upon which the characteristic X-ray lines are superimposed at specific wavelengths(or energies). This background must
be subtracted from the characteristic signal produced by the specimen to obtain a valid reading of intensity. X-ray continuum forms a
basic limitation to the ability to detect a characteristic 'line'
since it can be considered as X-ray noise. In addition the
continuum can cause fluorescence of X-rays from other atoms within
the specimen. It can however be of great use in quantitation since
it can provide us with a measure of specimen 'mass thickness'in
order to determine elemental concentrations.

Space here does not allow a description of quantitative procedures
for which the reader is referred to Hall[1].However a simple outline
may be given.

The X-ray signal collected from a given sample in a given time,
under given experimental conditions, and for a particular element,
may be considered to be proportional to the mass of that element.
If the mass thickness of the specimen is also measured by collection
of the white radiation(for example a region such as the hatched area
of fig 2),then a method of calculating elemental concentrations is
available. With crystal spectrometers, over a wide range of
intensities, the X-ray signal is exactly proportional to the number
of atoms of a particular element in the area being analysed. With
thick specimens precautions have to be taken regarding possible
absorption as the X-rays leave the specimen,but this is not troublesome in thin specimens.Standards having known weight fractions of
the elements concerned are used for calibration and such standards
need to be as close in nature to the specimen itself.Alternatively
very thin films of evaporated metals may be employed.

3. Instrumentation

Several configurations of electron optical systems are available,
many of which make use of X-ray analysis facilities. For biological
materials the methods of viewing the specimen fall into two main
categories. For thin specimens the image is viewed in transmission
and for bulk specimens by reflection.

For cytochemical work the instrumental arrangements of most
interest are :
(a)Scanning electron microscope + X-ray detection(SEM + X-ray);
(b)Scanning transmission electron microscope + X-ray detection
(STEM + X-ray);

(c) Transmission electron microscope + X-ray detection(TEM + X-ray);
and the X-ray detection systems most commonly used are : Wavelength
dispersive crystal spectrometers;and energy dispersive solid state
detectors.

Another instrument combining electron optics with X-ray microanalysis is the electron probe microanalyser(EPMA)but its usefulness
in cytochemical investigation is limited and it has been largely
superseded by the SEM & X-ray systems.

The two important parameters to be married together are the best
resolution for imaging the specimen and the highest sensitivity
available from the X-ray detector system.

"Wavelength dispersive crystal spectrometer"

Fig 3. illustrates the principle of wavelength crystal spectrometers. X-rays leaving the specimen may cover a range of wavelengths
from a range of elements excited by the electron beam.

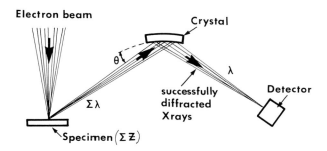

Fig 3.Operation of wavelength crystal spectrometers. The
wavelength of X-rays successfully reflected(diffracted)
depends on the angle of the crystal.

X-rays leave the specimen at all angles but a narrow cone passes
towards a specially curved crystal such that a fraction of this
signal is reflected into a detector. The fraction of the X-ray beam
which is reflected depends upon the ability of the crystal to
"diffract" a particular wavelength while excluding all other wavelengths.

This phenomenon is described by Bragg's law which states that for
crystals :

where n is an integer, λ is the X-ray wavelength successfully
diffracted, d is the lattice spacing of the crystal and θ is the angle
of incidence(and of reflection) of the X-ray beam at the crystal.

This law illustrates that given a crystal of known lattice
spacing d, and the angle of incidence θ , then the wavelength of the
diffracted X-rays can be calculated. In practice the crystal is
scanned through a range of angles until a 'peak' is noted in the
detector at which point the conditions of Bragg's law are satisfied
and the X-ray detected can be identified.

For good resolution and highly efficient diffraction of the X-ray
lines, the crystal must be of excellent quality with no flaws or
blemishes. The machinery that governs the position of the crystal

relative to the specimen and to the detector must be of the highest
accuracy to achieve reproducibility.

After being selectively diffracted by the crystal the **X**-rays of a
particular wavelength pass towards the spectrometer detector which
is usually a gas flow, or sealed,proportional counter.

The range of wavelengths that such a crystal can cover depends
upon the range of angles that it can be varied through. This limits
a crystal to a few angstrom units of wavelength corresponding to the
X-ray lines of few elements in the periodic table. To extend this
range a diffracting spectrometer is often equipped with a number of
crystals having different d values,or lattice spacings, so that
for the same Θ range the wavelength(λ) range is enlarged and so a
greater number of elements are covered.

"Energy dispersive analyser"

A limitation of the crystal spectrometer is that it is unable by
definition to detect and display all X-ray energies leaving the
specimen at once. The energy dispersive analyser,otherwise called
a solid state detector(SSD)provides an energy spectrum of all
elements analysed simultaneously.

A very great advantage of the solid state detector is that it can
be placed very close to the source of X-rays and so accept a wide
solid angle of radiation. This increases detection sensitivity
enormously and thus provides better statistical data. Such detectors
are discussed more fully elsewhere in this book, but they form such
an essential part of any comprehensive analytical system that they
will be included in this paper.

The type of system chosen for a particular application depends
entirely on the nature of the problem. When dealing with biological
materials,however, and except in the case of teeth or bone, the
elemental levels are frequently very low$(< 1\%)$ and special conditions
govern the choice of analysing system as well as electron optical
system.

"Energy resolution"

Energy resolving power of a detector may be defined as the ability
to separate two adjacent peaks in the energy spectrum. The resolution
is generally taken to be the full width of the peak at half the
maximum intensity(fwhm) and usually is quoted at the energy 5.9 KeV
in the spectrum.

Besides major X-ray lines, there are many other overlapping lines
in the energy spectrum when a number of elements are present
together in the specimen region being analysed. For example the K_α
line of phosphorus occurs at 2.01 KeV whereas the M_α line of osmium,
a frequently employed stain in electron microscopy,occurs at 1.91
KeV, only 100 eV away. Such overlaps occur frequently in biological
studies especially when heavy metals are employed to provide stains
or are used in cytochemical techniques. Table 1 illustrates some of
the major X-ray lines obtained from each element and shows how close
many of these lines can occur.
The ability to resolve two such lines depends on the relative peak
heights, i.e. relative concentrations of the elements. Heavy metal
stains are frequently present in concentrations much greater than
the naturally occurring elements and resolution requirements vary.

Solid state detectors currently commercially available have

Table 1.

Element	Characteristic line	eV	Element	Characteristic line	eV
Cu	L_α	930	Sn	L_α	3443
Zn	L_α	1011	Sb	L_α	3604
Na	K_α	1041	Ca	K_α	3691
Mg	K_α	1253	Sb	L_β	3843
Al	K_α	1486	Ti	K_α	4510
Si	K_α	1740	Cr	K_α	5414
Si	K_β	1835	Mn	K_α	5898
Os	M_α	1910	Fe	K_α	6403
P	K_α	2013	Ni	K_α	7478
Au	M_α	2123	Cu	K_α	8047
P	K_β	2139	Zn	K_α	8638
S	K_α	2307	Os	L_α	8841
Pb	M_α	2345	Cu	K_β	8905
Cl	K_α	2622	Os	L_α	8911
Ag	L_α	2984	Cu	K_β	8977
U	M_α	3170	Zn	K_β	9572
K	K_α	3314	Au	L_α	9628
U	M_β	3336	Au	L_α	9713

resolving powers of about 160 eV although this figure is being
extended downwards almost monthly as quality improves. The resolution can also be artificially increased(made better) by computer
techniques of spectrum stripping where the noise and statistical
fluctuation is smoothed out. Physical limitations however(thermal
and quantum noise) place a lower limit on the resolving power likely
to be achieved with this detector(possibly around 100 eV).

Crystal spectrometers function at a much higher resolution by the
nature of their operation. The technique of wavelength dispersion by
diffraction ensures that only those X-rays of a very precise wave-
length(or energy) reach the detector. Resolving powers better than
10 eV can be achieved routinely with such a system.

"Sensitivity"

The ability to resolve X-ray lines alone is insufficient unless
the detector system allows adequate X-ray counts to be accumulated
in reasonable analysis times. However, a distinction must be made
here between "count rate sensitivity" and "mass detection sensitivity".

The solid state detector has the great advantage of being able
to be placed very close to the specimen(sometimes just a few milli-
metres away). This vastly raises the count rate. Thus a greater
number of X-ray photons can be detected in the same time or the same
number in a shorter time. This means that a lower electron signal
can be employed to generate a greater X-ray signal thus allowing a
smaller probe diameter and reducing specimen damage. However, a
faster counting rate does not necessarily in itself increase mass
detection sensitivity since this depends critically on the ratio of
peak counts to background counts.

To make a direct comparison between a solid state detector and a
crystal spectrometer arrangement would not be helpful because of the
very wide variation in design. Either system may be used with the

electron optical arrangements described below and often they are used in conjunction. The wavelength dispersive crystal spectrometer and the solid state detector essentially provide the same information about the elemental nature of the specimen but present it in different ways.

The various combinations of electron optical systems with X-ray microanalytical detectors that are of most interest to cytochemistry are illustrated below.

"Scanning Electron Microscopy + X-ray analysis"

Fig 4. illustrates the arrangement by which crystal spectrometers or solid state detectors, or both, may be incorporated into a scanning electron microscope. The microscope may be operated for secondary emission or transmission imaging. A great advantage of the SEM is the ease with which detectors may be brought close in to the specimen for greater sensitivity.

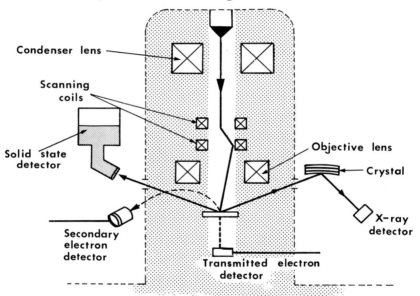

Fig 4. Schematic representation of a scanning electron microscope incorporating X-ray microanalysis.

The SEM allows X-ray information to be obtained in the form of a two dimensional elemental map across the specimen;as a line trace across a region of interest, or by static probe analysis.

Two dimensional analysis is performed by setting the detector to collect only the particular element of interest and then allowing the focused electron beam to scan the specimen surface. The X-ray signal from the detector is displayed on a cathode ray oscilloscope by synchronising with the electron beam raster.

For one dimensional analysis,i.e. a line trace, the output is simply fed into a chart recorder so that changes in elemental content can be monitored across that line in the specimen.

211

Most sensitive and quantitative analysis is obtained with a
static probe positioned over the region of interest and by counting
the X-ray emission over a period of time. For qualitative analysis
the crystal spectrometer is made to run through a range of wavelengths to cover the elements of interest. With the SSD, qualitative information is at once produced as an energy spectrum for the
static probe.

"Transmission electron microscopy + X-ray analysis"

A logical extension of transmission electron microscopy is to
combine high resolution imaging of ultrathin sections with high
efficiency X-ray microanalysis.

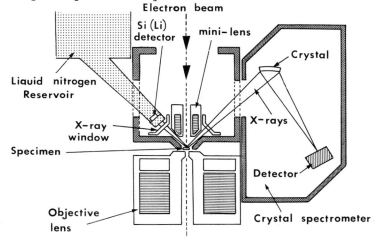

Fig 5. EMMA-the electron microscope microanalyser, combining
wavelength crystal spectrometers with a solid state
detector.

Fig 5. illustrates this combination in the integral instrument
EMMA (electron microscope microanalyser[2]) where the design has been
such as to avoid compromise between the high resolution electron
imaging and the efficiency of analysis. The departure from a
conventional electron microscope exists in the centre where,
immediately above the specimen an extra condenser lens is involved.
This small conical lens, called a "minilens", allows the illumination
to be focused to a small probe on the specimen. Probe diameters of
0.1 - 0.2 μm are commonly employed.

Because of the nature of the imaging the transmission image is
visible on the microscope screen even during analysis so that very
accurate location is possible. X-rays leaving the specimen enter
the detector systems (double crystal spectrometers and solid state
detector) via thin windows to allow the microscope to operate at
high vacuum and without introducing contamination into the specimen
region. Both detector systems can function simultaneously, and in
some arrangements the SSD can be made to slide in and out towards
the specimen to increase the solid angle of collection of X-rays.

Fig 6. illustrates the arrangement and demonstrates the great
advantage of this configuration whereby the ultrastructure can be
correlated with subcellular microanalysis for detailed cytochemical
studies.

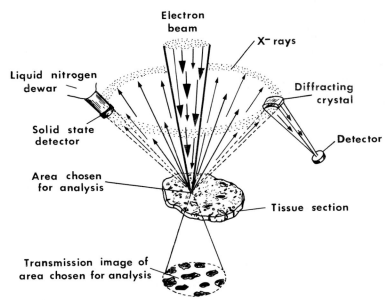

Fig 6. Schematic representation of X-ray microanalysis of thin sections with transmission electron microscopy.

The instrument is operated basically in the same mode for microanalysis as for normal transmission electron microscopy with the range of voltages from 40-100 KV suitable for most types of ultrathin specimen. Although beam deflectors are incorporated in the instrument in the condenser system the probe is not normally scanned as in the previous configuration but the instrument is used essentially as a static probe analyser. With typical ultrathin sections (< 1000 AU) the amount of material to be analysed is extremely small (generally < 10^{-15} gms) and such scanning systems would rarely yield useful information. Typical counting times are in the region of 100 seconds per point analysis.

There are many hundreds of transmission electron microscopes in all corners of the world which are capable of being converted to X-ray analysis instruments simply by attaching a suitable detector to the specimen region. This often requires some modification to the microscope column but quite a satisfactory arrangement can often be obtained without too much disturbance of the electron optics. If no minilens is incorporated into the condenser system the smallest available probe is that of the conventional microscope i.e. 1-2 μm. With a minilens however, probes down to 500 AU can be achieved but with quite low intensity beam currents.

Crystal spectrometers were the first type of detector to be incorporated into microscopes in this way but these have to some extent given way to solid state detectors with their rapid development in recent years. Such detectors are more versatile and flexible so that they can be interfaced between different electron optical instruments. There should be a great increase in the numbers of transmission electron microscopes having this attachment facility.

Again, like EMMA, this combination instrument finds its greatest use in static probe analysis and combines X-ray analysis with detailed morphological studies on ultrathin sections.

4. Specimen Preparation.

X-ray microanalysis has been performed on both hard and soft tissue in thick and thin forms as well as in isolated cells. For an interpretation of the analytical results the investigator must have full confidence in the integrity of the specimen, i.e.that elements that have been analysed shall be distributed within the specimen in the same way as in the original living tissue, or at least that any differences should be entirely understood.Different preparatory techniques will affect elements in the tissue in different ways depending whether they be free or bound. Soluble electrolytes, for example, are unlikely to remain in their in vivo sites when subjecting tissues to organic or aqueous solutions. The binding forces involved in linking elements to organic structures are very complex and little understood. Elements may exist also in tissue in both the bound form and as free ions and will be affected differently by different procedures.

Correlative techniques are necessary to establish the nature of these binding forces and the effects of various treatments upon them. For example, atomic absorption spectrometry is used regularly in the author's laboratory to determine the degree to which elements are free or bound in tissues subjected to conventional preparation involving solvents.

Thick specimens (> 2000 AU)

"Liquid methods"

Some standard histological methods may be quite suitable for some types of analyses provided one is confident that the final specimen is sufficiently similar to the original tissue. Such methods may involve the use of fixatives such as alcohol,aldehydes,formalin,or acetone; embedding media such as paraffin,styrene,methacrylate and low viscosity resins;clearing solutions such as Xylol,toluol, benzene,chloroform or cedarwood oil; and possibly flotation on water and staining for histological correlation or histochemical tests.

All or any of these procedures can possibly contaminate the specimen or cause elemental loss or redistribution. Several cases have been quoted where this has arisen and the losses have been intolerable, yet other examples, on different tissues and with analysis of different elements have proven otherwise.

"Freeze drying"

The commonly used method of rapidly freezing the tissue block, freeze drying, vacuum impregnation and sectioning critically reduces the risk of sample contamination, loss of soluble substances, displacement of cell constituents and chemical alteration of reactive groups.

However the possibility of changes occurs during the freeze drying procedure as the ice-vapour front moves through the tissue, and from the embedding medium where elements such as sulphur, chlorine,calcium and silicon may be present.

A method described by Läuchli et al[3] is of importance in this
context. Anhydrous ether is used to freeze substitute the tissue,
followed by infiltration with a low viscosity epoxy embedding resin[4].
Sections are floated on hexylene glycol instead of water to avoid
leaching of electrolytes.

"Cryomicrotomy"

The technique of cutting frozen sections and freeze drying them
ready for analysis is of value in order to be absolutely sure that
no risk is attached to elemental redistribution or loss.Histological
detail in these sections suffers from difficulties associated with
ice crystal formation, the hygroscopic nature of the specimen,
unevenness,compression artefacts, and folds. The morphological
information is nowhere near as good as with conventional methods but
in many cases is good enough for the spatial resolution of analysis.
With section thicknesses between 1-5 µ electron diffusion is considered to occur within a narrow cone,thus providing reasonable spatial
resolution with relatively large quantities of material for analysis
(compared with ultrathin sections).

"Air drying"

For the analysis of cell suspensions(e.g.blood,sperm,etc)
relatively simple air drying procedures can be adopted. Some care
has to be taken to obtain 'pure'cells by this method. For example,
human semen contains much debris in the seminal fluid which
contaminates the cells. Washing the cells is not satisfactory
because of elemental redistribution. Freeze drying of cell suspensions is an alternative for air drying.

"Mounting"

Specimens are mounted on flat substrates of materials that will
not interfere with the analysis(e.g. carbon,quartz,beryllium,nylon
films) and are thoroughly coated with a thin film(e.g. carbon or
aluminium) to prevent electrostatic charging or specimen damage from
the electron beam.

Thin specimens (< 2000 AU)

The same principles apply regarding elemental loss and
redistribution as described for thick specimens but the methods
involved in overcoming these difficulties vary somewhat. Since the
principle of performing analysis on ultrathin sections is that of
correlating ultrastructure with analysis, methods must be chosen
which will not affect the chemicals of the sample too much but will
also preserve the ultrastructure to a very high degree.Sensitivity
of analysis is all important because of the relatively small amount
of material compared with thick sections) but image resolution is
far superior.

"Liquid methods"

An important property of the fixative is that it should form
stable bonds which will hold the tissue molecules together during
dehydration and subsequent treatments. Giese[5] discusses the solubility of proteins,lipids,carbohydrates and conjugated compounds in
various solutions. The whole story of chemical fixation is at once
enormously complicated and little understood. Contamination may arise
from buffers containing phosphorus,sodium, or other elements.

Dehydrating agents like alcohols,acetone,gelatin, as well as being agents for replacing water, are nearly all strong organic solvents which can cause some extraction of cell constituents.Propylene oxide has been found to extract lipids, possibly even from fixed tissue. The various embedding media may contain contaminating elements,and might introduce errors into the analysis of concentrations depending upon their stability in the electron beam.

Sections need to be thin enough to allow good resolution in the image yet thick enough to allow sufficient quantity of elements for detection. They must have even thickness where quantitative analysis is to be performed. Flotation on a water bath is another possible source of artefact where subcellular constituents may be displaced from the sections.

"Staining"

Electron stains increase the possibility of obtaining cytochemical information from the sections but can interfere with the analysis. The heavy metals employed produce characteristic X-ray emissions which may mask lines of other elements being examined. In some cases it is a quantitative estimation of the stain material itself which is required. Heavy metals also make the detection of trace elements within the tissue more difficult by raising the background signal.

"Cryoultramicrotomy"

The use of wet chemicals may be avoided altogether by cutting frozen sections and freeze drying for analysis. Analysis has been mostly performed on those sections at room temperatures but efforts are being made to maintain them frozen and with the water still present during analysis in the microscope. The technique is very young at the present time and a great deal of work is yet to be done to improve the quality of ultrastructure obtained by this method.

Other methods of histochemistry using frozen fixed and freeze substituted tissue go some way to overcoming the losses involved in conventional methods.

"Mounting"

As with thick specimens, sections are mounted on supports that will not interfere with the analysis. Grids of carbon,beryllium and aluminium are commonly used. Sections are carbon coated to overcome electron beam damaging effects.

5. Applications

Despite all the gloomy predictions concerning specimen preparation,a surprisingly large number of cytochemical applications of X-ray microanalysis have been performed in the last few years.

Whereas the major number of applications in X-ray microanalysis as a whole have been on mineralised tissues such as teeth and bone, our major interest here is in subcellular analysis and we shall turn our attention to those applications in particular. Once again it is helpful to consider applications in terms of thick and thin specimens.

"Thick specimens (> 0.2 μm)"

The distribution of calcium,iodine and phosphorus has been studied in human thyroid glands by Robison et al[6] using a scanning instrument with a crystal spectrometer. Specimens were prepared for analysis by freezing the tissue in liquid nitrogen and cutting 12 μ sections on a cryostat. Other samples were fixed in formalin and cut frozen or embedded and sectioned. Scanning X-ray images indicated a heavy distribution of calcium in follicular cells and colloid material, while iodine was found uniformly distributed in the follicles of normal thyroids.

The accumulation of Ca,P & K in tobacco leaf guard cells was studied by Sawhney & Zelitch[7], again with a scanning device. Light stimulated leaves were freeze dried and examined whole. Qualitative X-ray distribution maps were obtained for the three elements and it was demonstrated that significant differences occurred in potassium concentration in the guard cells under varying light stimulation.

Galle[8] has used a combination instrument involving an electron probe microanalyser equipped with transmission electron imaging to study the accumulation of various elements in the kidney in certain pathological cases.

Some of the more commonly used histological staining methods can be used in conjunction with X-ray microanalysis where detectable elements are found to be associated with the stain. For example, Gomori's chrome-haematoxylin could be identified by analysis of chromium in the tissue.

Gallocyanin-chromalum is a chromium containing dye that reacts with nucleic acids and has been used for quantitative determinations[9]. Sims & Marshall have observed the distribution of chrome alum stain in 7 μ thick sections of the spinal cords of rats fixed in formalin and compared the X-ray scanning images of chromium in the micro-analyser with the distribution of coloured stain in the optical microscope[10]. The chromium was seen to be associated with nuclear chromatin, nucleoli and Nissl substance.

Since the need for an electron dense final reaction product is obviated by the use of X-ray microanalysis this has important consequences in enzyme histochemistry. Engel has examined 10 μ thick sections of fresh frozen human skeletal muscle to demonstrate alkaline phosphatase without going to the final reaction product of CoS[11]. Instead the distribution of calcium was observed by an X-ray scanning image after $CaHPO_4$ precipitation by the Gomori technique.

"Thin specimens (< 0.2 μm)"

The role of zinc and calcium in human and rat prostate is being studied in the author's laboratory using an EMMA instrument with ultrathin (< 1000 AU) sections. Specimens are fixed with potassium pyroantimonate and osmium tetroxide to retain cations in situ which are then analysed in each subcellular region quantitatively. Fig 7. illustrates a section through columnar cells of rat prostate acini fixed in this manner and unstained. The effect of various hormonal controls on Zn and Ca distributions can be successfully measured in nucleoli, nucleoplasm, secretory granules and other subcellular regions. The lumen is found to contain high levels of zinc with calcium, presumably ready for secretion into prostatic fluid.

Ultrathin section of rat prostate (Fig 7.) Zinc and calcium localised in nucleoli, nucleoplasm, granules and lumen.

7

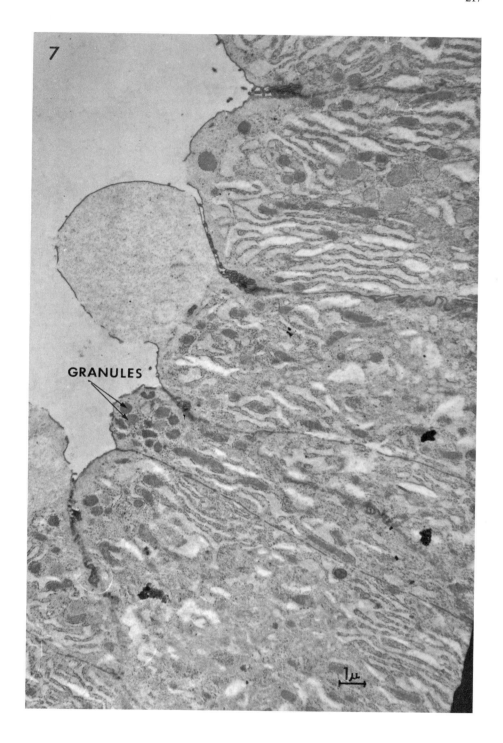

GRANULES

Lacy & Pettitt[12] have demonstrated the distribution of Si,P & S in Sertoli cells using EMMA and have shown that the lipid droplets contain sulphur and silicon with highest sulphur concentration in the small granules associated with the lipid. Maturation phase spermatids were also found to contain these 3 elements in normal rat tissue but to a much lower extent in animals treated with the spermatogenesis-inhibiting steroid 20α-hydroxy-pregn-4-en-3-one.

Ultrathin frozen sections of mouse pancreas have been examined with EMMA by Appleton[13] and were seen to contain the electrolytes Na and K discretely distributed between the subcellular organelles. Those elements would not have been detectable in specimens prepared by conventional wet methods.

Yarom & Chandler[14,15] studied the distribution of calcium in frog skeletal muscle treated with the pyroantimonate method of fixation. Fig 8. shows the transmission electron microscope image obtained in EMMA from the unstained specimen. The areas analysed are indicated by circles produced on the section deliberately by allowing contamination to build up. Calcium was found to be distributed between the triads, sarcolemma, nucleus and extracellular regions.

Chandler & Chou[16] studied the distribution of preservatives in wood, and their uptake by fungal hyphae infecting the wood,with EMMA. The preservative contained copper,chromium and arsenic each of which could be measured locally in the thin sections. Certain fungal cells were also cultured in a copper containing nutrient and prepared for electron microscopy by fixing and embedding. Fig 9.shows an ultra thin section of the cells. Accumulation of copper and sulphur were found to occur within the cell and in extracellular granules.

Fig 10 is a single human metaphase chromosome analysed with EMMA (Chandler, unpublished results) and found to contain significant levels of phosphorus,sulphur,zinc and calcium. The chromosomes were prepared for electron microscopy by application of cells to the surface of a Langmuir trough. The cells ruptured and nuclei and chromosomes were picked up on carbon coated grids. The phosphorus content is to be expected from nucleic acid content;calcium possibly is an enzyme cofactor(nuclease);sulphur is a protein constituent; and zinc may also be associated with nuclear enzymes.

Electron histochemistry is still limited compared to the available methods for light microscopy since it depends on the production of an electron scattering final reaction product. However X-ray microanalysis can be effectively used to extend the range of histochemical techniques by the localisation of individual elements that may constitute part of a stain that is not electron dense.

For example the ferric ferricyanide method for cystine containing proteins could be employed where the presence of iron would define disulphide groups only. Olszewska's technique[17] for disulphide bonds - the presence of selenium could be monitored by X-ray microanalysis in thin sections.

Fig 8. Frog skeletal muscle treated with pyroantimonate fixative for analysis.

Fig 9. Copper accumulation found subcellularly in Poria Monticola.

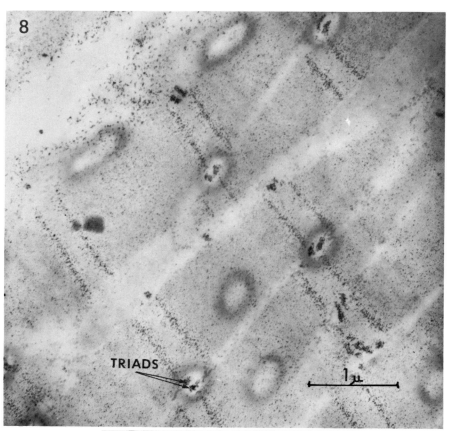

8

TRIADS

1μ

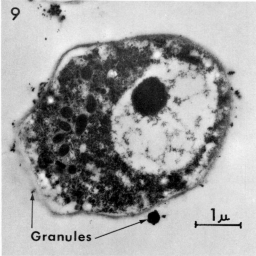

9

Granules

1μ

Low & Freeman[18] have studied chromium compounds as stains for nucleic acids in ultrathin sections for electron microscopy. X-ray microanalysis could well be applied together with those techniques for quantitative nucleic acid determinations _in situ_. Other methods for staining nucleic acids may also be used in microanalysis, involving the metals silver, lead, uranium, indium, bismuth or mercury.

An extremely important application is the labelling of steroids with elements detectable by microanalysis. Some steroids, notably estradiol, have an affinity for mercury. For example the 17β estradiol can be reacted with a mercurial salt to form the 4-mercuri-17β-estradiol.[19] Aromatic mercurials have been found to react with sulphydryl groups which in turn have been implicated in the binding of 17β estradiol by receptor protein. Thus mercury may be used in this way for affinity labelling studies of receptor protein in subcellular regions of thin sections. It is thought that strong mercaptide binding will occur at the receptors such that wet chemical preparatory methods may be employed.

For immunocytochemistry X-ray microanalysis may usefully be used for quantitative studies of protein in thin sections. Ferritin is easily measured by its iron content and in the method employing acid phosphatase the lead ions are readily analysed. However, the advantages of the peroxidase technique are so important for demonstrating antibody in tissue that methods of forming final reactive products containing detectable elements would be of great value (e.g. labelling DAB).

The above examples demonstrate the wide range of applications which X-ray microanalysis has in cytochemistry. The development of further methods of labelling cell components with detectable elements directly; via light optical stains; electron stains; by complexing with antibodies; and by labelling with enzymes, should make the technique of great importance in quantitative cytochemical work.

Acknowledgements

The author gratefully acknowledges the generous support of the Tenovus Organisation.

Fig 10.   A human metaphase chromosome analysed in EMMA and found to contain S,P,Ca,Zn.

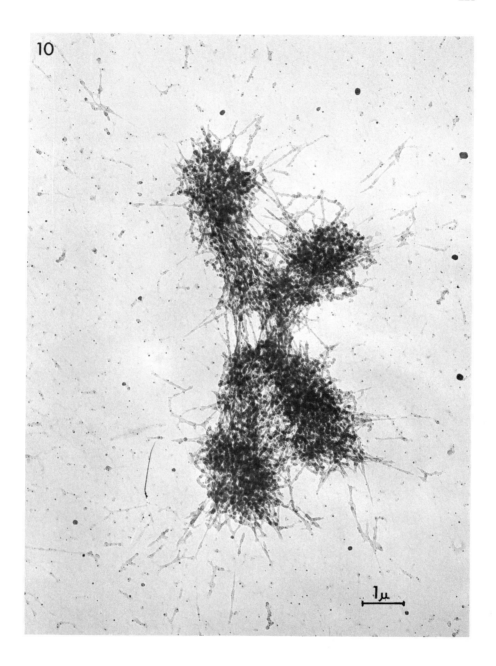

References

1. Hall,T.A. 1971 in "Physical Techniques in Biomedical Research"(Ed G.Oster)Vol.1A,2nd Ed.Academic Press, N.Y.

2. Chandler,J.A.1971. American Laboratory 3, 50

3. Läuchli,A., Spurr,A.R.,Wittkopp,R.W.,1970.Planta(Berl), 95, 341.

4. Spurr,A.R.1969 J.Ultrastr Res. 26,31

5. Giese,A.C. 1968 in"Cell Physiology" W.B.Saunders & Co., Philadelphia, U.S.A.

6. Robison,W.C.,Middlesworth,L.V.,Davis,D. 1971. J.Clin.Endocrin. Metab. 32, 6, 786.

7. Sawhney, B.L. & Zelitch,I. 1969. Plant Physiol 44, 1350.

8. Galle,P.,Berry,J.P.,Stuve,J. 1968. Proc.Nat.Conf.Electron Microscope Anal. 3rd Chicago, paper 41.

9. Einarson, L. 1951. Acta path,Scand. 28, 52.

10. Sims, R.T. & Marshall,D.J. 1966. Nature, 212, 1359.

11. Engel,W.K.,Resnick,J.S. & Martin,E. 1968. J.Histochem Cytochem. 16, 273.

12. Lacy,D. & Pettitt,A.J. 1972. Micron 3,1, 115.

13. Appleton,T.C. 1972. Micron 3,1, 101.

14. Yarom,R. & Chandler,J.A. 1972. 5th Europ.Cong.Elect.Mic. 138, Inst.Phys.Lond.

15. Yarom,R. & Chandler,J.A. 1973 (In preparation)

16. Chandler,J.A. & Chou, C.K. 1970. 7th Int.Conf.Elect.Mic. (Ed.P.Favard) Vol.1, 493, Soc.Francaise de Mic.Elect.

17. Olsewska,M.J.,Wronski,M.,Fortak,W. 1967. Folia Histochem. Cytochem. 5, 7.

18. Low, F.N. & Freeman,J.A. 1956. J.Biophys.Biochem.Cytol. 2, 626.

19. Chin, C.C. & Warren, J.C. 1968. J.Biol.Chem. 243,19, 5056.

Electron microscopy and Cytochemistry, eds. E. Wisse, W.Th. Daems, I. Molenaar and P. van Duijn.
© 1973, North-Holland Publishing Company - Amsterdam, The Netherlands.

MICROANALYSIS OF THIN SECTIONS IN THE TEM AND STEM
USING ENERGY-DISPERSIVE X-RAY ANALYSIS

John C. Russ

EDAX Laboratories
Raleigh, North Carolina 27612

1. Introduction

The electron microscope is a powerful and widely used tool in the
biological sciences because it offers the ability to study morphology on a fine scale. Many of the efforts to better understand biological processes involve relating morphology to function, and these
efforts frequently raise questions of chemical or elemental composition of structures in the sample. Accordingly, a large fraction of
the work in the electron microscopy of biological materials, whether
measured as the fraction of specimen preparation effort or the bulk
of literature, has involved the use of chemical stains. These chemicals produce electron-dense deposits within the sections that aid
recognition of certain structures and in some cases are presumed to
mark the location of specific elements or sites of chemical activity.

This type of interpretation is subject to a multitude of pitfalls,
false results, artifacts, and a lack of accuracy or reproducibility
due to variations in preparative techniques, fading or oxidation of
the stains, and human error. Quantitation based on deposit density
is at best crude and often impossible.

The biologist's interest in elemental analysis spans three categories: The analysis of naturally occurring elements important in
cell chemistry, the localization of elements associated with pathological conditions, and the detection of marker elements added
either in vivo or during preparation to identify sites of chemical
activity. Since most of the mass of the tissue is the light elements (C, H, O) the most interesting elements analytically are the
heavier ones, from about Na up in the periodic table. This is for-
tunate, because as we will see, these elements are best suited for
x-ray analysis.

2. Instrumentation - The Electron Microscope

The electron microscope produces a familiar image that can be
used to identify structures to be analyzed. At the same time the
interaction of the high energy electrons with the atoms of the sample produces x-rays that can be used to identify the elements present. When an incident electron knocks a bound electron out of its
orbital position in an atom in the specimen, the vacancy thus created is immediately filled by a higher energy electron. The energy
released by this process can be emitted as an x-ray photon, and the
energy of the photon is just the difference in energy levels of the
atom from which it came. Since different elements have unique energy levels, the x-rays are characteristic of the element from which
they come, and can be used to identify it.

The electron microscope can be either a conventional TEM, an SEM
equipped with transmission capability, or a combination instrument.
The TEM offers generally superior image resolution, higher accelerating voltages that provide superior excitation of many elements

(better x-ray production), and often have better specimen facilities and anti-contamination devices. The SEM, on the other hand, has provision for smaller incident beams, which affects the spatial resolution of the x-ray analysis, can produce useful images of much thicker sections, by scanning transmission electron microscopy, and is frequently easier to modify mechanically for analytical work because of the large specimen chamber. The new combination instruments are generally being designed with x-ray analysis in mind, and offer excellent performance in both modes of operation.

3. Instrumentation - X-ray Detection

The traditional method of measuring x-rays, as for example in the electron probe microanalyzer, makes use of the phenomenon of diffraction. Photons of a given energy E have an effective wavelength λ, and will diffract from a crystal when the geometric requirements of Bragg's law $n\lambda = 2d \sin \theta$ are obeyed (n is any integer). The angle θ is controlled by a mechanical linkage, while d is the atomic lattice spacing of the crystal and is chosen to cover a range of elements. Typically, three or four crystals can be used to cover the entire periodic table, one element at a time.

This type of spectrometer can be attached to either the TEM (e.g. EMMA-4) or SEM, and used for microanalysis of tissue sections. It is, however, largely being replaced by the newer energy-dispersive spectrometer. In these devices the x-ray photons enter a solid-state detector and lose their energy by a process that produces a small electrical pulse, one for each x-ray. Each pulse has a height that is proportional to the energy of the x-ray that produced it, and so the pulses can be amplified and measured electronically to identify all of the elements present, simultaneously. This feature of simultaneous analysis of all elements is one of the major advantages of the energy-dispersive method. The other is the fact that the detector can be placed very close to the sample to collect a large fraction of the x-rays, which are emitted in all directions. The high efficiency of the device enables the use of much lower beam currents and hence smaller beam diameters (and reduced specimen damage).

4. Detection Limits

Complicating the analysis is the fact that not all of the x-rays reaching the detector are characteristic x-rays from the atoms in the specimen. The deceleration of electrons in the sample also produces x-rays of all energies, up to the maximum electron energy. Since these x-rays can have any energy, they do not yield useful information on the elements present, and they make it more difficult to determine the presence of small amounts of the elements of interest. Since the counting of x-ray photons is a statistical process, we can define a peak counting rate of photons when the element is present, and a background counting rate when it is not.

The standard deviation in the number of counts tallied is just the square root of the number (Gaussian statistics) so that we can specify the requirement for determining that an element is really present. If the peak rises above the background by 2 standard deviations, we are 95% sure that the peak is real; 3 standard deviations give 99% confidence. In this way, we can describe the minimum concentration which will, for a given analyzing time and instrumental conditions (kV, current, etc.) be detectable.

When favorable conditions are used -- for example, a small but intense beam of high energy electrons, a thin sample to minimize background and proper collimation of the detector so that x-rays from the supporting structures cannot be seen -- the minimum concentration that can be detected is in the range of 100-500 ppm for most elements (in an organic matrix), for analyzing times of a few hundred seconds. Longer times improve the minimum detectable limit (in proportion to the square root of time), but are rarely practical because of specimen contamination or damage.

Since the total mass that can be analyzed in a thin section is very small -- 10^{-15} to 10^{-16} grams -- and so this limit of concentration would imply that an elemental mass of 10^{-19} grams or less could be detected. In practice, this has not been realized because the total signal from such a small mass is so low that the statistics are poor. Detection limits of 10^{-18} grams can be achieved [1], and the advent of new higher brightness electron guns, which will boost signal levels proportionately, should help to improve these values further.

5. Spatial Resolution

In addition to detecting small amounts of trace elements, we are concerned with pinpointing their location in the specimen, usually with reference to observable structural features. With the conventional TEM, the entire area being viewed is illuminated with electrons and hence emits x-rays -- there is no way to tell where in the area a particular element may be. Hence, it is necessary to restrict the incident beam to the structure or feature of interest. This is normally done by bringing the beam to crossover, and possibly using the stigmator to shape it to the feature to be analyzed. Several TEM's can now produce incident beams of under 1000 Å, and by using scanning attachments, beams of 100 Å and less can be achieved. SEM's can generally produce 100 Å as well.

The use of small incident beams does not, however, directly produce good x-ray spatial resolution. The electrons scatter within the material, and spread laterally so that the excited volume is flared into a cone shape, wider at the bottom than at the top. This means that x-rays come from an area somewhat larger than the incident beam diameter, and so the x-ray spatial resolution is poorer than the beam diameter, and poorer than the image resolution in the scanning transmission mode.

For conveniently prepared (embedded) sections, the diameter of the area being analyzed will be between 1/2 and 3/4 of the section thickness (thicker sections produce more transverse scattering of the incident beam [1]. The use of sections thinner than about 1000 Å is usually difficult because of heating damage and subsequent tearing under the intense beam, and because the total excited mass is so small that total count rates become very low, and analyzing times must be extended (which aggravates the specimen damage). For many purposes, sections about 1500 Å in thickness are desirable, and for this case, x-ray spatial resolution of about 700 Å has been demonstrated [2].

The use of frozen sections with the water ice still present should give similar results. Removing the ice by freeze-drying the sections offers several advantages -- the removal of mass reduces lateral electron scattering so that spatial resolution improves to a few hundred angstroms; it also effectively concentrates the elements of interest as the water is removed, and by reducing the total mass, reduces the background x-ray intensity so that smaller peaks can be distinguished, improving the limits of detection. The disadvantages of the method are that the image is frequently difficult to interpret, and the drying process can cause gross motion of elements to nearby surfaces. Much of the emphasis in this field has been on the specimen preparation, since the conventional methods certainly alter drastically the content and location of many trace elements. The most promising methods involve rapid freezing, with or without subsequent chemical treatment or drying.

6. Quantitation

Since the sections used in transmission electron microscopy are thin, it is generally safe to assume that negligible x-ray absorption takes place in the sample. Somewhat thicker sections may be encountered in scanning transmission electron microscopy, but as was pointed out before the best spatial resolution is obtained with thin sections, and they are hence preferred. The assumptions of linearity to be described certainly hold for most specimens up to 0.5 μm thick, and give moderately accurate approximations beyond that (3).

In the absence of absorption or other interelement effects, the x-ray intensity measured for an element will be linearly proportional to the number of atoms of the element present, and hence to the mass of the element. To relate this to concentration information (normally expressed in weight percent) we must know the total excited mass, which depends on the section density and thickness. Fortunately, the total excited mass can be determined from the continuum x-ray intensity described before. The total background radiation is linearly proportional to the total mass that is decelerating the electrons. This means that the ratio of net peak intensity to total background intensity is proportional to the elemental concentration (4).

The constant of proportionality is normally determined for each element of interest from standard samples, and under standard conditions of analysis. However, since the methods of specimen preparation in use result in the addition and/or removal of mass from the sample, and in addition a substantial loss of organic mass occurs under electron bombardment, it is generally difficult to relate elemental concentrations in the section to those in the original tissue. For this reason, it is simpler to work with elemental concentration ratios. Frequently, it is these ratios themselves that are interesting (e.g. Ca/P or Na/K ratios), while in other cases, it may be possible to relate all elements of interest to one element that is known or can be determined by addition methods. We frequently find it convenient to use the Cl content of embedding media, for instance, to quantitate the other elements present.

Since the peak-to-background ratio for each element is proportional to its concentration, and the background is the same since it is coming from the same specimen, the elemental concentration ratios are proportional to their net intensities

$$c_1/c_2 = K_{12} (I_1/I_2)$$

where the constant of proportionality K_{12} can be determined from a standard sample (5). Since the constant depends on analyzing conditions, it would need to be measured under the same kV, etc. as the specimens.

However, it is possible to eliminate this empirical constant and calculate the proportionality directly from fundamental constants. For two elements, the ratio of measured intensity values is proportional to their respective number of atoms (which can be converted to weight concentrations with the relative atomic weights) by a series of probabilities: (1) The probability that an incident electron ionizes an atom of the element; (2) the probability that the atom emits an x-ray, and that the x-ray is of the emission line being counted; and (3) the probability that the emitted x-ray is detected and counted. These terms are just the ionization cross-section, the fluorescent yield, and the spectrometer efficiency, respectively.

The first two terms are straightforward and can be calculated knowing the electron accelerating voltage. The spectrometer efficiency for a wavelength-dispersive (diffractive) spectrometer is difficult to measure, or express analytically, but for an energy-dispersive detector, it is given by an expression of the form

$$T = \delta \cdot e^{-C_1/E^3}(1 - e^{-C_2/E^3})$$

where δ is the solid angle (a machine constant which will thus cancel out of all ratio terms), and C_1 and C_2 are parameters that describe the response of the detector as a function of energy E. The constant C_1 is approximately 0.58 for an entrance window of 7 μm beryllium foil, and C_2 is approximately 18700 for a 3 mm. thick detector (6).

With these expressions, it is possible to directly and quickly compute relative elemental concentrations, or to prepare a set of tables to use (at a given kV) to relate intensities to concentrations.

7. Summary

Most of the electron microscopy performed by biological researchers has been basically morphological. Efforts to obtain biochemical information using stains that deposit an electron-dense marker at a presumed site of chemical activity in the cell are generally plagued by ambiguity and variability, in addition to the problems of specimen preparation. Recent progress toward obtaining in-situ biochemical data on sections of biological material has centered on the use of energy-dispersive x-ray analytical equipment mounted on the TEM or STEM. These systems detect, measure, and count the x-rays emitted from the atoms of the sample that are excited by the electron beam. The identification of the elements present is specific, and detection of local concentrations of a few hundred parts per million or a concentrated mass of 10^{-18} grams, is possible under favorable conditions.

The spatial resolution of the method is limited by several factors. Many TEM's cannot reduce their beam crossover at the specimen to less than about 0.5 μm, in which case the area of the specimen analyzed is at least this big. The STEM, and now several of the newer TEM's, produce incident electron beams of 200 Å and below; in this case, the limitation is due to transverse scattering of electrons in the thin section. In a conventionally embedded and sectioned specimen 1500 Å thick, an area about 800 Å in diameter is excited. Thinner sections reduce this value, but at the expense of total signal, so that analyzing times of several minutes are needed. The use of freeze-dried sections improves the spatial resolution to a few hundred angstroms and also effectively concentrates the nonvolatile elements. Many questions remain to be answered regarding the desirable methods of preparing sections for this type of analysis, the goals being preservation of recognizable structure and prevention of elemental migration.

Quantitation of elemental concentrations is meaningful only in terms of ratios of one element to another, since an unknown amount of mass has been added or removed in preparation. The equations to compute elemental ratios directly have been derived for the case of a thin section (i.e. thin enough that negligible interaction takes place between the generated x-rays and the atoms of the specimen). This condition is satisfied in nearly all cases for tissue sections up to thicknesses exceeding 0.5 μm. Tests indicate that the results are generally accurate within $\pm 10\%$. The only information required from the operator is the electron accelerating voltage. This level of accuracy is greatly superior to that offered by more primitive analytical methods.

The biological researcher can use this new technique and instrumentation to analyze several different biochemical problems: (1) the analysis and localization of naturally occurring, biochemically important elements such as electrolytes, enzyme cofactors, etc.; (2) the identification of pathological elements and their sites of accumulation; (3) the detection and localization of a new class of marker chemicals. Such chemical stains can be made far more specific since they need not produce an electron dense deposit, but simply contain one or more elements in the atomic number range 9-92 not naturally present in the sample.

References

1. Russ, J.C.,1972, Resolution and Sensitivity of X-ray Microanalysis in Biological Sections by Scanning and Conventional Transmission Electron Microscopy, 5th Annual SEM Symposium, IITRI, Chicago.

2. Russ, J.C., 1972, Energy-Dispersive Analysis of Biological Thin Sections in the Electron Microscope, Proc. 5th European Cong. on Electron Microscopy, Manchester.

3. Russ, J.C., 1972, Obtaining Quantitative X-ray Analytical Results from Thin Sections in the Electron Microscope, 12th Annual Mtg., American Society for Cell Biology, St. Louis.

4. Hall, T.A. and Werba, P., 1968, The Measurement of Total Mass Per Unit Area and Elemental Weight-Fractions Along Line Scans in Thin Specimens, 5th Int'l Congress on X-ray Optics and Microanalysis, p. 93.

5. Duncumb, P., 1968, Journal de Microscopie 7, p. 581.

6. Russ, J.C., 1973, Microanalysis of Thin Sections, Coatings and Rough Surfaces, SEM Symposium, IITRI, Chicago.

Electron microscopy and Cytochemistry, eds. E. Wisse, W.Th. Daems, I. Molenaar and P. van Duijn.
© 1973, North-Holland Publishing Company - Amsterdam, The Netherlands.

CRYOULTRAMICROTOMY, POSSIBLE APPLICATIONS

IN CYTOCHEMISTRY

T.C.Appleton

Physiological Laboratory, Cambridge CB2 3EG

1. Introduction

Biological materials which have been subjected to fixation, dehydration embedding and staining techniques are ideally suited to morphological examination, where the main interest is in ultrastructural detail. Such techniques have reached an advanced stage, and can be used on a routine basis. Such techniques, however are not always suitable for chemical analysis which may require the use of fresh, unfixed, or unembedded materials. The 'art' of cutting ultrathin frozen sections has been developed in many laboratories on a 'semi' routine basis. Furthermore the retrieval rate is sufficiently high to be able to use such sections in a variety of different ways - e.g. histochemistry, X-ray microanalysis, and perhaps autoradiography.

Biological material, whether it be in the form of tissue, single cell, suspensions of cells or homogenates, can be frozen rapidly and cut at thicknesses down to 40 nm, although for X-ray microanalysis or autoradiography the quantity material available for 'analysis' is too small at thicknesses below 100 nm. Sections may be cut in a variety of ways:

(a) unfixed and unembedded, onto <u>dry glass knives</u>[1,2,3,4]

(b) fixed or unfixed, encapsulated in gelatine or albumen, <u>onto DMSO</u>[5]

(c) unfixed and unembedded, <u>onto organic solvents such as Cyclohexene</u>[6,7]

A full review of techniques has been given by Christensen[3], and Bernhard[5].

Frozen sections should not be used as short cut to ultrastructural morphology, but as a means of obtaining chemical information about biological systems which may not be possible after the tissues have been subjected to conventional fixation, dehydration and embedding procedures. The morphological preservation is distinctly different from that seen in conventional microscopy and may require some experience in interpretation. Staining with Osmium vapours[4] may increase the contrast to some extent but may render the sections unsuitable for microanalysis or interfere with autoradiographic processes. The morphology however is sufficient to localize the chemical information to specific identifiable cells and their organelles, and may even provide new morphological information on the structure of living systems[2] complementing results obtained by other techniques such as freeze-etching, freeze-fracturing, and X-ray or electron deffraction.

The purpose of this communication is to describe advances in X-ray micro-analysis of naturally occuring diffusible ions such as Na, Mg, K, Ca, P, S and Cl. It is necessary to avoid contact with all fluids to avoid the possibility of diffusion of such elements - sections must be cut DRY: and we must continually and honestly ask ourselves the question "Can we trust the frozen section?"

2. Materials and Methods

Small pieces (1 mm^3 or less) of fresh tissue were frozen onto the tips of brass or silver specimen holders by immersion in either liquid nitrogen or iso-pentane cooled with liquid nitrogen. Freezing is extremely rapid and ice-crystal damage minimal provided the tissue samples are kept very small and the tissue is immersed in the quenching medium before the specimen holders. Tissues on specimen holders may be stored in liquid nitrogen until required: experience has shown that no deterioration in morphology occurs when tissues are stored in liquid nitrogen for periods up to 2 years.

Sections were cut in an ultramicrotome cryostat, capable of maintaining a chamber temperature of -80°C, at a thickness of 100 nm. Ribbons of sections were pulled away from the edge of DRY glass knives (45° inclined at 10° to the vertical) with a fine vacuum delivered through a flattened serum needle[1,2]. The procedure is summarized in figure 1.

The ribbons of ultrathin frozen sections are suspended in air at right angles to the knife edge (Fig. 1C) in a similar configuration to conventional sectioning of embedded material where the surface tension of the flotation bath pulls the sections away from the knife edge. In the case of dry frozen sections there is no flotation medium; the vacuum (generated by a vibrator pump) provides the necessary tension to pull sections away from the knife edge. A full description of the technique has been described by Appleton[1,2].

The sections are freeze-dried in the cryostat at -80°C and at atmospheric pressure: drying of 100 nm sections is complete in 1½ to 2 hours. The sections are warmed to ambient temperatures in the presence of dry nitrogen and transferred under dry nitrogen to the analytical electron microscope (AEI EMMA-4).

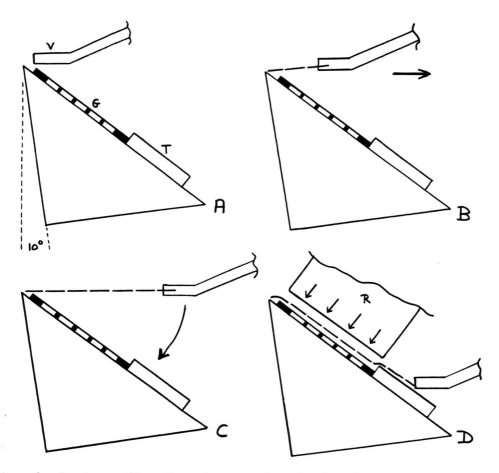

Figure 1. Drawing a ribbon of sections away from the edge of a DRY glass knife with a vacuum (**v**) delivered through a flattened serum needle. 'T' represents a shelf of PVC tape which supports the formvar coated grid 'G'. A ribbon of sections is drawn away from the knife edge (A,B,C) until it is sufficiently long to traverse the grid. The vacuum device is lowered until the sections lie across the grid. Sections can be pressed into contact with grid by light pressure with polished copper rod (R), according to the method described by Christensen. See 1(D).

2. Results and Discussion

Frozen sections prepared in the manner described can be used for X-ray micro-analysis of diffusible electrolytes and show compartments with different levels of elements such as Na, Mg, Si, P, S, Cl, K, Ca, Fe and Cr. Analysis has been undertaken using crystal diffracting spectrometers (where the characteristic elements are diffracted according to their characteristic wavelength) or energy dispersive silicon detectors (where a complete spectrum of elements present are displayed according to the energies of the X-rays detected). A more complete appraisal of the theory and advantages of the two systems has been given by Beaman & Isasi[8] at the physical level and by Weavers[9] at the biological level. Chandler[10] and Russ[11] have provided useful background information on X-ray micro-analysis in this symposium.

One of the criticisms that can be levied against the use of frozen sections for the localization of diffusible material is that the very act of sectioning may induce a momentary thawing as the knife passes through the frozen block. Thornburg and Mengers[12] considered that for a cutting resistance of 20g/cm of block width the energy dissipated per centimeter of knife travel would be 1.96×10^4 ergs; allowing for the heat of fusion of ice this would melt a layer of pure ice approximately 60 nm thick. Their calculations were based at a time when only thick sectioning (i.e. circa 5 um) was possible - this would mean that 60% of a 100 nm section had been subjected to momentary thawing during the cutting phase: their figures could be considerably better for smaller blocks, thinner sections and lower cutting temperatures. Hodson and Marshall[7] suggest that a rise of $100^{o}C$ could be produced and that part of this energy is dissipated into the sect-

Figure 2. Frozen section (100 nm) of NaCl in carboxy methyl cellulose, after Freeze-drying. No Na detectible within 120 nm resolution in any of the clear areas such as areas within the circles. Insert shows high magnification of NaCl crystals at end of feather structures.

Figure 3. Frozen section as in Fig.2 which had been allowed to melt for less than 30 seconds. Appearance of crystals and feathers changes to 'snow flakes'. There were no clear areas in such a section where sodium was not detected.

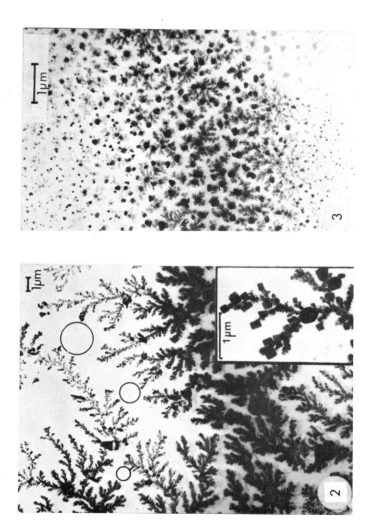

ion and part into the block. If their figures were correct there would be little point in using frozen sections cut at temperatures above -100°C for analysis of diffusible elctrolytes, and that sections cut below this temperature (Appleton[1,2], and Christensen[3]) would show no differential levels of such electrolytes. Ribbons of ultrathin frozen sections cut as described do show considerable differences in the levels of elements such as Na, Mg, S, Cl, K, Ca, P in compartments within the cell and in immediately adjoining areas at a spatial resolution of 120-150 nm. Yet the fact that ribbons are produced suggests that some momentary thawing probably takes place and accounts for the adhesion of sections into ribbons. Perhaps refreezing is so rapid that redistribution (within the levels of resolution now obtainable) does not take place. Redistribution in the order of 10 nm would not be detected with a spatial resolution of 100 nm. Alternatively it may be that the major part of energy dissipation during cutting is at the point where the block strikes the knife and that this thawing zone along the edge accounts for adhesion of sections into a ribbon. Clearly junction areas between sections would be unsuitable for analysis.

Figures 2 and 3 indicate that in a non biological system there was no redistribution of sodium due to a thawing phase during sectioning, when the section of NaCl in carboxy methyl cellulose (concentration of NaCl at 100 mM/1) was freeze-dried. However when the section was allowed to melt the structure changed from the feather-like appearance in Fig.2 to the 'snow-flake' appearance in Fig.3. Clearly melting had caused a redistribution of sodium. This probably represents the worst possible case because the concentration of sodium used was some ten time greater than in the cell of most tissues, and there are no membranes or organelles

---

Figure 4. 'Ganglion' cell in 'dry' frozen section of Polymorphus minutus: the dense granules are not seen in fixed and embedded preparations.

Figure 5. Peak levels of Calcium and Phosphorus in marked areas of cell from fig. 4. The values shown represent the peak-background corrected for differences in mass.

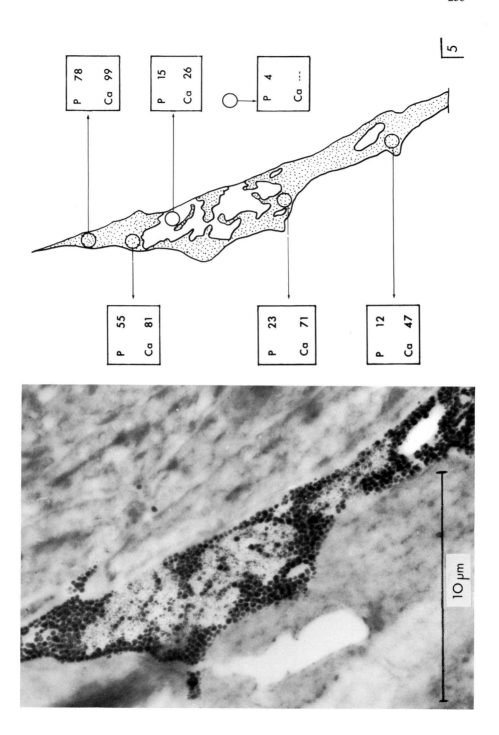

to interfere with diffusion.

Evidence against diffusion during cutting was also obtained by analysing areas within a ganglion cell of a parasitic helminth worm (Polymorphus minutus). (Fig. 4 & 5). Frozen sections, unstained, showed these cells to be filled with dense granules: conventionally prepared material showed no such granulation. Analysis by wave-length dispersion showed different levels of Calcium and phosphorus from one end of the cell to the other and that the levels of the two elements were, at least partly, independent of each other.

Analyses of areas around desmosomes in the spinosum layer of rat skin are shown in Figs. 6 & 7 and in Table 1. The spectra shown were obtained using an Ortec silicon detector; elements are displayed as a single spectrum according to the energies of the X-rays collected. There is some difference between the spectra of junctional areas and the areas over the semi-dense diffuse areas on either side of the junction (distances between junction and diffuse area analysed was 0.25 um) - Fig.6. There was an even more marked difference between the spectra of junctional and dense areas compared with adjacent cytoplasm (0.5 um from junction and 0.25 um from diffuse area).

Figure 7 and Table 1 show the levels of elements in three different desmosomes and three areas of immediately adjacent cytoplasm in rat skin. There is considerable difference between the desmosomal and cytoplasmic counts. The significance of the presence of Silicon in the desmosomes is not clear. It is unlikely that contamination from vacuum pump oils used in the microscope, or contamination from the glass knife, could account for such differing levels of silicon within areas which are separated by a few microns (less than 20). Silicon appears to be present in nearly every frozen section analysed.

Figure 6. Spectra obtained by energy dispersive analysis of three types of areas around desmosomes in rat skin: junctional, diffuse dense, and cytoplasm.

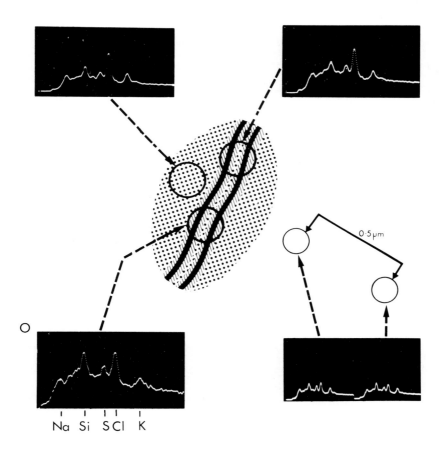

Na Si S Cl K

Figure 6

TABLE 1

Integrated peak counts of desmosomes (junction) and cytoplasmic areas shown as spectra in Fig.7.   Corrected for background and formvar.

Note: Peak to background ratios all greater than 5:1

Specimen		Elements							
		Na	Mg	Si	P	S	Cl	K	Ca
Desmosome	F	1206	–	431	–	757	5759	2025	–
"	G	3051	–	1255	7401	3279	8616	4594	894
"	J	752	391	3548	11101	2980	6674	6742	974
MEAN		2384	130	1744	6167	2339	7016	5120	623
S.E. ±		836	319	1319	2942	1010	3771	2193	129
Cytoplasm	A	1406	–	–	–	787	7292	2284	129
"	B	803	–	–	–	1121	3536	1107	383
"	C	834	–	–	–	424	4748	1406	–
MEAN		1014	–	–	–	777	5192	1599	170
S.E. ±		399	–	–	–	286	1566	499	124

4. Conclusions

The results presented suggest that cryoultramicrotomy has an important role to play in our understanding of the chemical content of cells and their organelles. But have we in any way answered the question posed in the introduction? "Can we trust the frozen section?" Results obtained with a physical model involving sodium chloride in Carboxy methyl cellulose suggest that ions as mobile as sodium are retained and that any redistribution which may occur during momentary thawing at the cutting phase may not be a problem within the spatial resolution (circa

Figure 7. Spectra obtained by energy dispersive analysis over three desmosomes, three adjacent areas of cytoplasm, and neighbouring area of formvar membrane. Computed corrected counts shown in Table 1: figures at bottom left corner indicate vertical scale of spectra.

239

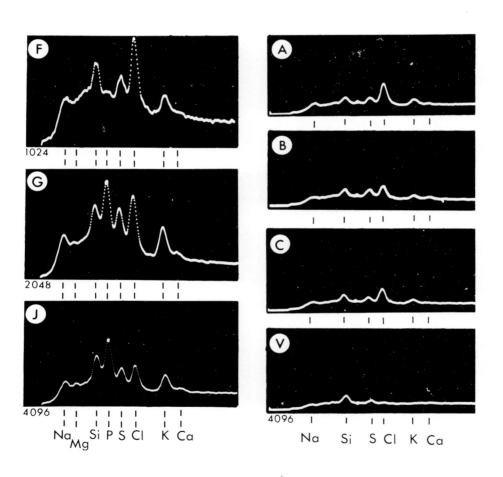

Figure 7

100 nm) obtainable at the present time. Results with biological material show that different levels of diffusible electrolytes can be detected within the cell to within a resolution of 250 nm. Results reported elsewhere [1,2] have shown that different compartments can be retained in ultrathin frozen sections of mouse pancreas within a resolution of 120 nm. If redistribution does occur it must be below the levels detectible at the present time.

The results shown also suggest that the sensitivity of detection is within normal physiological levels. Estimates have shown that with elements such as sodium, magnesium, potassium and calcium, concentrations in the order of 1-5 mM can be detected with peak to background levels greater than 2:1.

Clearly one needs to check the reliability of the method even further and continually ask oneself the question "Can we trust the frozen section?" We need also to check carefully whether the freezing process itself can be trusted. Is there any possibility of redistribution of diffusible elctrolytes during the quenching procedure? Many questions still need to be answered; the prospects of getting meaningful physiological information are good.

5. References

1. Appleton T.C., 1972, J.de Microscopie 13. 144

2. Appleton T.C., 1973, J. of Microscopy. In Press

3. Christensen A.K., 1971, J. Cell Biol. 51, 772

4. Werner G., Morgenstern E. & Neumann K., 1973, this symposium

5. Berhard W., 1971, J. Cell Biol. 49, 731

6. Hodson S., & Marshall J., 1970, J. of Microscopy 91, 105

7. Hodson S., & Marshall J., 1972, J. of Microscopy 95, 459

8. Beaman D., & Isasi J.A., 1971, Electron Probe Microanalysis. Amer. Soc. Testing & Materials. Special Publ. 506

9. Weavers B.A., 1971, Micron 2, 390

10. Chandler J., 1973, this symposium

11. Russ J., 1973, this symposium

12. Thornburg W., & Mengers P.E., 1957, J.Histochem. Cytochem. 5, 47

6. Acknowledgements

This work was generously supported by a grant from the Medical Research Council (Grant No G.970/341/B.). The Emma-4 analytical electron microscope was supported by a grant from the Wellcome Trust.

STAINING REACTIONS

Electron microscopy and Cytochemistry, eds. E. Wisse, W.Th. Daems, I. Molenaar and P. van Duijn.
© 1973, North-Holland Publishing Company - Amsterdam, The Netherlands.

STAINING OF INTRACELLULAR GLYCOPROTEINS

A. Rambourg

Département de Biologie

Centre d'Etudes Nucléaires - Saclay - France

Glycoproteins mays be defined as branched macromolecules that contain short oligosaccharide side chains attached to a protein backbone. The carbohydrate side chains, which are composed of two to fifteen monosaccharide units, may consist of two to six different types of sugars : amino sugars (D-glucosamine and D-galactosamine), neutral sugars (D-galactose, D-mannose and L-fucose) and amino sugar acids : the sialic acids (N-acetyl, N-glycolyl, and sometimes O-substituted neuraminic acid). The hexosamines, invariably in the N-acetyl form are usually attached to the protein backbone by means of o-glycosyl linkages involving seryl or threonyl residues or by glycosidic ester bonds involving glutamic or aspartic acid residues. Among the neutral sugars, D-mannose frequently acts as a branching point for an additional side chain whereas L-fucose is normally found at the end of the chains. A large number of the carbohydrate side chains are terminated by sialyl residues. The presence of terminal sialic acids bearing carboxyl groups of low pK (pK = 2.6 for N-acetyl neuraminic acid) confers a high degree of negative charge on some glycoproteins which can thereby be detected by means of cationic dyes. Furthermore, in all these macromolecules, adjacent hydroxyl groups are always available for periodic acid oxidation in terminal sialyl and/or fucosyl residues so that glycoproteins may also be revealed by techniques derived from the periodic acid-Schiff technique or techniques involving the use of phosphotungstic acid at low pH.

1. Cationic dyes

Cationic dyes such as alcian blue, colloidal iron or colloidal thorium are positively charged molecules which, consequently, are able to bind and precipitate polyanions. Their binding is electrostatic and therefore depends on the pH of the staining medium and on the pK values of the anionic groups present in the tissues [51]. When used below pH 2.5, they may react with the carboxyl and sulphate groups of mucopolysaccharides or with the sialic acids of glycoproteins so that the chemical specificity of the reaction should always be controlled by blocking experiments (methylation, sulfation, saponification) or enzymic digestions (e.g. neuraminidase). In the electron microscope, the use of cationic dyes is seriously impaired by the lack of penetration of the reagents and, although some results have been obtained by staining frozen sections with colloidal iron [6, 51] or treating methacrylate or epoxy ultrathin sections with colloidal thorium [43, 3, 54] or colloidal iron [54] the technique has been so far restricted to studies of cell surface glycoproteins (see reviews in Martinez-Palomo (1970), or Rambourg (1971)).

2. Techniques derived from the periodic acid-Schiff technique

a) The periodic acid-Silver methenamine (PA-Silver) technique

The use of silver methenamine after chromic acid oxidation as an histochemical test for the demonstration of complex carbohydrates in the light microscope was introduced by Gomori in 1946. Later on, the technique was adapted to the electron microscope but the oxidation was done in periodic acid instead of chromic acid [7, 16, 5, 22, 23, 24, 29, 49, 52, 53, 14, 27, 55, 8, 37, 50, 34]. Like the periodic acid-Schiff technique, the PA-Silver reaction is carried out in two steps :

1) In a first step, periodic acid oxidizes adjacent glycol and α-amino alcohol groups, thereby transforming them into aldehydic groups.

2) In a second step, these aldehydic groups reduce the silver tetramine contained in the silver methenamine reagent with release of free silver.

After glutaraldehyde and to a lesser extent after paraformaldehyde fixation, many structures are stained in the absence of periodic acid oxidation. Such staining, which is referred to as "unspecific" has given rise to considerable skepticism as regards to the specificity of silver technique for the detection of carbohydrates [20, 5, 11]. However, when a large survey of the unspecifically stained tissues was done in the light microscope (Rambourg and Hernandez, unpublished), the main finding was the complete lack of staining of carbohydrate containing structures. Hence, the unspecific staining, once it is recognized by examination of unoxidized control sections, should not interfere with the detection of carbohydrates done by silver staining after periodic acid oxidation. One of the main advantages of periodic acid is that the oxidative process does not go further than the aldehydic stage. Thus, regions which stain with silver or Schiff, exclusively after periodic acid oxidation are likely to contain aldehydic groups bound to carbohydrates and particularly to glycoprotein molecules [9]. Such is the case for thyroid colloid, mucus in digestive and respiratory tracts, glycoprotein secretions of pituitary or salivary gland cells, acrosome of spermatide, basement membranes, cartilage matrix and so on. Furthermore, glycogen, which is strongly stained with the PA-Schiff technique does not react with silver methenamine after periodic acid oxidation [20, 37] and most lipids are extracted during histological processing [15, 17, 18]. It may therefore be assumed that once unspecific staining has been detected by examination of control sections, glycoproteins are responsible for PA-Silver staining [37, 41]. In contrast to the cationic dyes, periodic acid and silver methenamine penetrate easily through thin sections of epoxy embedded material so that the technique may be used for the detection of extra and intracellular glycoproteins.

b) The periodic acid chromic acid-silver methenamine (PA-CrA-Silver) technique

Mowry (1959) [30], working with the light microscope, showed that the insertion of a chromic acid step between periodic acid oxidation and silver methenamine increased the contrast and specificity of the PA-Silver technique. This technique was therefore adapted to the electron microscope by Hernandez et al. (1968) [13] and slightly modified by Rambourg (1970, unpublished). As for PA-Silver, glutaraldehyde fixed tissues are embedded in Epon or Vestopal. Pale gold thin sections are placed on uncoated stainless steel or platinum grids and floated for 15 minutes on a 1 °/₀ periodic acid solution in 50 °/₀ aceton. They are rinsed for 15 minutes in distilled water and then oxidized for 15 minutes on a 10 °/₀ chromic acid solution in 50 °/₀ aceton. After three short rinses on distilled water, the sections are stored overnight on a last bath of distilled water.

As for PA-Silver, staining is achieved by floating the oxidized sections on two baths of silver methenamine as indicated in Rambourg (1967) [37].

In the electron microscope, the PA-CrA-Silver technique furnishes the same overall pattern as PA-Silver, but the non-specific staining observed after PA-Silver is minimized, whereas glycogen which did not react with the latter technique becomes intensely stained. The relative lack of unspecific staining after the double oxidation procedure has been attributed to chromic acid oxidation of various reducing groups [41], since in contrast to periodic acid oxidation, chromic acid oxidation is not restricted to adjacent glycols and/or α-amino alcohol groups.

The mechanism of chromic acid oxidation for the demonstration of carbohydrates with the Schiff reagent (Bauer's technique) is less known than that of periodic oxidation. It is generally accepted, however, that chromic acid like periodic acid transforms glycols into aldehydic groups, but that chromic acid oxidation, if prolonged, may convert these aldehydic groups into acidic groups which no longer react with the Schiff reagent. After a 5 minutes oxidation with 10 °/₀ chromic acid, the stainability of carbohydrates with the Schiff reagent is usually weaker than after periodic acid oxidation. In contrast, their reactivity to silver methenamine is greater. It might be argued that some glycol groups have been oxidized into carboxyl groups which do not react any longer with the Schiff reagent but are perhaps responsible for the increase in silver staining. This possibility, however, was eliminated since treatment of the oxidized sections by chlorous acid, which irreversibly transforms aldehydes into carboxyl groups, abolished both Schiff and Silver staining. The PAS and PA-CrA-Silver reactions were also prevented by acetylation of unoxidized sections. It was then concluded that after the double oxidation procedure, the aldehydic groups engendered by periodic acid oxidation of glycoproteins were reorientated in a different fashion or that the loss of some aldehydic groups by transformation into carboxyl groups was compensated by the appaearance of new aldehydic groups located in other parts of the molecule [41].

c) The periodic acid-thiocarbohydrazide-silver proteinate (or osmium tetroxide) (PA-TCH-Silver) technique

Introduced in histochemistry by Seligman et al. (1965) [46] and Thiery (1967) [53], this technique utilizes the reducing properties of thiocarbohydrazide (or thiosemicarbazide) for demonstrating the aldehydic groups engendered by periodic acid oxidation. Tissue sections are first oxidized in periodic acid which converts adjacent hydroxyl or α-amino alcohol groups into aldehydes. These aldehydes are then condensed with thiocarbohydrazide (or thiosemicarbazide) to yield thiocarbohydrazones which are powerful reducing agents. Thus, after exposure to osmium vapors[46] or silver proteinate[53] these thiocarbohydrazones are finally revealed in the electron microscope as osmium or silver deposits at the reactive sites.

In the light microscope, after carnoy or formaldehyde fixation, the PA-TCH-OsO$_4$ technique reveals the same structures as the PAS technique except that collagen and reticulin fibers which are PAS positive remain unstained[46].

In the electron microscope, the type of carbohydrate revealed by the PA-TCH-Silver technique depends on the time of exposure to TCH which may vary from 30-40 minutes for demonstrating glycogen to 48-72 hours for the detection of glycoproteins[53]. As with PA-Silver, many structures are unspecifically stained after glutaraldehyde fixation. They may be detected in control unoxidized sections. Unspecific staining is minimized after osmium fixation. It should be kept in mind, however, that osmic acid may oxidize proteins[12] and unsaturated lipids[17, 18, 19, 44, 45] so that not only carbohydrates but also proteins and lipids may react with TCH after osmium fixation and periodic acid oxidation. Furthermore, it has been suggested that osmic acid may react with adjacent cis-hydroxyl groups[28, 17, 18, 19, 45] and, indeed, some regions which were positive after glutaraldehyde fixation and periodic acid oxidation became unreactive when tissues were fixed or postfixed in osmium tetroxide (e.g. the cell surface in ref. 53). Hence, although the resolution of the technique may appear better, the chemical specificity of the reaction should always be tested on control sections in any interpretation of the results obtained after osmium fixation or postfixation.

d) Other techniques

Besides silver methenamine or thiocarbohydrazide, other reagents have been proposed to demonstrate dialdehydes engendered by periodate oxidation of carbohydrates. For instance, Thiery (1967)[53] revealed tissue aldehydes by coupling the electron dense phosphotungstic acid stain to the final product of the

periodic acid-Schiff reaction. An alternative approach was to condense aldehydes
with compounds such as penta or monofluorophenylhydrazine [4, 48]. The disavantages of these techniques, however, are the low contrast of the PA-Schiff-PTA
technique and the difficulty in obtaining chemicals for the phenylhydrazine technique. More recently, Ainsworth et al. (1972) [1] have demonstrated periodate
engendered dialdehydes by floating periodic acid oxidized tissue sections on a
chelated alkaline bismuth subnitrate solution. Like the silver methenamine
reagent, this bismuth subnitrate solution is a mild oxidizing agent which oxidizes
aldehydes and is itself reduced with release of free bismuth at the reactive sites.
According to the authors, the advantages of this new electron miscroscopic
technique over existing methods would be the stability of the staining solution,
the scarcity of non specific background staining and the high resolution allowed
by the very fine precipitate of the reaction product.

3. Phosphotungstic acid at low pH : the chromic acid or hydrochloric acid-
phosphotungstic acid (CrA or Hcl-PTA) technique

Although the mechanism of this technique is still a matter of controversy,
there is some evidence that when unfixed or aldehyde fixed tissues are embedded
in hydrophilic resins, aqueous solutions of phosphotungstic acid (PTA) may
detect carbohydrate macromolecules in thin sections [32, 38, 24]. Lowering the
pH of phosphotungstic acid by dissolving 1 °/. PTA in a strong mineral acid such
as chromic, hydrochloric or sulphuric acid [25, 33, 38] increases the contrast
and specificity of the technique to such extent that in a survey of a large variety
of rat tissues [39, 40] all the PTA stained structures were also stained by the
PA-Schiff technique. When staining by Hcl-PTA was combined with radioautography after fucose-^{3}H injection to label glycoproteins, the label was exclusively
localized on PTA-stained structures [42]. It was then assumed, since glycogen
and most lipids were extracted during embedding into hydrophilic resins, that
glycoproteins were responsible for PTA staining [41].

This opinion has been criticized by Silverman and Glick [47] as well as by
Quintarelli et al. [35, 36]. According to Glick and Scott [10] ..." There is no
direct experimental evidence that a free polysaccharide is stained with PTA,
despite observations that sites where polysaccharides are known to be present
show positive staining ... " No doubt, substances such as proteins may be

selectively stained by PTA once it is used in anhydrous conditions [2, 31] but the claim that PTA can only react with amino groups should also explain why sections of glycol methacrylate embedded agarose (a polymer of galactose) are intensely stained with Hcl-PTA [41]. Furthermore, deamination of rat tissues [25, 31] or methylation to block carboxyl groups and hydrolyse sulphate groups do not modify PTA staining in aqueous solutions [25]. In contrast, sulfation or acetylation to block hydroxyl groups abolishes the reaction [25, 39]. The reactivity of PTA stained structures is not modified after periodic acid oxidation which converts hydroxyl into aldehydic groups [31] but disappears when aldehydic groups are transformed into carboxyl groups by an unduly prolonged chromic acid oxidation (Rembourg 1968, unpublished). Presumably, then, hydroxyl groups as well as aldehydic groups are involved in this reaction whose specificity can be compared to that of the PA-Schiff technique [25, 31, 41].

In conclusion, several techniques, mainly derived from the periodic acid-Schiff technique are now available to detect intracellular glycoproteins at the ultrastructural level with a good specificity. However, since in these techniques, the fixatives routinely used in electron microscopy, frequently introduced unspecific staining, the chemical specificity of the reaction should always be tested by studies of control sections as well as by blocking experiments or enzymic digestions. As will be seen later in this symposium, the use of carbohydrate binding proteins such as concanavalin A may appear promising from the point of view of providing reagents of known saccharide specificity. Yet, due to lack of penetration of the reagents inside cells or tissue sections the use of this promising method has been so far limited to the detection of cell surface glycoproteins.

References

1. Ainsworth, S.K., Ito, S., and Karnovsky, M., 1972. J. Histochem. Cytochem. 20, 995.

2. Bloom, F.E., and Aghajanian, G.K., 1968. J. Ultrastruct. Res. 22, 361.

3. Berlin, J.D., 1967. J. Cell. Biol., 32, 760.

4. Bradbury, S., and Stoward, P.J., 1967. Histochemie, 11, 71.

5. Churg, J., Moutner, W., and Grishman, E., 1958. J. Biophys. Biochem. Cytol., 4, 841.

252

6. Curran, R.C., Clark, A.E., and Lovell D., 1965. J. Anat., 99, 427.

7. Dettmer, N., and Schwarz, W., 1954. Z. Wiss Mikrosk., 61, 423.

8. Follenius, E., and Doerr-Schott, J., 1966. C.R. Acad. Sci., Ser. D., 262, 912.

9. Glegg, R.E., Clermont, Y., and Leblond, C.P., 1952. Stain Technol., 27, 277.

10. Glick, D., and Scott, J.E., 1970. J. Histochem. Cytochem., 18, 455.

11. Goldblatt, P.J., and Trump, B.F., 1965. Stain Technol., 40, 105.

12. Hake, T., 1965. Lab. Invest., 14, 1208.

13. Hernandez, W., Rambourg, A., and Leblond, C.P., 1968. J. Histochem. Cytochem., 16, 507.

14. Hollmann, K.H., 1965. J. Microsc. (Paris), 4, 701.

15. Idelman, S., 1965. Histochemie, 5, 18.

16. Jones, B.D., 1957. Am. J. Pathol., 33, 313.

17. Korn, E.D., 1966. Science, 153, 1491.

18. Korn, E.D., 1966. Biochim. Biophys. Acta, 116, 317.

19. Korn, E.D., 1967. J. Cell. Biol., 34, 627.

20. Lillie, R.D., 1954. J. Histochem. Cytochem., 2, 127.

21. Marinozzi, V., 1960. Electron Microsc. Proc. 2nd Eur. Reg. Conf., Delft, 2, 626.

22. Marinozzi, V., 1961. J. Biophys. Biochem. Cytol., 9, 121.

23. Marinozzi, V., 1963. J. Roy. Microsc. Soc., 81, 141.

24. Marinozzi, V., 1967. J. Microsc. (Paris), 6, 68 A.

25. Marinozzi, V., 1968. Electron Microsc., Proc. 4th Eur. Reg. Conf., Rome, 2, 55.

26. Martinez-Palomo, A., 1970. Int. Rev. Cytol., 29, 29.

27. Marx, R., and Mölbert, E., 1965. J. Microsc. (Paris), 4, 799.

28. Milas, N.A., Trepagnier, J.H., Nolan, J.T., and Iliopulos, M.I., 1959. J. Amer. Chem. Soc., 81, 4730.

29. Movat, H.Z., 1961. Am. J. Clin. Pathol., 35, 528.

30. Mowry, R.W., 1959. J. Histochem. Cytochem., 7, 288.

31. Palladini, G., Lauro, G., and Basile, A., 1970. Histochemie, 24, 315.

32. Pease, D.C., 1966. J. Ultrasruct. Res., 15, 555.

33. Pease, D.C., 1968. 26th Annual Meeting of Electron Microscopists of America, p. 36.

34. Pickett-Heaps, J.D., 1967. J. Histochem. Cytochem., 15, 442.

35. Quintarelli, G., Zito, R., and Cifonelli, J.A., 1971. J. Histochem. Cytochem., 19, 641.

36. Quintarelli, G., Cifonelli, J.A., and Zito, R., 1971. J. Histochem. Cytochem., 19, 648.

37. Rambourg, A., 1967. J. Histochem. Cytochem., 15, 409.

38. Rambourg, A., 1967. C.R. Acad. Sci., Ser. D, 265, 1426.

39. Rambourg, A., 1968. Electron Microsc., Proc. 4th Eur. Reg. Conf., Rome, 2, 57.

40. Rambourg, A., 1969. J. Microsc. (Paris), 8, 325.

41. Rambourg, A., 1971. Int. Rev. Cytol., 31, 57.

42. Rambourg, A., Bennett, G., Kopriwa, B., and Leblond, C.P., 1971, J. Microsc. (Paris), 11, 163.

43. Revel, J.P., 1964. J. Microsc. (Paris), 3, 535.

44. Riemersma, J.C., 1963. J. Histochem. Cytochem., 11, 436.

45. Riemersma, J.C., 1968. Biochem., Biophys. Acta, 152, 718.

46. Seligman, A.M., Hankers, J.S., Wasserkrug, H., Dmochowski, H., and Katzoff, L., 1965. J. Histochem. Cytochem., 13, 629.

47. Silverman, L., and Glick, D., 1969. J. Cell. Biol., 40, 761.

48. Stoward, P.J., and Bradbury, S., 1968. Histochemie, 15, 93.

49. Suzuki, T., and Sekiyama, S., 1961. J. Electronmicr., 10, 36.

50. Swift, J.A., and Saxton, C.A., 1967. J. Ultrastruct. Res., 17, 23.

51. Szirmai, J.A., 1963. J. Histochem. Cytochem., 11, 24.

52. Thiery, J.P., 1964. Electron Microsc., Proc. 3rd Eur. Reg. Conf., Prague, B, 209.

53. Thiery, J.P., 1967. J. Microsc., Paris, 6, 987.

54. Thiery, J.P., 1970. Electron Microsc., Proc. 7th Int. Congr., Grenoble, 1, 577.

55. Van Heyningen, H., 1965, J. Histochem. Cytochem., 9, 234.

56. Wetzel, M.G., Wetzel, B.K., and Spicer, S.S., 1966. J. Cell. Biol. 30, 299.

Electron microscopy and Cytochemistry, eds. E. Wisse, W.Th. Daems, I. Molenaar and P. van Duijn.
© 1973, North-Holland Publishing Company - Amsterdam, The Netherlands.

PHOSPHOTUNGSTIC ACID STAINING AND THE QUANTITATIVE STEREOLOGY OF
SYNAPSES.

G. Vrensen and D. de Groot.
Mental Hospital "Endegeest", Oegstgeest, (The Netherlands).

1. Introduction

Quantification of synapses is of particular interest for
neurobiological studies dealing with dynamic aspects of synaptic
functioning. Using quantitative stereological principles [4,6] a
number of relevant parameters can be obtained by simple countings;
e.g. number per area, surface area of synaptic contact zones per
volume and mean length. The countings are facilitated and object-
ified using a selective staining technique. Ethanolic phosphotungstic acid (E-PTA) selectively stains synaptic contact zones [1,2]
optimally resulting in opaque contact zones with presynaptic dense
projections, interlines and postsynaptic bands (fig. 5 p, i and b)
in a nearly unstained background. In a recent study the present
authors [5] have shown that E-PTA stained contact zones are quantitatively equivalent to synaptic membrane thickenings in OsO_4-fixed
material. The routine procedure [2] has some disadvantages that are
especially inconvenient when systematic analyses over larger areas
of tissue are to be carried out.

2. Results and Discussion

A more or less selective staining of synaptic contact zones is
not only achieved with 1% PTA in absolute ethanol [2] but also with
1% PTA in distilled water (pH~2) or in buffer solutions at
pH < 4(A-PTA), (figs. 1 and 2). Applied to 1 mm^3 blocks of aldehyde-
fixed cerebral cortex the staining is optimal up to 100-200 μm
from the outer faces of the blocks. In E-PTA material the opacity
of the neuropilic background gradually increases from outside to
inside. The centre of the blocks is not stained, (fig. 3). A-PTA
treatment results in a marked background contrast, especially of
mitochondria, over the entire stained area, (fig. 1).

These observations indicate that two factors likely influence
the uniformity and level of contrast of the PTA procedures:
1. PTA slowly penetrates the tissue
2. ethanol extracts phospholipids from cellular membranes in

aldehyde fixed tissue. This assumption explains the difference between the E-PTA and A-PTA contrasting effect. Proteins attached to membranes will be liberated when the phospholipids are partially or totally extracted thus preventing localized staining with PTA. The paramembranous proteins and glycoproteins of synaptic junctions are highly organized [1,3] and likely less vulnerable to extraction.

Based on these conclusions we have tried to improve the E-PTA procedure by using thin slices of tissue (50, 100 and 200 μm) and by raising the temperature of the staining solution to 40, 60 and 80° C. Tissue slices of 50 and 100 μm are uniformly stained over their entire thickness, (fig. 4). In 200 μm slices a rim of less optimally stained tissue is present in the middle of the slice. This indicates that the penetration of PTA is indeed very slow and that the thickness of the slices used is very critical. Raising the temperature results in a decrease of the contrast of the neuropilic background, (figs. 5 and 6). This again points to an extraction effect of ethanol. At 80° C the ultrastructure of the junctions has changed. The dense projections and interlines are less clearly outlined. This implies that for qualitative studies this material is less suitable. For systematic quantitative studies it had first to be proved that treatment at elevated temperatures does not influence the parameters previously mentioned. The table summarizes the results of two independent experiments. It can be concluded that temperatures up to 60° C have no systematic effect on the quantitative parameters. At 80° C a slight decrease in number and mean length is present, especially in tissue that has been kept for long time in the buffer solution (experiment 2).

All figures are taken from visual cortex of the rabbit, perfusion-fixed with aldehydes and treated as indicated. To avoid secondary contrast differences micrographs are taken and negatives enlarged under identical lighting conditions. The bars indicate 1 μm.

Fig. 1. A-PTA, 40° C, 1 mm³ block. Note the marked staining of background and especially the mitochondria (→).

Fig. 2. E-PTA, 40° C, 1 mm³ block. The picture is taken from an area intermediate between optimally stained and unstained.

Fig. 3. A-PTA, 40° C, 1 mm³ block. This figure illustrates the absence of staining reaction in the centre of the blocks.

Fig. 4. E-PTA, 80° C, 50 μm slice. Low power micrograph illustrating the homogenous staining over large areas.

Figs. 5 and 6. E-PTA, 100 μm slices, 20 and 80° C, resp. Note the optimal staining with 100 μm slices and the slight decrease in background staining at the higher temperature.

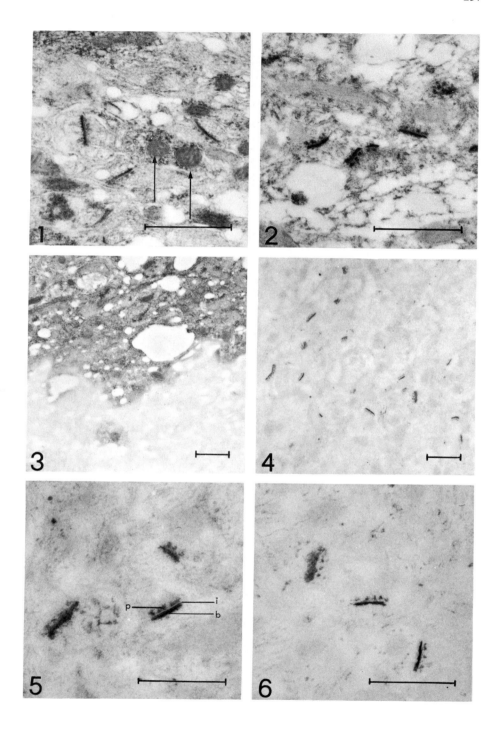

Table E-PTA STAINING AT DIFFERENT TEMPERATURES

Comparison of quantitative parameters

T	Experiment 1			Experiment 2		
	N_A	S_V	L	N_A	S_V	L
20°C	279.0 ± 17.5^x	85.0 ± 6.0	0·240	253.5 ± 12.0	97.1 ± 6.9	0·301
40°C	268.2 ± 14.4	79.1 ± 4.6	0·232	267.1 ± 15.9	100.2 ± 8.4	0·294
60°C	284.3 ± 17.2	76.4 ± 5.2	0·212	261.9 ± 21.4	102.6 ± 7.2	0·308
80°C	235.4 ± 13.7	69.9 ± 3.5	0·235	223.9 ± 16.1	79.8 ± 6.0	0·279

N_A = number per 1000 μm^2

S_V = surface area in μm^2 per 1000 μm^3

L = mean length in μm

x = mean of 20 pictures and standard error of the mean.

3. Conclusions

Synaptic contact zones are stained by PTA in aquaeous and ethanolic solutions, E-PTA being the more selective. The cause of this selectivity and the inhomogenous staining in small blocks of tissue is discussed. E-PTA staining of 50-100 μm slices at 60° C greatly enhances the selectivity and uniformity of the staining reaction. The significance and reliability of this improved procedure for the quantitative stereology of synapses is discussed.

References

1. Bloom, F.E., 1972, in: Structure and function of synapses, eds. G.D. Pappas and D.P. Purpura (Raven Press, New York) p. 101.
2. Bloom, F.E. and Aghajanian, G.K., 1968, J. Ultrastruct. Res. 22, 361.
3. Pappas, G.D. and Waxman, S.G., 1972, in: Structure and function of synapses, eds. G.D. Pappas and D.P. Purpura (Raven Press) p.1.
4. Underwood, E.E., 1973, Quantitative stereology (Addison-Wesley, Reading).
5. Vrensen, G. and de Groot, D., 1973, Brain Res. 58.
6. Weibel, E.R. and Elias, H., 1967, in: Quantitative methods in morphology, eds. E.R. Weibel and H. Elias (Springer, Berlin). p.89

Electron microscopy and Cytochemistry, eds. E. Wisse, W.Th. Daems, I. Molenaar and P. van Duijn.
© 1973, North-Holland Publishing Company - Amsterdam, The Netherlands.

SELECTIVE GLYCOGEN CONTRAST BY HEXAVALENT OSMIUM OXIDE COMPOUNDS.

W.C. de Bruijn and P. den Breejen.
E.M. Department, Pathol.Lab., Medical Faculty, Erasmus University
P.O. Box 1738, Rotterdam, The Netherlands.

1. INTRODUCTION.

Since the introduction of aldehyde fixatives in the electron microscopic histotechnique, $Os^{VIII}O_4$ was used, either as double fixative for substances not fixed by the primary fixation or for contrasting primary fixed organelles. Modification of such OsO_4 fixatives by the addition of complex cyanides (e.g. $K_3Fe(CN)_6$, $K_4Fe(CN)_6$, $K_3Co(CN)_6$, $K_2Ru(CN)_6$ or $K_4Os(CN)_6$) resulted in selectively contrasted glycogen and membranes, whereas the ribosomes were uncontrasted[1,2].

2. RESULTS LEADING TO THE USE OF HEXAVALENT OSMIUM OXIDE COMPOUNDS.

When such (complex) cyanides were not added to the OsO_4 double fixative, but OsO_4 double fixed, tissue blocks were incubated in cyanide containing solutions, it was noticed that:

a) KCN-treatment completely depleted the tissue from its original contrast obtained from the OsO_4 double fixation treatment (fig.1).

b) $K_3Fe(CN)_6$ or $K_4Fe(CN)_6$-treatment did the same, but in the periphery of the blocks a new selective contrast was created in the glycogen and membranes similar to that of the modified fixatives.

c) these differences in glycogen contrast were manifested by P.T.A-staining of the ultrathin sections (fig. 2).

From these experiments it was deduced that due to the reduction of the OsO_4 by the aldehyde-fixed tissue, lower osmium oxides were involved in the selective glycogen contrasting reaction by the complex cyanides.

3. TISSUE CONTRAST BY DOUBLE FIXATION WITH HEXAVALENT OSMIUM OXIDE COMPOUNDS.

Among the known hexavalent osmium oxide compounds, $K_2OsO_4 \cdot 2 H_2O$ and $OsO_3 \cdot 2$ Pyridine are water soluble. Application of such solutions to the aldehyde-fixed tissue blocks resulted in an identical contrast distribution:

a) In the untreated ultrathin sections the glycogen areas were the least electron dense parts of the cells (fig.3).

b) P.T.A.-staining selectively enhanced the glycogen contrast(fig.4).

From these results it was concluded:

1st. The presence of the osmate in the glycogen areas is not sufficient to create positively[4] contrasted glycogen.

2nd. The presence of osmate is selectively visualized by the P.T.A.-staining, resulting in positively contrasted glycogen.

3rd. The accumulation reaction,performed here by the P.T.A.-staining of the ultrathin section, might be similar to the reaction of $K_4Fe(CN)_6$ in the periphery of the by OsO_4-fixed tissue blocks, described above.

4. TISSUE CONTRAST BY HEXAVALENT OSMIUM OXIDE COMPOUNDS, MODIFIED BY THE ADDITION OF $K_4Fe(CN)_6$.

0.05 M $K_4Fe(CN)_6$ was added to the original K_2OsO_4.2 H_2O or OsO_3.2 Pyridine solutions and applied to aldehyde-fixed tissue blocks. In the untreated ultrathin sections of these blocks, contrasted glycogen was observed. The ribosomes were not contrasted, whereas the membranes were contrasted in the tissue blocks treated by the combination K_2OsO_4.2 H_2O plus $K_4Fe(CN)_6$, but were not contrasted in the tissue blocks treated by the combination OsO_3.2 Pyridine plus $K_4Fe(CN)_6$ (Fig. 5 and 6). From these results it was concluded:

1st. Hexavalent osmium oxides are selectively bound to a ligand present in the aldehyde-fixed tissue glycogen.

2nd. Accumulation of a heavy metal complex, (possibly composed of hexavalent osmium oxide and a complex cyanide, e.g. $K_4Fe(CN)_6$), at this selectively bound hexavalent osmium oxide makes the glycogen appear in positive contrast[4] in untreated ultrathin sections.

5. STAINING SECTIONS BY THE K_2OsO_4.2 H_2O plus $K_4Fe(CN)_6$ COMBINATION.

Staining ultrathin sections of glutaraldehyde-fixed,epon embedded tissues resulted in selectively contrasted glycogen, when treated by this combination. This result offered an experimental approach for the approval, that really a large complex of osmate and ferrocyanide was built up at the glycogen sides, converting the negative contrast gradually into positive. Instead of staining the ultrathin sections by the combination of K_2OsO_4.2 H_2O plus $K_4Fe(CN)_6$, several separated ultrathin sections were intermittently stained by single 0.05 M K_2OsO_4. 2 H_2O and/or 0.05 M $K_4Fe(CN)_6$ solutions, with a rinse in distilled water in between these reactions.

Fig.1 KCN removed the contrast from the OsO_4 double fixation, untreated section. Fig.2 P.T.A.-staining; glycogen contrast similar to that of aldehyde fixed tissue. Fig.3 Absence of contrast in the glycogen areas after K_2OsO_4. 2 H_2O double fixation. Untreated section. Fig.4 In tissues double fixed by OsO_3.2 Pyridine, P.T.A.-staining detected the lower osmiums oxides present in the glycogen areas. Fig.5 Selective glycogen contrast after double fixation by K_2OsO_4 plus $K_4Fe(CN)_6$. Fig.6 Selective glycogen contrast after double fixation by OsO_3.2 Py. plus $K_4Fe(CN)_6$. Fig.7 Glycogen contrast in aldehyde fixed tissue after staining with two intermittent solutions of K_2OsO_4.2 H_2O and $K_4Fe(CN)_6$. Fig.8 Glycogen contrast after four such intermittent treatments.

(Bars indicate 0,5 μ)

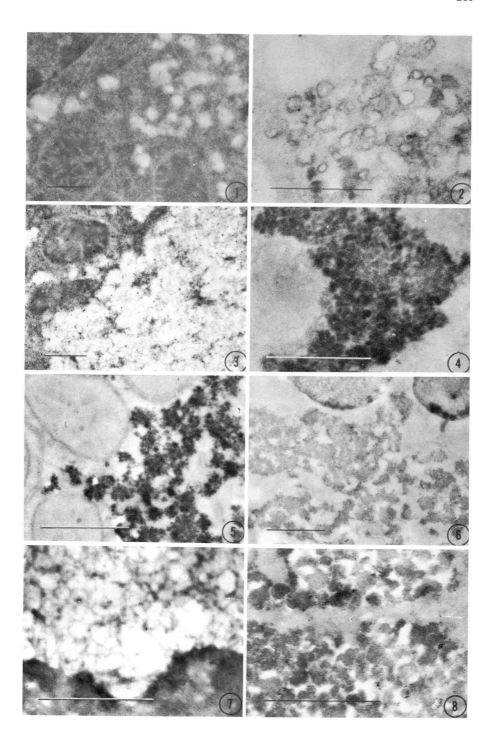

No glycogen contrast was observed in the ultrathin sections treated
once with either $K_2OsO_4 \cdot 2\ H_2O$ or $K_4Fe(CN)_6$, neither when the sections
were treated by the osmate solution, rinsed carefully and subsequently were treated by the $K_4Fe(CN)_6$ solution. Nor when sections were
again treated by osmate or even got an additional second treatment
with $K_4Fe(CN)_6$ (Fig.7). Some glycogen contrast was observed in the
sections treated intermittently three times by osmate and three times by $K_4Fe(CN)_6$, whereas full glycogen contrast was obtained following four such intermittent treatments (Fig.8).

6. DISCUSSION:

The presence of osmium oxides in the glycogen areas of the tissue
is detected by a P.T.A.-treatment of the ultrathin section. In tissues fixed by glutaraldehyde or in tissues double fixed by OsO_4 and
subsequently treated by KCN prior to embedding, a different type of
glycogen contrast was observed (compare Fig. 2 with Fig. 4).

The presence of (hexavalent)osmium oxide compounds in the glycogen areas is also detected by $K_4Fe(CN)_6$ in such solutions. Schematically the reactions can be summarized as follows:

COMPLEX CYANIDES + $Os^{VI}O_2^{2-}(OH)_4 \rightleftarrows$ OSMATE/CYANIDE COMPLEXES
ALDEHYDE-FIXED GLYCOGEN +

CONTRASTED GLYCOGEN COMPLEXES.

Criegée et al.[3] and Lott and Symons[5] produced evidence that in
aqueous solutions, $K_2OsO_4 \cdot 2\ H_2O$ had to be considered as $OsO_2^{2-}(OH)_4$.
For the binding of $K_4Fe(CN)_6$ to the osmate in the osmate/cyanide
complex formation, two alternatives were available in the literature.
One produced by Criegée et al.[3] explaining the structure of the $OsO_3 \cdot 2$
Pyridine and the other proposed by Saxena[6] which can be written as:

$$\left\{ \left[Os^{VI}O_2(OH)_2 \cdot 2 \left\{ Fe^{II}(CN)_6 \right\} \cdot Os^{VI}O_2(OH)_2 \right]^{8-} \right\}_n ?$$

As $OsO_3 \cdot 2$ Pyridine together with $K_4Fe(CN)_6$ is able to contrast the
tissue glycogen, we prefer the complex structure as proposed by
Saxena; although the possibility, that the Pyridine is replaced by
the $K_4Fe(CN)_6$ during this complex formation instead of the
$K_4Fe(CN)_6$ being added to the osmate in the Pyridine-osmate complex,
is not investigated.

7. REFERENCES.
1. Bruijn, W.C.de, Proc.Eur.Reg.Conf.4th,1968, Vol. II p.65 (1968).
2. Bruijn, W.C.de, J. Ultrastruct.Res. 42, 29-50 (1973)
3. Criegée, R.,Marchand,B.and Wannowius,H. Analen der Chemie 550,
   99-133 (1942)
4. McGee-Russell,S.M. and de Bruijn, W.C. in Cell Structure and its
   interpretation, S.M. McGee-Russell and K.F.A. Ross ed. p.115-133
   (E. Arnold Ltd. London 1968)
5. Lott, K.A.K. and Symons, M.C.R., J.Amer.Chem.Soc.X, 973-976 (1960)
6. Saxena, O.G. Microchemic J. 12, 609-611 (1967).

Electron microscopy and Cytochemistry, eds. E. Wisse, W.Th. Daems, I. Molenaar and P. van Duijn.
© 1973, North-Holland Publishing Company - Amsterdam, The Netherlands.

ON THE STAINING OF GLYCOGEN FOR ELECTRON MICROSCOPY WITH POLYACIDS
OF TUNGSTEN AND MOLYBDENUM.

I. DIRECT STAINING OF SECTIONS OF OSMIUM FIXED AND EPON EMBEDDED
MOUSE LIVER WITH AQUEOUS SOLUTIONS OF PHOSPHOTUNGSTIC ACID (PTA)

H.A.R.Schade

Institut für Biophysik und Elektronenmikroskopie
Universität Düsseldorf

1. Summary

Thin sections of osmium fixed mouse liver with or without prefixation with an aldehyde mixture are stained with aqueous PTA solutions with or without periodic acid pretreatment. Staining is followed or not by HCl washing as well as immediate blotting dry or subsequent water washing resulting in positive or respectively negative
intracellular glycogen staining.

2. Introduction

The suggestion[5,6,10,14,15,16] to consider PTA as having "a special affinity for polysaccharides"[6] has caused an elaborate controversy[1,3,5,6,9,11,12,13,15,16,19,20] although the specificity claimed
for PTA has since been demonstrated for many cell types and organelles[6,1,16]. However, the material dealt with in most cases com-
prises glycoproteins or mucopolysaccharides usually fixed with aldehydes and embedded in some type of methacrylate. So hardly any results have been shown in the context of this controversy for the
section staining of a simple polysaccharide, e.g. a homoglycan such
as glycogen within differently fixed and embedded material.

Earlier staining experiments with PTA or phosphomolybdic acid
(PMA) for glycogen in osmium fixed and methacrylate embedded tissue
have shown inconsistent results: either definite positive or more
or less negative staining or none at all as well as differences between PTA and PMA[4,7,8,22,23,24]. However, experiments conducted in
this laboratory[17,18] have mostly shown intracellular negative staining of glycogen both with PTA and PMA and also after periodic acid
treatment, these results on first sight being compatible with those
obtained by block staining with aqueous PTA[21].

Because of this situation a thorough re-investigation of section
staining of glycogen in osmium fixed and epon embedded material
with PTA was undertaken.

3. Materials and methods

Mouse liver was fixed with OsO_4 (1g%) with or without prefixation
in a formaldehyde (1g%) / acrolein (1Vol%) mixture and embedded in

epon. Thin sections on formvar coated grids (Cu or Au) were stained
from 1 to 30 min. with aqueous solutions of PTA (e.g. 2g%) with or
without pretreatment (20 min.) with periodic acid PA (o.o5M). Staining was followed or not by washing with HCl (5N, 1N or 0.1N) from
1 to 5 min. In both cases grids were either blotted dry immediately
or washed with water from 5 to 20 min.

4. Results

Positive staining of glycogen was obtained in all cases except
after washing with water which caused negative staining of glycogen
along with an extremely low general contrast in these sections while
washing with 0.1N HCl gave more or less transitory results. Besides
glycogen only collagen was stained significantly in positively
stained sections whereas ribosomes could not be detected. The same
was the case in negatively stained sections. The only difference between these two cases, disregarding the fact of different glycogen
staining, seemed to be a gradual destaining of membranes in positively and of restaining of these structures in negatively stained sections.

5. Discussion

Positive staining of glycogen with aqueous PTA appears to be an
advantageous method to differentiate between glycogen and ribosomes
histologically. Positive staining furthermore appears to be a simpler
and more selective method histochemically than the PA-Tollens-(PAT)
or the PA-Nylander-(PAN) reactions which are straight analogs of the
PA-Schiff-(PAS) reaction, for staining with PTA causes far less
background staining than the two step staining reaction with basic
solutions of the Tollens-Ag- and the Nylander-Bi-reagents.

Following the criteria established for qualitative analytical
procedures[2] the reaction considered may be valued as a selective
reaction for the detection of compounds containing 1,2-vic-hydroxyl-
functions since their detection is interfered to only a very small
extent by other material present and only in sections not pretreated
with PA, if compared e.g. with the PAT- or the PAN-reaction. It is,
however,not a specific reaction for such compounds under the circumstances described because specificity of a reaction requires no
interference at all whatever other material may be present[2]. And

Electron micrographs of sections of osmium fixed and epon embedded
mouse liver. Fig. 1. Stained with PTA. Fig. 2. Stained with PTA and
washed with water. Fig. 3. Treated with PA, stained with PTA.
Fig. 4. Treated with PA, stained with PTA and washed with water.

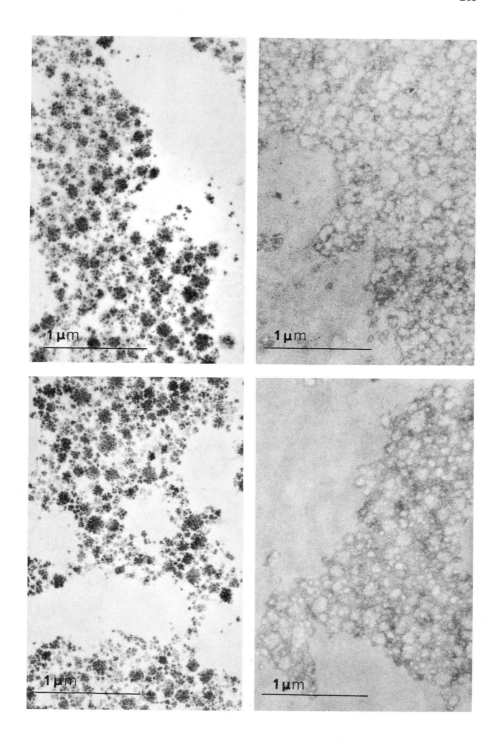

266

osmicated membraneous material does at least interfere to some extent, i.e. in sections not treated long enough with PTA.

The gradual destaining of such material by PTA treatment along with a noticeable increase of density of PTA stained glycogen with time, above all, however, the fact that even PA treated glycogen, i.e. its polyaldehyde is positively PTA stained leads to the conclusion that PTA exerts an oxidative effect on the tissue; PTA removes reduced osmium from the tissue thus destaining membranes and transforms 1,2-vic-hydroxyl-functions into aldehyde-functions which then combine with some, possibly a reduced PTA-compound not to be defined at present to form positively stained glycogen. The same is the case - mutatis mutandis - for collagen.

Evidence for this hypothesis[9] will be presented e.g. in the form of light microscopical histochemical work (PTA-Schiff-reaction) in subsequent papers which will also deal with the histochemistry of the observed negative staining of glycogen in thin sections.

References

1. Bloom,F.E. and Aghajanian,G.K.,J.Ultrastruct.Res.22,361-375(1968)

2. Feigl,F.,Chemistry of specific, selective and sensitive reactions (1949)

3. Glick,D. and Scott,J.E.,J.Histochem.Cytochem.18,455(1970).

4. Marinozzi,V.,J.Roy.Micr.Soc.81,141-154 (1963).

5. Marinozzi,V.,J.Microscopie 6,68a (1967).

6. Marinozzi,V.,Electron Micr. (Bocciarelli,ed.)Vol.2,55-56 (1968).

7. Marinozzi,V. and Gautier,A.,C.R.Acad.Sc.Paris D 253,1180-82(1961).

8. Mercer,H.E.,J.Roy.Micr.Soc.81,179-186 (1963).

9. Palladini,G.,Lauro,G.et Basile,A.,Histochemie 24,315-321(1970).

10. Pease,D.C.,J.Ultrastruct.Res.15,555-588(1966).

11. Pease,D.C.,J.Histochem.Cytochem.18,455-458(1970).

12. Quintarelli,G. et al.,J.Histochem.Cytochem.19,648-653(1971).

13. Quintarelli,G. et al.,J.Histochem.Cytochem.19,641-647(1971).

14. Rambourg,A.,C.R.Acad.Sc.Paris D 265, 1426-28(1967).

15. Rambourg,A.,Electron Micr. (Bocciarelli,ed.)Vol.2,57-58(1968).

16. Rambourg,A.,J.Microscopie 8,325-342(1969).

17. Schade,H.A.R.,Thesis, Universität Düsseldorf 1969.

18. Schade,H.A.R.,Symp.Gschft.Histochem. Düsseldorf 1971, in print.

19. Scott,J.E.,J.Histochem.Cytochem.19,689-691(1971).

20. Scott,J.E. and Glick,D.,J.Histochem.Cytochem.19,63-64(1971).

21. Silverman,L. and Glick,D.,J.Cell Biol.40,761-767(1969).

22. Swift,H. and Rasch,E.RCA Sci.Instr.News 3,1-5(1958).

23. Swift,H. and Rasch,E.,J.Histochem.Cytochem.6,391-392(1958).

24. Watson,M.L.,J.Biochem.Biophys.4,475-478(1958).

Electron microscopy and Cytochemistry, eds. E. Wisse, W.Th. Daems, I. Molenaar and P. van Duijn.
© 1973, North-Holland Publishing Company - Amsterdam, The Netherlands.

ELECTRON MICROSCOPICAL DETECTION OF DNA

G. Moyne

Laboratoire de Microscopie Electronique, Institut de
Recherches Scientifiques sur le Cancer. Villejuif (France)

Summary

A new electron microscopical reaction for localization of DNA has been
developed. It involves revelation of the classical Schiff's reagent used in
Feulgen's reaction by thallium ethylate, an electron stain specific for hydroxyl
groups. To avoid side reactions, the tissues are acetylated prior to application
of Schiff's reagent.

The reaction allows localization of DNA in chromatin and also in the cores
of DNA viruses and in bacteria. It is suppressed by DNase digestion of formaldehyde-fixed blocks.

Introduction

The Feulgen reaction is well known as one of the most specific methods
available to the light microscopist. Unfortunately Schiff's reagent combined to
apurinic acid produced during Feulgen's reaction is not electron dense. It can
be revealed at the electron microscope level, however,by PTA[1,5] but the use of
this stain entails a loss of specificity with reference to the Feulgen or PAS
reactions. We find that a more specific revelation can be achieved through the
use of thallium ethylate, a hydroxyl-binding reagent[3]. In order to avoid staining of other hydroxyl-containing cell structures, the tissues are acetylated
prior to treatment with Schiff's reagent. Several methods have been developed,
varying with the moment of embedding : before or after hydrolysis or acetylation
or both[4].

Material and methods

- Rat tissues ; normal and virus infected tissue cultures as well as
  bacteria have been studied.
- For routine procedure, tissues are fixed in 1.6 % glutaraldehyde in
  pH 7.3 phosphate buffer for 1 h at 4°C. After thorough washing, the

blocks are dehydrated in acetone and acetylated overnight at 45°C in a mixture of pyridine-acetic anhydride (60/40 v/v). After Epon embedding, the sections mounted on gold grids are floated for 20 to 30 mins. on 5 N HCl at room temperature and rinsed. They are subsequently floated on Schiff's reagent for 30 mins., rinsed and dried. They are finally stained with thallium ethylate in ethanol (1 mg/ml) adding one drop of water to speed up the reaction as described[3].

- DNA detection can also be carried out entirely on sections of glutaraldehyde-fixed, GMA-embedded tissues[4].
- Bacteria have been fixed according to Granboulan and Leduc[2].

Results

In order to ascertain the mechanism of the reaction, various steps have been omitted, except final thallium ethylate staining :

- Untreated sections display a general staining due to thallium ethylate.
- Acetylated, unhydrolyzed, sections always present a weak staining.
- Acetylated and hydrolyzed controls, untreated by Schiff's reagent are not stained ; Schiff's reagent is needed for obtention of the complete reaction.
- Some blocks have been fixed in formaldehyde and treated by DNase before DNA staining : the reaction is then negative.
- The complete reaction is characterized by staining of DNA-containing structures. In the tissues which we studied, chromatin (**figs. 1 and 2**), cores of DNA viruses and bacterial DNA were stained. Nucleoli, RNA viruses and cytoplasm had no contrast.

Discussion

The fact that a light staining can be seen on unhydrolyzed sections shows that the reaction is not strictly Feulgen-like. This could be expected if one considers that the specificity of the reaction is not based on Schiff's reagent but on removal of all hydroxyl groups , except those eventually created by combination of Schiff's reagent with apurinic acid[6]. This drawback is compensated

---

C12TSV5 cells (Hamster cells transformed by SV 40 virus). DNA staining.

Fig. 1. Staining of the chromatin (chr) is evident. Contrast of the nucleoli (nu) and cytoplasm (cy) is low. Nuclear pores (⟶) are devoid of stained material.

Fig. 2. Nucleolus. Perinucleolar and intranucleolar (▶) chromatin is well contrasted. Notice the fibrillar appearance of nucleoplasm (Np) as compared with nucleolar body (nu).

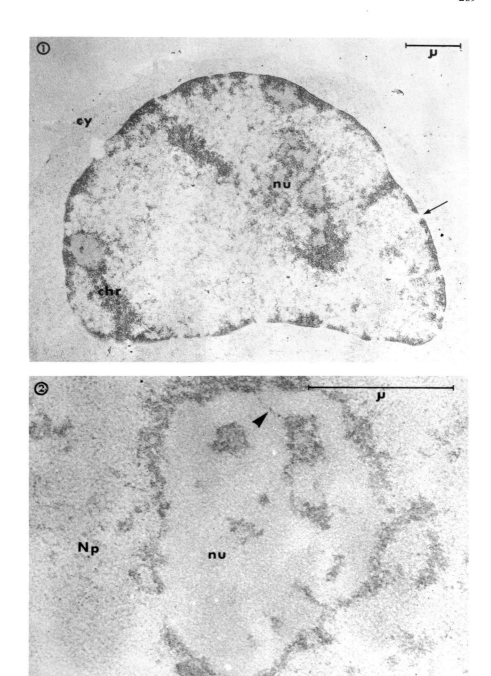

by the observation that acetylated and hydrolyzed controls, untreated with Schiff's reagent are not stained.

We can finally assume that a very preferential, if not specific, staining of DNA is achieved since the complete reaction stains only structures known to contain DNA and is suppressed by DNase.

References

1. Gautier, A. and Schreyer, M., 1970, 7ème Cong. Intern. Micr. Electr. Favard, P. ed. vol. I pp 559-560 (Société Française de Microscopie Electronique).
2. Granboulan, P. and Leduc, E.H., 1967, J. Ultrastr. Res. 20, 111.
3. Mentré, P., 1972, J. Microscopie 14, 251.
4. Moyne, G., 1973, J. Ultrastr. Res. in press.
5. Thiéry, J.P., 1967, J. Microscopie 6, 987.
6. Wieland, H. and Scheuing, G., 1921, Ber. Deut. Chem. Ges. 54, 2527.

Electron microscopy and Cytochemistry, eds. E. Wisse, W.Th. Daems, I. Molenaar and P. van Duijn.
© 1973, North-Holland Publishing Company - Amsterdam, The Netherlands.

"FEULGEN-TYPE" AND "PAS-TYPE" REACTIONS
AT THE ULTRASTRUCTURAL LEVEL*

A.Gautier, R.Cogliati, M.Schreyer and J.Fakan

Center for Electron Microscopy,

The University of Lausanne (Switzerland)

Introduction

We have previously shown that it is possible to detect DNA on ultrathin sections of fixed and embedded tissues by the Feulgen reaction
or Feulgen-type reactions[3]. However, only a low contrast was observed
with the Schiff's reagent itself and irregular results were obtained
with 12 tested Schiff-type reagents[4], thus prohibiting routine work.
Since an inorganic polyammine, Ruthenium Red, showed similar properties to a Schiff-type reagent[2], we tried parallel experiments with
other polyammines of heavy metals. Best results were obtained with a
recently synthesized[1] osmium ammine. This reagent may be used not only
to detect DNA, but also polysaccharides.

The Reagent

Osmium Ammine (OA) is a black crystalline powder, whose chemical
composition is not yet exactly determined. This reagent is easy to
prepare as described previously[1] and shows highly reproducible staining properties. It is weakly soluble in water ($\leqslant 0.1\%$ p/v). OA has an
absorption spectrum in UV with peaks at 322 and 238 nm and an additional shoulder at 380 nm. This spectrum has some similarities to the
corresponding spectra of mixed-ammine-chloro complexes of osmium**.

Material and Methods

Ciliates, various mammalian tissues and tissue cultures infected
with DNA-viruses were used in preliminary experiments. The fixation
involved the usual procedures with aldehyde, aldehyde-OsO_4, OsO_4 or
$KMnO_4$ buffered solutions. Epon or Araldite was used for embedding.
Mild acid hydrolysis was obtained usually by floating ultrathin sec-

*   Research supported by a grant from the Swiss National Foundation
     for Scientific Research.

**   Thanks are due to Prof. A.LUDI, the University of Berne (Switzer-
     land) for these analyses.

271

tions on 5N HCl solution at 20°C for 15 min. HCl- or cold-PCA-hydroly-
ses on blocks gave similar results. Periodic acid (PA) treatment was
achieved by floating sections on a 1% solution at 20°C for 15-45 min.
Sections were then floated on a solution of the reagent dissolved in
water and previously treated by a 10 min. SO_2-bubbling. The concentration, the temperature and the duration for this staining reaction have to be adjusted for a given specimen to the desired intensity and
grain fineness. In order to test the Feulgen-type character of the reaction[5], the Oster and Mulinos histochemical procedure with hot Aniline-HCl was used (as well as controls with NaCl-HCl). These reactions
were carried out after hydrolysis and before staining, following modifications suggested by Peters and Giese[7].

Results

Feulgen-type reactions in case of aldehydic fixation.

In mammalian nuclei, intensity of chromatin is clearly enhanced by
such a procedure (Fig.1). The stain may be prepared quite strong with
rough grain (for quick determinations at low magnifications) or very
fine (with grains approximatively 1 nm in diameter). DNA may be detected by this reaction in cell nuclei, in ciliate mitochondria and in
DNA-viral nucleoids (Fig.2). No staining occurs when the reagent is
not pretreated with SO_2, or if the tissue is not hydrolysed, or in the
case where the aldehydic groups of hydrolyzed DNA are blocked by the
Aniline-HCl procedure.

PAS-type reactions in case of aldehydic fixation.

In mammalian tissues, every component showing definite PAS-positive
reaction under the light microscope is also strongly characterized at
the ultrastructural level by the PA-OA staining procedure; liver glycogen (Fig.3), glycocalix of various cells (Fig.4) or mucous droplets
in intestinal goblet cells may be quoted as examples. In liver glyco-

---

Fig.1. Rat exocrine pancreas cell. Glutaraldehyde/Epon/HCl/[OA 0.05%
+ SO_2]$_{\underline{60}}$'20°C

Fig.2. Shope fibroma viruses in an infected culture cell (R.Scherrer,
Villejuif). Glut./Epon/HCl/[OA 0.1% + SO_2]$_{\underline{90}}$'60°C

Fig.3. Dog hepatocyte. Glut./Araldite/PA$_{\underline{45}}$'/[OA 0.5% + SO_2]$_{\underline{60}}$'20°C.

Fig.4. Mouse intestinal epithelial brush border. Glut./Epon/PA$_{\underline{45}}$'/
[OA 0.1% + SO_2]$_{\underline{90}}$'20°C.

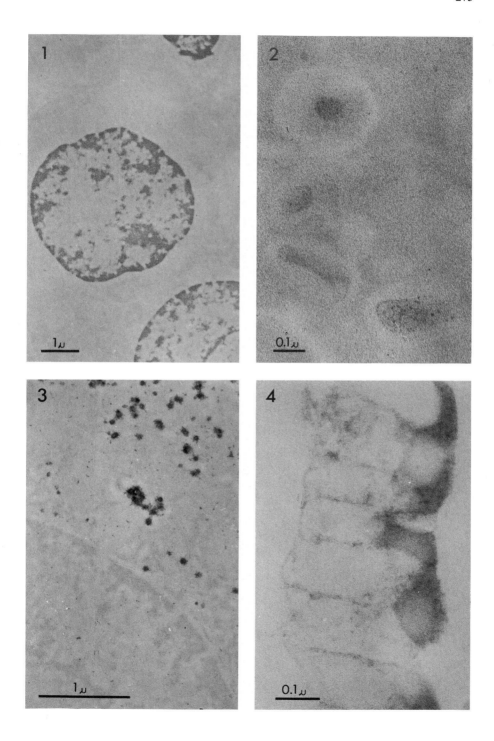

274

gen the fine structure may be easily resolved even to their minute γ-particles.

Feulgen-type and PAS-type reactions with other fixatives.

The applicability of the present reaction is not restricted to aldehyde fixed specimens. After osmic or aldehyde-osmic fixations, the reaction is indeed less specific, but still preferential for DNA or polysaccharides. As examples, we detect ciliate mitochondrial DNA even after OsO_4 fixation, or an increased intensity of glycogen staining in liver fixed with glutaraldehyde-OsO_4. Even after $KMnO_4$ fixation, DNA may be clearly identified in the Feulgen-type ultrastructural OA-reaction in cell nuclei and in DNA-viral nucleoids.

OA as stain for extracellular acidic mucopolysaccharidic moieties.

If 0.16% OA (not treated with SO_2 !) is added to glutaraldehyde and OsO_4 fixative solutions, like Ruthenium Red in the Luft's procedure[6], a similar reaction occurs: the glycocalix is strongly stained. The fine structure of this precisely localised reaction is the same as that obtained with Ruthenium Red, as may be seen, for instance, on the brush borders of intestinal epithelial cells. The mechanism of both reactions, i.e. the chelation of extracellular acidic mucopolysaccharide moieties may be analogous, both being seemingly polynuclear ionic complexes between oxygen-bridged heavy metal atoms and ammonia.

References

1. Cogliati, R. and Gautier, A., 1973, C.R.Acad.Sc. (Paris), in press.
2. Gautier, A. and Schreyer, M., 1970, Proc.7th.internat.Conf.E.M. 1, 599.
3. Gautier, A., Schreyer, M., Cogliati, R. and Fakanova, J., 1971, Experientia (Basel) 27, 735.
4. Gautier, A., Schreyer, M., Cogliati, R. and Fakan, J., 1972, J. Micr. 14, 48a.
5. Kasten, F.H., 1960, Internat.Rev.Cytol. 10, 1.
6. Luft, J.H., 1971, Anat.Rec. 171, 347.
7. Peters, D. and Giese, H., 1970, Proc.7th internat.Conf.E.M. 1, 557.

Electron microscopy and Cytochemistry, eds. E. Wisse, W.Th. Daems, I. Molenaar and P. van Duijn.
© 1973, North-Holland Publishing Company - Amsterdam, The Netherlands.

INTRACELLULAR AND SURFACE COAT STAINING IN KERATINIZED AND
NON-KERATINIZED ORAL EPITHELIA USING THE PERIODIC ACID -
THIOCARBOHYDRAZIDE - SILVER PROTEIN TECHNIQUE (PA-TCH)

A.F. Hayward
Department of Oral Anatomy
Royal Dental Hospital of London

1. Summary

The membrane-coating granules of most keratinized squamous
epithelia studied and of all non-keratinized epithelia stain with
PA-TCH showing the probable presence of glycoproteins. Discharge of
the granules from the distal border of the cells is followed by the
appearance of a PA-TCH-positive coat on the distal cell surfaces.
Certain less-densely keratinized epithelia display no staining
reaction of this type.

2. Introduction

Membrane-coating granules (MCGs)[1] are found in most, if not all
keratinized and non-keratinized squamous epithelia. As a keratino-
cyte differentiates, in the stratum spinosum the MCGs first increase
in numbers predominantly along the distal borders of the cells.
Subsequently in the stratum granulosum, as the granule contents are
discharged into the intercellular spaces [2,3], they disappear from
the cytoplasm completely.

In keratinized epithelia, MCGs contain lysosomal hydrolases [5,6]
and internal lamellations, possibly composed of phospholipids [4].
MCGs of non-keratinized epithelia display similar characteristics
but have non-lamellated contents.

Many MCGs stain with the periodic acid thiocarbohydrazide
reaction [7,8] probably indicating the presence of glycoproteins.
Variations in this staining have been investigated in a range of
epithelia.

3. Material and methods

Keratinized and non-keratinized oral mucosa were taken from
6 species including man. Tissues fixed in Karnovsky's aldehyde
solution [9] with and without post-osmication were embedded in
Araldite. Thin sections were stained with Thiéry's PA-TCH method[10]

using processing times and control methods described previously[8].
Sections of post-fixed tissues were first treated with hydrogen
peroxide.

3. Results

MCGs generally stained positively with PA-TCH (Figs. 1,2).
Control reactions were compatible with the presence of glycoproteins.
The only exceptions were found in the less densely keratinized
epithelia characteristic of parts of the rodent oral cavity where
no staining was ever detected. The intensity of staining varied
and was greatest in human gingival epithelium. MCGs of non-
keratinized epithelia generally stained intensely.

After the discharge of MCGs from the cytoplasm, a PA-TCH reaction,
not encountered in deeper cell layers occurred along the distal cell
membranes (Figs 3,4). Some reaction product was dispersed through
the intercellular spaces. Cell surface coat staining persisted
throughout the stratum corneum and to the oral surfaces of non-
keratinized mucosae. The characteristics of the reaction differed
slightly after post-osmication.

4. Discussion

The precise functions of MCGs are still unknown. They can be
categorized as primary lysosomes and as such would be expected to
contain glycoproteins since some of the hydrolytic enzymes are
glycosylated[11]. MCGs are unusual among lysosomes in their discharge
from the cells. Their stainable component seems to be related to
their own limiting membrane and subsequently provides a cell surface
coat for the more distal keratinocytes. The hydrolytic enzyme
activity might be related to cell desquamation. The secretion of
glycoprotein into the intercellular spaces and as a cell coat may be
a factor in the overall permeability of the epithelium and might
also play a part in cell shedding by altering the character of the
distal cell membrane.

Fig. 1. Electron micrograph of membrane-coating granules of keratinized human gingiva, processed with post-osmication and PA-TCH
showing staining of lamellae. Fig. 2. In MCGs processed without
osmication the staining is peripheral. Fig. 3. Surface coat staining
of stratum granulosum of keratinized mucosa after MCG discharge.
Fig. 4. Surface coat staining in distal layers of non-keratinized
buccal mucosa.

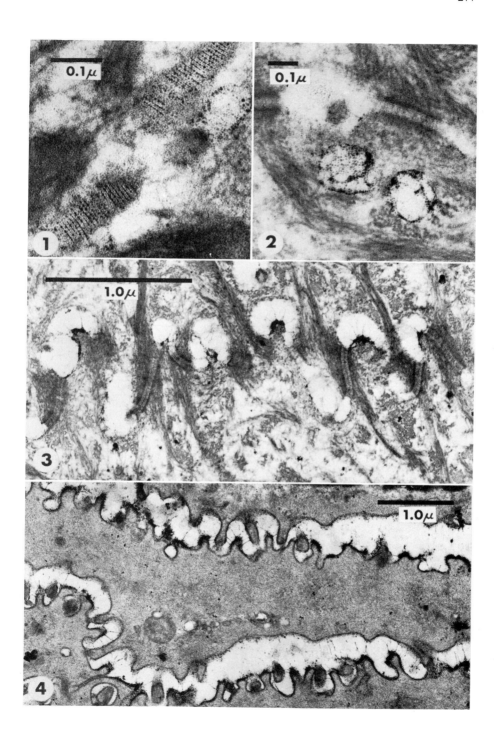

278

References

1. Matoltsy, A.G. and Parakkal, P.F., 1965, J. Cell Biol. 24, 297.

2. Farbman, A.I., 1964, J. Cell Biol. 21, 491.

3. Martinez, I.R. and Peters, A., 1971, Am. J. Anat. 130, 93.

4. Hashimoto, K., 1971, Arch. Dermatol. Forsch. 240, 349.

5. Wolff, K. and Schreiner, E., 1970, Arch. Dermatol. 101, 276.

6. Olson, R.L., Nordquist, R.E. and Everett, M.A., 1969. Arch. klin. exp. Dermatol. 234, 15.

7. Hayward, A.F., 1973, Archs. oral Biol. 18, 67.

8. Hayward, A.F. and Hackemann, M.M.A., 1973, J. Ultrastruct. Res. in press.

9. Karnovsky, M.J., 1965, J. Cell Biol. 27, 137A.

10. Thiéry, J-P., 1967, J. Microsc. (Paris) 6, 987.

11. Spiro, R.G., 1969, New England J. Med. 281, 1043.

CYTOCHEMISTRY OF THE LIPID DROPLETS CONTAINING VITAMIN A IN THE LIVER

K. Wake [1)]

Department of Anatomy, Osaka City University Medical School

Abenoku, Osaka, 545 Japan

1. Summary

Gold chloride reaction has been studied cytochemically in normal and hyper-vitaminosis A rat livers. Metallic gold deposition has been observed on the surface of the lipid droplets containing vitamin A in the stellate cells. These lipid droplets have developed in the multivesicular bodies (MVBs). Specifity of this reaction has also been examined.

2. Introduction

In a previous paper[1], the author has reported that vitamin A is not stored in the phagocytic Kupffer cells, but in the stellate cells first described by Kupffer[2], using the gold impregnation method. These stellate-shaped cells are located in the space of Disse, and are identical with the "fat-storing cells". The lipid droplets in their cytoplasm contain vitamin A, and react intensely with gold chloride. In the present paper, this gold reaction is shown to be applicable to demonstrate vitamin A, and is applied to ultracytochemistry in order to examine the mechanism of vitamin A-uptake and storage in the stellate cells.

3. Materials and methods

Sprague-Dawley **rats**, fed laboratory chow, received vitamin A acetate (Eisai Co.Ltd., Tokyo) in four subcutaneous injections (Total dosage, 1,320,000 IU/kg.) with two days' intervals. The methods adopted for light and fluorescence micro-scopy are the same as described in an earlier paper[1]. For electron microscopy 1mm thick liver blocks were fixed in 6% glutaraldehyde in 0.1M phosphate buffer, at pH 7.4, for two hours. After washing in distilled water, the tissue was incubated in 0.02% gold chloride solution, at pH 2.8, for 16 hours in darkness at room temperature. A short rinse in unbuffered 1% osmium tetroxide before incubation is recommended for p revention of fine structure destruction. The

---

1) Present address: Department of Anatomy, Justus Liebig-University,

63 Giessen, Germany

tissue was post-fixed for two hours in unbuffered 2% osmium tetroxide, dehydrated and embedded in Epon 812.

4. Results

Intensity of the gold reaction in the stellate cells increased remarkably after the administration of excess vitamin A. Metallic gold deposition occurs on the surface of the lipid droplets in the stellate cells (Figs. 2 and 3), as well as of the chylomicrons in the phagocytic Kupffer cells. These lipid droplets impart vitamin A fluorescence. This reaction occurs only below pH 4.2, increasing with the reduction of pH value and is strong at pH 2.6. No reaction occurs after irradiation with ultraviolet light, as a result of destruction of vitamin A. Vitamin A acetate, palmitate and cod liver oil react with gold chloride in vitro to produce a violet colour with precipitation. There exist two kinds of gold-reactive lipid droplets in the stellate cells (Fig. 4): The Type I is surrounded by a unit membrane derived from the limiting membrane of the MVB. The Type II exists in the matrix of the cytoplasm which seems to have developed from the Type I after losing the unit membrane. Various developing steps of lipid droplets from the MVB to those without a unit membrane have been observed.

5. Discussion

The present gold reaction occurs only on the surface of the lipid droplets which impart vitamin A fluorescence, and seems to be a reduction reaction of gold chloride by vitamin A. This method is also applicable for proper fixation of these lipid droplets. It is suggested that retinyl ester is hydrolysed in the space of

Fig.1. A fluorescence micrograph showing vitamin A fluorescence from the lipid droplets of the stellate cells of a hypervitaminosis A rat liver.

Figs. 2 and 3. Electron micrographs of the stellate cells (SC) from the same rat after gold reaction. Note the well-preserved lipid droplets (L) with a reduced gold precipitation on their surface. The reaction fails to occur at the lipid droplets (arrows) in the parenchymal cells (HC). S, sinusoid; ER, granular reticulum.

Fig. 4. Some developmental stages of the vitamin A-lipid droplets in the stellate cells. Note the small lipid droplets (∗) in the matrix of the multivesicular bodies (MVB). The lipid droplets (L_1) are surrounded by unit membrane derived from the limiting membrane of the MVB. The mature lipid droplets (L_2) have no unit membrane.

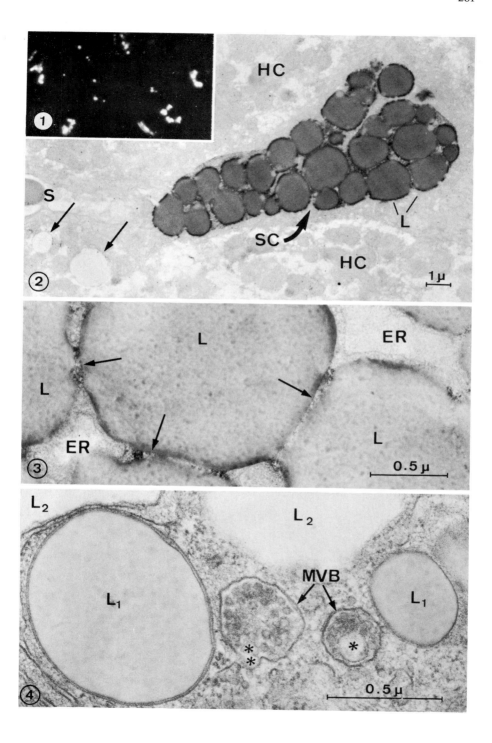

282

Disse, and retinol crosses the cytoplasmic membrane of the stellate cells. Re-
esterification of retinol at pH 4.5[3] seems to occur in the MVBs in the stellate cells

References

1. Wake, K. Am. J. Anat. 132: 429-462 (1971)

2. von Kupffer, C. Arch. mikr. Anat. 12: 353-358 (1876)

3. Futterman, S. and Andrews, J.S. J. Biol. Chem. 239: 4077-4080 (1964)

Electron microscopy and Cytochemistry, eds. E. Wisse, W.Th. Daems, I. Molenaar and P. van Duijn.

THE USE OF PHYTOAGGLUTININS COUPLED WITH PEROXIDASE FOR THE LOCALIZATION
OF SPECIFIC CELL SURFACE RECEPTORS

W. Bernhard

Institut de Recherches sur le Cancer, VILLEJUIF (France)

1. Introduction

There already exist several cytochemical techniques, developed in the
past 10 years, which allow the visualization of the polysaccharide compounds of
the cell coat or even those inside the cell. These most useful procedures are
carried out on fixed cells, and are generally based on the transformation of
glycol groups into aldehydes, on the presence of sulfuric esters or carboxyl
groups[22,29,30,31,38].

Avrameas[4] first proposed an entirely different procedure for visualising
sugar residues, primarily on the cell coat, but also in certain cases inside the
cytoplasm, by means of phytoagglutinins (lectins). The lectins have two main
advantages :
1) they possess a very high degree of specificity in the immunological sense for
   certain sugar residues ;
2) they may be employed on living cells without any artificial changes of the
   molecular configuration of the sugars.

The development of this technique is closely linked with cancer research.
In 1963, Aub and collaborators[3] prepared a crude extract of wheat germ which
had the property of a lipase. This extract was shown to agglutinate more strongly
certain cancer cells than normal cells. Later, Burger and Goldberg (1967)[11]
further purified this product and obtained a crystallizable phytoagglutinin
which was called WGA (wheat germ agglutinin). In the same year, Agrawal and
Goldstein[1] purified a comparable extract from Jack Beans (Canavalia ensiformis)
which was called Concanavalin A (Con A). This lectin also strongly agglutinated
certain cancer cell strains, whereas the homologous normal cells could be agglutinated only after a mild predigestion of the cells with a protease. A series of

papers by Burger, Sachs and their collaborators[9,10,19,20,34] confirmed these findings and lead to the hypothesis that cancer cells tend to have more free surface receptors for phytoagglutinins than normal cells, which possess the same receptors but in a masked or cryptic form. Subsequently, a great number of investigators tried to distinguish normal from malignant cells of various origin with this simple procedure. It became clear quite rapidly that although the original observation could be confirmed for certain cell strains, increased agglutination could not considered to be a general feature for all cancer strains[29], Furthermore, normal embryonic cells also were found to agglutinate very strongly [37]. On the other hand, the hypothesis of an increased lectin binding in tumor cells could not confirmed[2,12,28]. There was an increasing need for ultrastructural methods allowing the direct visualization of the cell surface receptors specific for the different types of lectins.

2. The Con A-peroxidase method

An obvious way to realize this was to couple one of the lectins currently used in agglutination studies with a marker which is easily visible in the electron microscope. Horseradish peroxidase seemed an excellent candidate as it leads to a reaction product visible both in the light[4] and in the electron microscope[7]. When brought in contact with concanavalin A, it does not need a special coupling agent, as by mere chance this type of enzyme is a glycoprotein, containing 18 % of sugar moieties, among which the three sugar residues reacting predominantly with the acceptor sites of the lectin are present : α-D-glucopyranosyle, α-D-mannopyranosyle, and β-D-fructofuranosyle[15]. Under suitable conditions, peroxidase is fixed stereospecifically on the free acceptor site (s) of the lectin molecule which may occur as monomer, dimer or tetramer. The principle of the reaction is illustrated in Fig. 1. The Con A molecules are fixed on the surface of a living cell culture by the corresponding specific sugar residues which are present in the glycocalyx. After removal of the unfixed lectin, peroxidase is added and becomes fixed on the free acceptor sites of the bivalent or polyvalent lectin. The catalytic group of the enzyme is then revealed in the classical way by means of the Graham-Karnovsky reaction with DAB[16], leading to an asmiophilic precipitate visible in the electron microscope (fig. 1). The specificity of this reaction can be demonstrated by the addition either before or after treatment of the cells of a sugar which blocks the free acceptor sites of Con A. The most specific inhibitor is α-methyl-D-mannoside, added in high concentration (0,1 M). A glucose solution (0,2 M) can also be used for this test, but

it is clearly less specific. Under the experimental conditions indicated below,
no non-specific binding of peroxidase takes place at the cell surface (figs. 2,
3, 4).

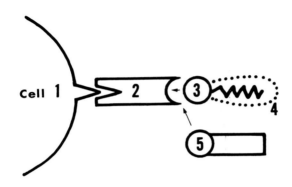

Fig. 1. Mechanism of the Con A-peroxidase reaction. 1- Cell surface with specific
carbohydrate residues ; 2- Concanavalin A with 2 or more receptor sites ;
3- Horseradish peroxidase ; 4- Electron dense precipitate after DAB reaction ; 5- Competitive sugar for control.

3. Standard procedure for the Con A-peroxidase reaction for cell cultures

1. Remove supernatant, wash gently the living cells 3 times 1 min. with
   5 ml of PBS (Ca^{++} and Mg^{++} free according to Dulbecco).
2. Add Con A 50 μg/ml. Incubate 15 min. at 20°C (or at any temperature
   desirable for the experiment).
3. Wash 3 x 1 min. in PBS.
4. Add Peroxidase 50 μg/ml. Incubate 15 min. at 20°C.
5. Wash 3 x 1 min. in PBS.
6. Fix in 1,6 % of glutaraldehyde (or formaldehyde) in PBS for 15 min.

7. Carry out the Gramham-Karnovsky reaction with DAB, 0,5 mg/ml in 0,1 M Tris buffer, at pH 7,4 after addition of one drop of H_2O_2 per 5 ml of medium immediatly before use.

8. Wash 3 x 1 min. in PBS.

9. Postfix 1h in 2 % osmium tetroxide.

10. Dehydrate and embed in situ within the culture bottles or the Petri dishes using Epon.

11. After polymerization, break the glass or cut the plastic bottom. The embedded cell layer can be easily detached by heating the glass-plastic fragment in hot water at 90°C for 3-5 min.

12. Sectionning is done preferentially with diamond knives in the longitudinal sense of the spread cells, in order to visualize the cell coat reaction.

13. Counterstaining is better omitted, but in order to enhance slightly the contrast, brief staining (1-3 min) in lead citrate may be useful.

This technique may be modified in various ways : the treatment with Con A and peroxidase may be shorter, the concentration of the lectin and the enzyme may be lower (down to 5 µg/ml). The whole reaction may be carried out on cells previously fixed in an aldehyde. However, it seems very important to carry out all reactions in situ, including the embedding and polymerization, in order to avoid irregular loss of the reaction product at the cell surface. The reaction may also be carried out on cell suspensions, but irregular loss of the DAB-osmium precipitate may occur during centrifugation.

We have repeatedly tried to stain the cell surface of compact tissues, such as liver, kidney, solid tumors, etc... but the results have been very disappointing because the Con A-peroxidase complex does not penetrate well into the intercellular spaces. A certain degree of penetration (one or two cell layers) may be obtains if the times of treatment are very much prolonged (up to 24hrs), but non-specific diffusion may then lead to false localizations.

As far as the penetration into the cytoplasm is concerned it may be considerable inside pinocytotic vesicles, phagocytic vacuoles or the Golgi complex, provided however that the reaction has been carried out in vivo. Therefore, this

---

Fig. 2. Intestinal mucosa of the rat. Con A-peroxidase reaction on cross-sectionned microvilli.

Fig. 3. Con A-peroxidase reaction on the surface of a Burkitt tumor cell. No counterstain.

Fig. 4. Burkitt tumor cell unlabelled (control for fig. 3). Addition of 0,2 M of α-methyl-D-mannoside to the solution of Con A and peroxidase.

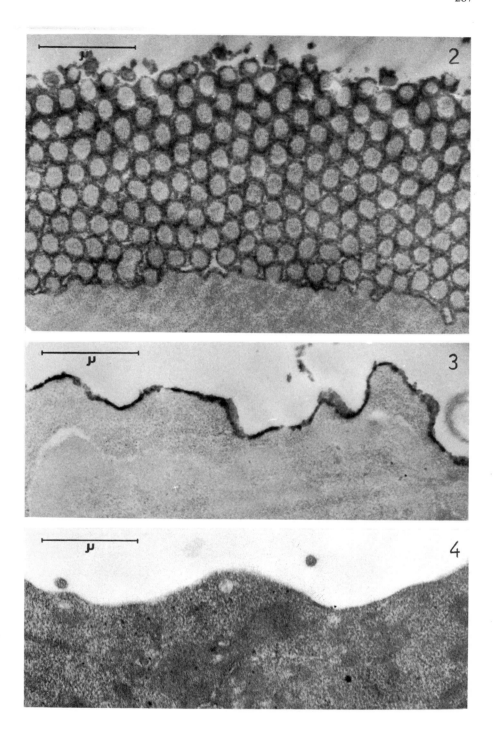

intracellular penetration reveals an active process of the living cell. Penetration does not occur if the cells are fixed with glutaraldehyde prior to cytochemical treatment, but it does occur after mild formaldehyde treatment.

4. The Con A-ferritin method

Nicolson and Singer[25,27] developed a method different from ours using ferritin as a marker instead of peroxidase. The marker of course does not react spontaneously with the enzyme, but has to be coupled chemically by means of glutaraldehyde according to the procedure described by Avrameas[4]. The published electron micrographs convincingly demonstrate binding on areas of the cell coat of erythrocytes and normal and transformed cells in the areas of the specific receptors. It may be too early to judge the respective advantages and disadvantages of the peroxidase and ferritin method. Both seem useful for different purposes : the peroxidase-DAB-osmium precipitate is easily visible at low magnification which allows a rapid screening of many cells. The ferritin may be more useful for very precise localization of the receptors, and probably also, for quantitation.

5. The coupling of other lectins with peroxidase

The great interest in the use of lectins for cell surface studies is based on their individual specificities for different sugar residues. Unlike Con A, which predominantly reacts with glycose, mannose and fructose residues, WGA binds to N-acetyl-D-glucosamine[11], soybean extract to N-acetyl-D-galactosamine[35], the extract of Ricinus communis to other galactose residues[26]. There is, furthermore, a fucose-binding protein[21]. Other specificities may be expected for hitherto poorly explored phytoagglutinins from various sources (Tomita et al.)[39]. In all these cases, peroxidase has to be coupled chemically   by means of glutaraldehyde or other bifunctional agents, because the sugars present in this enzyme do not react with lectins other than Con A.

Figs. 5,6. Cell surface reaction of normal hamster embryo cells of 2 different secondary cultures. Continuous precipitation line with some labelled pinocytotic vesicles.

Fig. 7. Cell surface of hamster embryo culture transformed in vitro with SV 40 virus. Very weak and irregular labelling of the cell coat.

Fig. 8. Cell surface of a hamster tumor cell line induced in vivo with SV 40 virus. Irregular label. Figures of surface invagination.

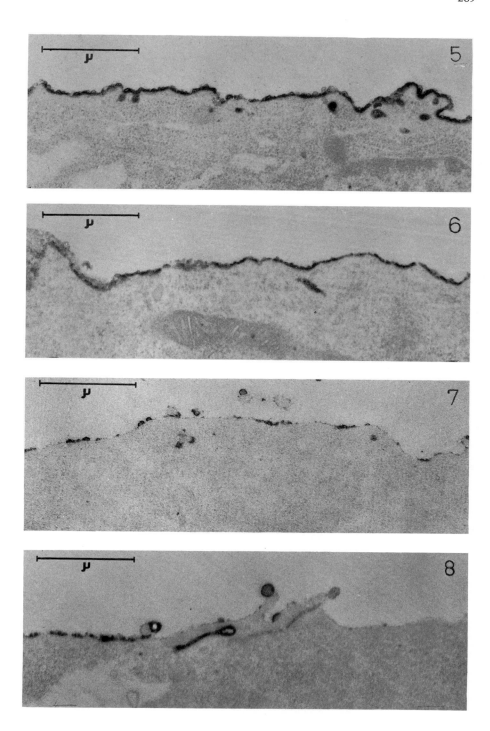

In our laboratory, Huet and Garrido[17] coupled WGA with peroxidase by means of glutaraldehyde. When applied to the study of both normal and tumor cells, the cell coat was stained in a manner similar to that of Con A (Garrido et al. 1974). A sample of soybean extract was also coupled by the same method to the same enzyme. Preliminary results indicated that the cell coat was equally well stained (C. Huet, unpublished).

6. A few results obtained with the Con A-peroxidase method

Most of our applications dealt with the possible difference between the cell surfaces of normal and virus-transformed cells in culture. Martinez-Palomo et al.[23] first demonstrated differences in the distribution of the reaction product on the glycocalyx of normal hamster fibroblasts and a malignant hamster cell line transformed by Polyoma virus. The cell coat of the normal cells appears to have a practically continuous precipitation line about 400 Å thick, whereas approximatively one third of the tumor cells showed a discontinuous marking with a more patchy distribution of the reaction product. These observations were extended and confirmed by Bretton et al.[8] using hamster cells transformed by SV 40 virus (figs. 5-8). The same tendency of irregular distribution of the cell surface reaction was observed, although it was not possible to distinguish with certainty individual normal cells from individual cancer cells. No difference could be shown in a preliminary experiment with rat fibroblasts, transformed with the same virus. In another paper, Rowlatt et al.[33] confirmed the more irregular distribution of the precipitate on hamster cells transformed with Adenovirus 12, but not in those cells which were transformed by the Simian Adenovirus SV 7. Furthermore, prefixed tumor cells showed the same continuous reaction line than the controls. From all these observations, it was concluded that in certain tumor cell lines, differences of the distribution of the receptor sites could be observed as compared with normal cells, but this observation had a quantitative character and could not be generalized for all the observed cases. Furthermore, this difference was visible only if the reaction was carried out in vivo, during about 40 min. Therefore, slight differences in cell surface mobility could thus be demonstrated by this method, revealing probably underlying metabolic differences. The Singer-Nicolson model[36] representing the cell membrane as a fluid mosaic is particularly useful for the interpretation of these results.

Further work has been carried out concerning the action of enzymes on the cell coat. Using low concentrations of proteases, hyaluronidase, neuraminidase, phospholipase C and, in addition, EDTA, Huet and Herzberg[18] observed only a slight action on the cell coat with the Con A-peroxidase methods with the exception, however, of trypsin and EDTA which induced after a very brief treatment considerable invagination of the cell coat, leading to disappearance of the surface precipitate. Similar disappearance of the cell surface reaction was demonstrated by Barat and Avrameas[5] on normal human lymphocytes. At zero incubation time a positive reaction was found on the entire cell membrane. Already after 15 min. of incubation with Con A, discontinuous labelling of the surface was the rule. The reaction product could be found inside cytoplasmic vesicles. After 2h, the surface had practically lost all label, but this latter was concentrated in the Golgi area. Similar observations are reported by Huet (this symposium) on normal and transformed cells where the disappearance of the surface label is a function of time and temperature. Both factors can be used to reveal differences in membrane fluidity.

Another study has been successfully undertaken by Garrido[13] to demonstrate differences of the receptor distribution during the mitotic cycle.

Finally, it should be mentioned that glycoproteins present in the tissue culture medium may be adsorbed secondarily onto the cell surface or the walls of the glass bottle, which may be a source of error in the interpretation of the results[32].

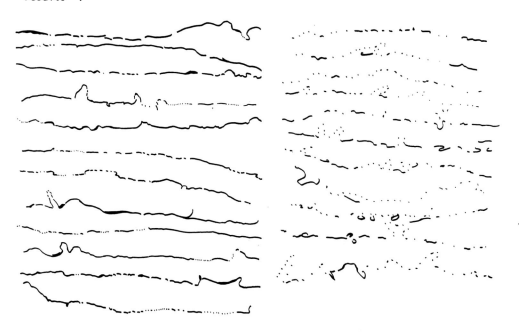

Fig. 9. "Glycograms" : Con A-peroxidase reaction of the cell coat of normal and malignant cells (hamster fibroblasts) (left), and homologous cells transformed by SV 40 virus (right). Considerable irregularities of the reaction on the malignant cells.

7. Conclusion

Phytoagglutinins coupled with an electron microscopically visible marker and, in particular, the Con A-peroxidase method have 3 main advantages compared to the other available procedures for revealing polysaccharides :
1) the reaction is extremely specific for certain sugar residues ;
2) the reactions may be carried out on the living cell without denaturation of the receptors ;
3) it can be applied to the study of the dynamic behaviour of the cell surface under various physiological conditions or after action of drugs and enzymes.

The main disadvantage is the poor penetration of the lectin between the cells of tissues or inside the cytoplasm. However, it may be hoped that the use of ultrathin sections will overcome these difficulties and enable the localization of intracellular sugar residues.

References

1. Agrawal, B.B.L. and Goldstein, I.J., 1967, Biochim. Biophys. Acta 133, 376.

2. Arndt-Jovin, D.J. and Berg, P., 1961, J. Virol. 8, 716.

3. Aub, J.C., Tieslau, C. and Lankester, A., 1963, Proc. Nat. Acad. Sci. USA 50, 613.

4. Avrameas, S., 1970, C.R. Acad. Sci. Paris 270, 2205.

5. Barat, N. and Avrameas, S., 1973, Exptl. Cell Res. 76, 451.

6. Benedetti, E.L. and Emmelot, P., 1967, J. Cell Sci. 2, 499.

7. Bernhard, W. and Avrameas, S., 1971, Exptl. Cell Res. 64, 232.

8. Bretton, R., Wicker, R. and Bernhard, W. 1972, Int. J. Cancer, 10, 397.

9. Burger, M.M., 1969, Proc. Nat. Acad. Sci. USA 62, 994.

10. Burger, M.M., 1970, Changes in the chemical architecture of transformed cell surfaces, p. 107. In : Permeability and Function in Biological Membranes (L. Bolis, A. Katchalsky, R. Keynes and W. Loewenstein, eds), North Holland, Amsterdam.

11. Burger, M.M. and Goldberg, A.R., 1967, Proc. Nat. Acad. Sci. USA 57, 369.

12. Cline, M.J. and Livingston, D.C., 1971, Nature (Lond.) 232, 155.

13. Garrido, J., manuscript in preparation.

14. Garrido, J., Burglen, M.J., Samolyk, D., Wicker, R. and Bernhard, W., 1974, Cancer Res. submitted for publication.

15. Goldstein, I.J., So, L.L., Young, Y. and Callies, Q.C.I., 1969, J. Immunol. 103, 695.

16. Graham, R.C. and Karnovsky, M., 1966, J. Histochem. Cytochem. 14, 291.

17. Huet, C. and Garrido, J., 1972, Exptl. Cell Res. 75, 523.

18. Huet, C. and Herzberg, M., 1973, J. Ultrastr. Res. 42, 186.

19. Inbar, M. and Sachs, L., 1969, Proc. Nat. Acad. Sci. (Wash.) 63, 1418.

20. Inbar, M. and Sachs, L., 1969, Nature (Lond.) 223, 710.

21. Inbar, M., Vlodavsky, I. and Sachs, L., 1972, Biochim. Biophys. Acta 255, 703.

22. Marinozzi, V.J., 1961, Biophys. Biochem. Cytol. 9, 121.

23. Martinez-Palomo, A., Wicker, R. and Bernhard, W., 1972, Int. J. Cancer 9, 676.

24. Moore, E.G. and Temin, H.M., 1971, Nature (Lond.) 231, 117.

25. Nicolson, G.L., 1971, Nature New Biology 233, 244.

26. Nicolson, G.L. and Blaustein, J., 1972, Biochim. Biophys. Acta 266, 543.

27. Nicolson, G.L. and Singer, S.J., 1971, Proc. Nat. Acad. Sci. USA 68, 942.

28. Ozanne, B. and Sambrook, J., 1971, Nature (Lond.) 232, 156.

29. Pease, J.C., 1966, J. Ultrastr. Res. 15, 555.

30. Rambourg, A., 1969, J. Microscop. 8, 325.

31. Rambourg, A., Hernandez, W. and Leblond, C.P., 1966, J. Cell Biol. 40, 395.

32. Rowlatt, C. and Wicker, R., 1972, J. Ultrastr. Res. 40, 145.

33. Rowlatt, C., Wicker, R. and Bernhard, W., 1973, Int. J. Cancer, 11, 314.

34. Sela, B., Lis, H., Sharon, N. and Sachs, L., 1970, J. Membrane Biol., 3, 267.

35. Sela, B., Lis, H., Sharon, N. and Sachs, L., 1971, Biochim. Biophys. Acta 249, 564.

36. Singer, S.J. and Nicolson, G.L., 1972, Science 175, 720.

37. Sivak, A. and Moscona, A.A., 1971, Science 173, 264.

38. Thiéry, J.P., 1967, J. Microscop. 6, 987.

39. Tomita, M., Osawa, T., Sakura, Y. and Ukita, T., 1970, Int. J. Cancer 6, 283.

Electron microscopy and Cytochemistry, eds. E. Wisse, W.Th. Daems, I. Molenaar and P. van Duijn.
© 1973, North-Holland Publishing Company - Amsterdam, The Netherlands.

EFFECTS OF ENZYMES, TIME AND TEMPERATURE ON CELL SURFACE LABELLING WITH
CONCANAVALIN AND PEROXIDASE

Ch. Huet

Institut de Recherches sur le Cancer, Villejuif (France)

Summary

Previous studies have shown that trypsin and EDTA can induce cell surface
movements in normal cells. By tracing endocytotic deplacements of Concanavalin A
bound surface molecules, we have observed that without special treatment, membrane flow is about twice as rapid in certain transformed cells as compared with
their homologous normal controls. Variations of temperature induce changes in the
Con A conformation : at low temperature the equilibrium dimer-tetramer is shifted
towards the dimer. This latter form probably cannot induce agglutination.

Introduction

The labelling of the cell surface by means of lectins coupled with peroxidase is now a well established procedure in E.M. cytochemistry[1,2]. We have been
using these techniques for studying cell surface movements and the action of
some agents on these movements in hamster cells cultured in monolayers.

Effects of enzymes

The effects on the cell coat of low concentrations of EDTA and of various
enzymes (trypsin, pronase, hyaluronidase, neuraminidase, phospholipase C) acting
"in situ" was processed on hamster embryo cell line[3]. Most of the enzymes had no
effect on the cell coat, but trypsin and EDTA induced movements of the plasma
membrane. Even after a brief exposure to trypsin 0.005 % for 3 to 5 minutes, the
Concanavalin peroxidase reaction occured entirely within endoplasmic invaginations of the plasma membrane, whereas the cell surface was devoid of it (see
fig. a).

Effect of time

With a slight modification of the original technique, we incubated cells in
PBS at 37°C after they had been treated with both Con A and PO. These incubations
were carried out for 15, 30, 60 and 240 minutes. Under such conditions it was
possible to demonstrate that previously published results interpreted as morphological differences between normal and transformed cells may have been in fact

the result of metabolic phenomena[4,5,6]. In other words, our experiments showed
that membrane flow was much greater in certain transformed cells (Cl_2TSV_5), since
all the Con A-PO labelled material disappeared from the cell surface within 15
to 30 minutes. Measurements of radioactive Con A showed that most of the Con A
penetrated the cells and that shedding of labelled material into the medium was
minimal. These movements were totaly blocked by aldehydes or dinitrophenol ;
some microvesicles were still seen after cytochalasin B action. The effect of
dinitrophenol suggests that some coupled energetic reactions are involved (see
figs. b, c, d).

Effect of temperature

Agglutination of the cells, Con A-PO staining and radioactive Ni-Con
binding were tested at several temperature, 0°C, 5°C, 25°C and 37°C. At 0°, the
transformed cells were not agglutining and the cytochemical reaction was very
weak. On the other hand, at 37°C, agglutination was very quickly completed, all
the cells becoming involved in clumps, and the cytochemical reaction gave a more
highly positive response. If the cells were prefixed with 2.5 % glutaraldehyde
at 0°C but labelled at 37°C, a heavy specific precipitate on the cell surface
was obtained. This demonstrate that the above effect at 0°C was the result of
the cytochemical reaction and was not due to modifications of the cell surface.

Because the cells were labelled for only 15 minutes with Con A, it was
necessary to measure the kinetics of the lectin fixation at the various temperatures studied. In order to eliminate any biological mechanism, we carried out
the reaction on an artificial substrate, namely Petri dishes incubated for 48
hours with complete culture medium but without cells. In such conditions, Ni-Con
A fixation was about 1.6 to 1.8 times higher at 37°C than at 0°C. Analytical
ultracentrifugation, performed under similar conditions of temperature and lectin concentrations gave us sedimentation coefficients of 3.8 and 5.8 respectively for the low and high temperatures. It is known that the tetrameric form of
the Con A (MW 110.000) has a coefficient of 5.8 and that the dimer (MW 55.000)
has one of 3.8[7].

Under such conditions, there is no evidence for any temperature sensitive
sites on the cell surface membrane, but it is clear that absence of non agglutination of cells at low temperature can only be due to the transition of the

a) Trypsin treated cell "in situ" : little staining of the surface.
b, c, d) Effect of time : b) secondary hamster embryos without reincubation and
(c) after 15' PBS ; (d) transformed cell after 15' PBS.
e, f) Effect of temperature : (e) labelling at 0°C ; (f) fixation at 0°C
and labelling at 37°C.

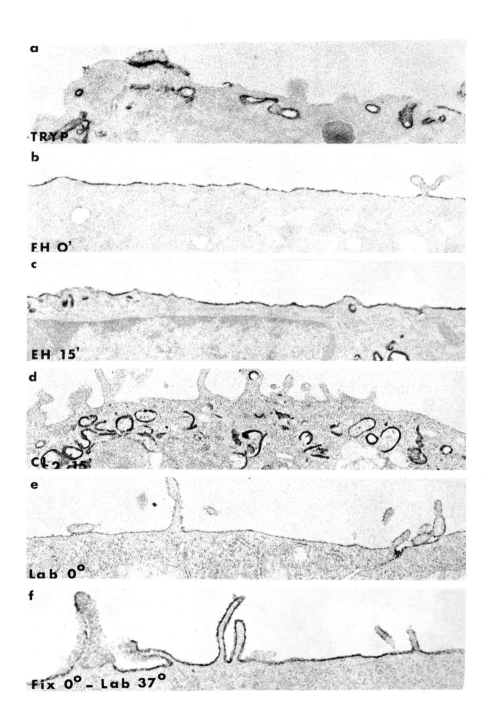

a TRYP

b EH 0'

c EH 15'

d CL₂ 15'

e Lab 0°

f Fix 0° – Lab 37°

lectin (see figs. E, f).

Conclusions

The surface membrane appears to have a high mobility in relation to the underlying cytoplasm. These movements are easily induced by trypsin or EDTA which are commonly used before agglutination experiments. This membrane flow is in our experimental system much higher in transformed cells than in the homologous normal ones. DNP inhibition of this membrane flow seems to indicate that some energetic reaction, are involved, perhaps localized in the membrane itself, and this process is uncoupled by the inhibitor. By lowering the temperature, the surface membrane movements are also inhibited but in addition, a temperature dependant change in Con A conformation occurs. It seems reasonable to think that two conditions are required for the living cells to agglutinate by Con A : first, the lectin must be present in its tetrameric conformation and second, active membrane movements may play an important role.

References

1. Bernhard, W. and Avrameas, S., 1971, Exptl. Cell Res., 64, 232.
2. Huet, Ch. and Garrido, J., 1972, Exptl. Cell Res., 75, 523.
3. Huet, Ch. and Herzberg, M., 1973, J. Ultrastr. Res., 42, 186.
4. Bretton, R., Wicker, R. and Bernhard, W., 1972, Int. J. Cancer 10, 397.
5. Rowlatt, Ch., Wicker, R. and Bernhard, W., 1973, Int. J. Cancer 11, 314.
6. Martinez-Palomo, A., Wicker, R. and Bernhard, W., 1972, Int. J. Cancer 9, 672.
7. McKenzi G., Sawyer, W. and Nichol, L., 1973, B.B.A. 263, 283.

Electron microscopy and Cytochemistry, eds. E. Wisse, W.Th. Daems, I. Molenaar and P. van Duijn.
© 1973, North-Holland Publishing Company - Amsterdam, The Netherlands.

CORRELATION BETWEEN CYTOCHEMICAL DETECTION BY HORSE-RADISH PEROXIDASE
AND BINDING OF CONCANAVALINE A ON NORMAL AND TRANSFORMED MOUSE
FIBROBLASTS.

J.H.M. Temmink and J.G.N.M. Collard

Departments of Electron Microscopy and Experimental Cytology,
The Netherlands Cancer Institute, Amsterdam, The Netherlands.

Introduction.

Mucoproteins are important constituents of the cell plasma-membrane
implicated in the growth control of animal cells in vitro. The plant
lectin concanavaline A (Con A) binds specific sugars on the plasma
membrane and it causes an increased agglutinability in cells that show
loss of density-dependent inhibition of growth in vitro. In addition
it can be made visible in the electron microscope by cytochemical
techniques. For those reasons Con A was used in a comparative study
on differences in the structure of the plasma-membrane between normal
(3T3) and transformed (SV-3T3) mouse fibroblasts. The differences in
amount of cytochemically detectable Con A bound to cells under different conditions were compared with differences in the amount of cell-
bound Con A found in binding experiments under the same conditions and
with differences in the agglutinability of similarly treated cells.

Materials and methods.

All cells were grown to confluence in Dulbecco's modified Eagle's
medium supplemented with 10% newborn calf serum. For electron microscopic cytochemistry the cells, grown on carbon-coated cover slips,
were first treated with Con A and then with horse-radish peroxidase
and 3-3'-diaminobenzidine + H_2O_2 (Per-DAB) as described by Bretton
et al[1]. Some cells were prefixed in situ for 15 min with 2% glutaraldehyde. In control experiments 0.1 M α-methyl-D-glucoside was added
to the Con A, proving in all instances that the Con A binding was
specific. In some experiments cells were treated with 10^{-3}% trypsin
at 20°C for 5 min prior to the addition of Con A or to prefixation;
in other experiments cells were exposed during the last 12 h of growth
to 10^{-5}M cycloheximide added to the medium. The methods used in the
binding and agglutination experiments will be described elsewhere.

Results.

The results of binding and agglutination experiments with Con A
on 3T3 and SV-3T3 cells indicate that amount of cell-bound ^{3}H-Con A
per cm^2 of cell surface and degree of agglutination are correlated
(Figs. 1a and 1b). There also is a correlation between amount of cell-
bound ^{3}H-Con A and amount of detectable Con A by peroxidase (Figs. 1a,
2a and 2b), thus suggesting a correlation between detectability of
cell-bound Con A and agglutinability by Con A. However, further experiments showed that the above correlations between binding and agglutinability, between binding and detectability, and between detectability and agglutinability are not causal relations: Table 1 summarizes
results of experiments where treatment of 3T3 cells with trypsin or
cycloheximide increased the agglutinability of these cells without
concomitantly increasing the amount of bound ^{3}H-Con A. After exposure of similarly treated cells to Per-DAB not more Con A was detected
on these cells with increased agglutinability (Figs. 2a, 2c, and 2d).
Absence of a causal relation between amount of cell-bound ^{3}H-Con A
and amount of detected Con A was shown by comparing non-prefixed and
prefixed 3T3 and SV-3T3 cells: Although prefixed cells bind approx.

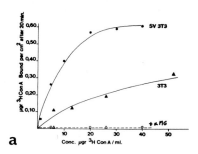
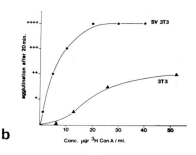

Figure 1. a). Concentration-dependent binding of ^{3}H-Con A by normal
          and transformed fibroblasts (3T3 and SV-3T3).
       b). Agglutination of normal and transformed fibroblasts
          (3T3 and SV-3T3) by different concentrations of Con A.

Table 1. Agglutination and binding of [3] - Con A on 3T3 fibroblasts after treatment with cycloheximide and trypsin.

Conc. µgr [3]H – Con A per ml	Treatment	Agglutination after 20 min	µgr [3]H – Con A bound per cm^2
12,5	cycloheximide	+	0,13
12,5	– –	(+)	0,12
25,0	cycloheximide	+ + (+)	0,20
25,0	– –	+	0,19
50,0	cycloheximide	+ + + +	0,31
50,0	– –	+ +	0,33
24,0	trypsin	+ +	0,15
24,0	– –	+	0,16
48,0	trypsin	+ + + +	0,28
48,0	– –	+ +	0,30

approx. 25% less ^{3}H-Con A than non-prefixed cells (not shown), the
amount of Con A-specific Per-DAB reaction product is much greater on
prefixed cells (Figs. 2a, 2b, 2e, and 2f).

Discussion.

    The results of these experiments leave two important questions:
1). Why does an increase in Con A-specific Per-DAB reaction product
not necessarily mean increased binding of Con A, as implicitly assumed
by other investigators[1,2,3,4]. 2). If increased binding of Con A by
treated 3T3 cells does not explain the greater agglutinability of
these cells, what other mechanism is involved? A number of possible
aspects may be considered and some have been summarized lately[5].
Cytochemical investigations will undoubtedly help to solve the problem
raised, but the present work shows how careful one should be while
interpreting specific cytochemical reaction patterns.

References.

1. Bretton,R., Wicker,R. and Bernhard,W.,1972, Int. J. Cancer 10, 397.
2. Martinez-Palomo,A., Wicker,R.and Bernhard,W.,1972,   „   9, 676.
3. Rowlatt,C., Wicker,R. and Bernhard,W., 1973,   „   11, 314.
4. Huet,Ch. and Herzberg,M., 1973, Ultrastruct. Res. 42, 186.
5. Burger,M.M., 1973, Federation Proc. 32, 91.

Figure 2. Normal and transformed fibroblasts (3T3 and SV-3T3) treated
          with Per-DAB after addition of Con A.
          a). Non-prefixed 3T3; b). Non-prefixed SV-3T3; c). Trypsin-
          treated, non-prefixed 3T3; d). Cycloheximide-treated, non-
          prefixed 3T3; e). Prefixed 3T3; f). Prefixed SV-3T3.

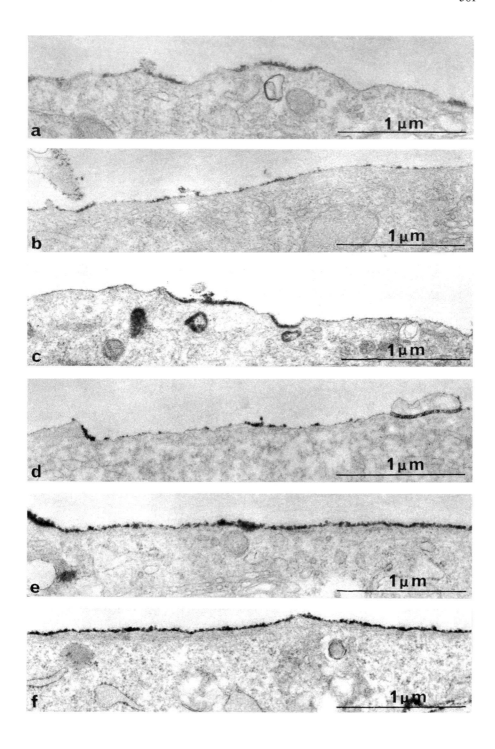

Electron microscopy and Cytochemistry, eds. E. Wisse, W.Th. Daems, I. Molenaar and P. van Duijn.
© 1973, North-Holland Publishing Company - Amsterdam, The Netherlands.

PREPARATION AND PERFORMANCE OF GOLD-LABELLED CONCANAVALIN A FOR THE LOCATION OF SPECIFICALLY REACTIVE SITES IN WALLS OF S. FAECALIS 8191.

J. M. Garland

Department of Microbiological Chemistry, School of Chemistry, The University
Newcastle upon Tyne, NE1 7RU

Summary

Colloidal gold has been used to coat Concanavalin A. Gold-Con. A has reactions similar but not identical to normal Con. A, and results using gold-Con. A as a topographical marker for α-glucoside or α-mannoside groups therefore require careful interpretation. Gold-Con. A reactive sites apparently specific for Con. A reactivity occur dispersed over the outer surface of S. faecalis 8191 walls and may relate to the position of teichoic acid.

1.   Introduction

Concanavalin A (Con. A) reacts specifically with α-glucoside or α-mannoside residues and has been coupled with Ferritin (1) or Peroxidase (2) as a histochemical marker. Many kinds of specifically active protein are adsorbed electrostatically to colloidal gold (3), and gold-labelled Con. A has been used to locate surface mannan in yeast cells (4, 5). The ribitol teichoic acid in S. faecalis 8191 walls is glucosylated (6), and isolated wall preparations react specifically with Con. A (7). Factors affecting the performance of gold Con. A as a marker for this teichoic acid have been investigated.

2.   Preparation of gold sol

All glassware must be siliconised. 60 ml. of 0.025% chloro-auric acid (BDH Chemicals, Poole, Dorset) in distilled water is reduced with 1 ml. of a 20% saturated solution of phosphorus in ether after the addition of 1 ml. 0.18N K_2CO_3. After stirring 5 min at room temperature, the solution is rapidly boiled until a light-red colour is produced, and aerated after cooling to remove residual ether and phosphorus. The final sol has a pH of 3.5, is stable, and can be adjusted to the required pH with additional K_2CO_3. Age of the phosphorus solution critically affects the quality of the sol, and pH control is extremely sensitive to the amount of phosphorus added, presumably through the formation of phosphoric acid. Gold sol is aggregated by low concentrations of salts, reflected in the loss of absorbance at 520 nm. and a visible change from red to blue. Salt-aggregation is inhibited by adsorbed protein.

3.   Protein colloid

Soluble Con. A (Sigma Chemicals Ltd) was used to coat gold sol at pH 7, and concentrated by centrifugation (18,000 g x 30 min 35,000 g x 45 min 78,000 g x 45 min, 3 active pellets). Colloid was saturated at approximately 150 μg protein/ml. colloid (Fig.1).

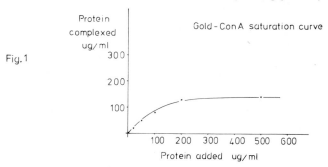

Fig.1

4.   Gold-Con. A Reactions

Colloidal gold is protected against salt-precipitation by levels of protein insufficient to display activity. Fig. 2 shows the precipitation reactions of saturated colloid; specific precipitation requires salts, and non-specific precipitation is caused by divalent cations. α-Methylglucoside (αMeG), a specific inhibitor of Con. A precipitation reactions through the formation of a soluble complex,was used as an inhibitor of gold-Con. A precipitation.

Fig. 2.

	Precipitin	Solution/Buffer	Precipitation
1	None	0.85% w/v NaCl	-
2	Glycogen	H_2O	-
3	Glycogen	0.85% w/v NaCl	+
4	Glycogen	0.85% w/v NaCl + 0.1M αMeG*	-
5	Glycogen	0.85% w/v NaCl + Peroxidase 500 μg/ml	-
6	Glycogen	0.1M Acetate/1M NaCl/ 10^{-3}M Ca, Mn, Mg	+
7	None	"	+

* α-Methyl Glucoside

Plate 1. Gold-Con. A particle labelled with Peroxidase and negatively stained.

Plate 2. Gold-Peroxidase particles similarly stained.

Plate 3. <u>S. faecalis</u> 8191 wall preparation, exposed to 1/10 dilution Gold-Con. A.

Plate 4. Same as 3, in presence of 0.1M αMeG.

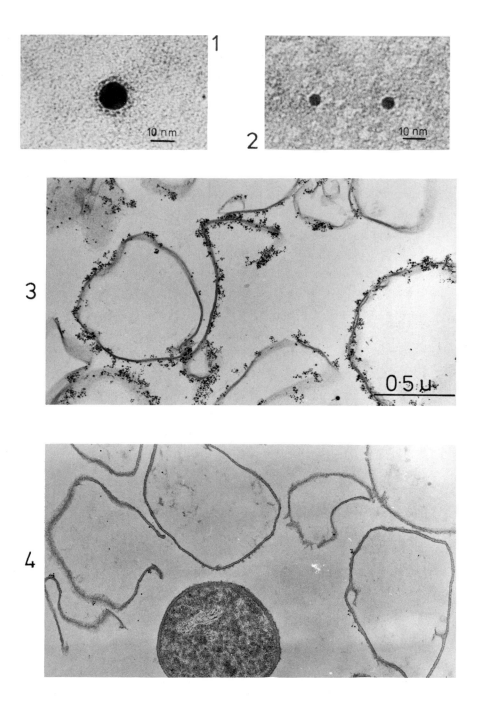

Gold-Con. A is adsorbed to Sephadexes G-50 and G200 but at less than 25% efficiency of unlabelled Con. A; elution occurs with αMethyl Glucoside but non-specific retention occurs and divalent cations cause disruption of the protein-colloid and eventual precipitation of gold.

5.   Electron Microscopy

An impure preparation of walls, derived from S. faecalis 8191 by passage through a French Press followed by extensive washing, was used to locate surface groups reacting with gold-Con. A. After staining and washing, samples were fixed in Osmium (8), stained with uranyl acetate and embedded for sectioning. Sections were post-stained with lead citrate.

Images of gold-Con. A secondarily labelled with Peroxidase and negatively stained with uranyl acetate show a clustering of density around the gold particles (Plate 1). Such images differed from those of gold-Peroxidase alone (Plate 2).

Plates 3 and 4 show walls exposed to gold-Con. A with or without the presence of αMeG as specific inhibitor. Walls are labelled on the outsides, and to a lesser extent on the insides and in the region of septa. Labelling is largely inhibited by αMeG. Walls exposed to gold colloid alone show very little labelling, and compare with those exposed to unsaturated but salt-protected colloid (Plates 5 and 6). Preparations exposed to 0.1M periodate which may be expected to cause scission of the teichoic acid polymer and which have lost nearly 60% of phosphate on an equivalent mass basis, do not show reactivity with gold-Con. A (Plate 7). A comparison with Peroxidase-Diaminobenzidine labelling after Con. A exposure has given similar results for untreated walls, but an opposite result for Periodate treated walls; explanations for this anomaly are under study.

References

1.   Nicholson, G. & Singer, S. J. (1971) Proc. Nat. Acad. Sci. 68, 942.

2.   Bernhard, W. & Avrameas, S. (1971) S. Exp. Cell Res. 64, 232.

3.   Faulk, W. P. & Taylor, E. M. (1971) Immunochemistry 8, 1051.

4.   Bauer, H., Horisberger, M., Bush, D. A. & Sigarlakie, E. (1972) Arch. Mikrobiol. 87, 202.

5.   Gerber, H., Horisberger, M. & Bauer, H. (1973) Infection and Immunity, 7, 487.

6.   Miller, F. Ph.D. Thesis, University of Newcastle upon Tyne (1969).

7.   Archibald, A. & Coapes, H. (1971) Biochem. J. 123, 665.

8.   Highton, P. J. (1969) J. Ultrastruct. Res. 26, 130.

Plate 5. Water-suspension of walls exposed to uncomplexed colloid.

Plate 6. Same as 3, using undersaturated colloid.

Plate 7. Wall preparation oxidised with 0.1M Sodium Periodate before exposure to Gold-Con. A.

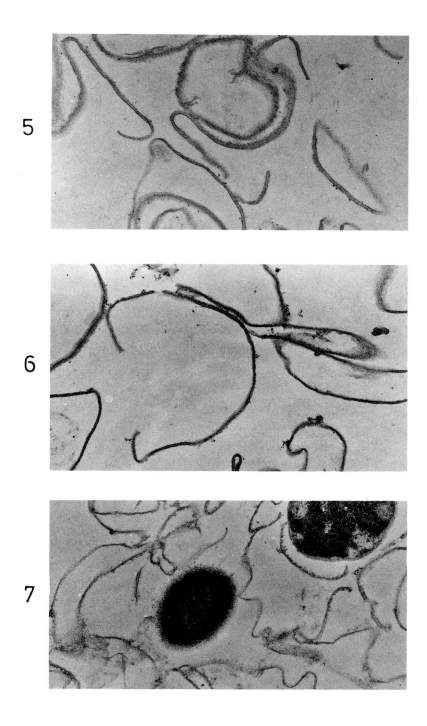

Electron microscopy and Cytochemistry, eds. E. Wisse, W.Th. Daems, I. Molenaar and P. van Duijn.
© 1973, North-Holland Publishing Company - Amsterdam, The Netherlands.

CYTOCHEMICAL STUDIES ON THE SURFACE COAT OF PARAMECIUM AURELIA

Aleksandra Przełęcka and Elżbieta Wyroba

Department of Cell Biology, Nencki Institute of Experimental
Biology, Warsaw, Poland

1. Summary

Surface coat of Paramecium aurelia binds ruthenium red and concanavalin A. This ability is abolished or changed to different degree after the previous treatment of cells with neuraminidase,
pronase, trypsin and lipid solvents.

2. Introduction

Paramecium aurelia is known to posses surface a _gen proper -
ties /1/. Such properties in cells of higher organisms are found
to be associated with the surface coat of the plasma membrane, and
particularly with the presence of sugar residues therein /2/. Hence
it might be supposed that the plasma membrane of Paramecium aurelia is covered with a similar surface coat responsible for the above properties.

3. Material and methods

Paramecium aurelia, Ciliata /stock 51, syngen IV/ cultured on
lettuce medium inoculated with Enterobacter aerogenes was used.Ru-
thenium red /RR/ was applied /Luft, 1971/ in 0.15 M cacodylate
buffer. Concanvalin A /Con A/ was applied on glutaraldehyde fixed
cells /Bernhard, 1971/. The cells were treated with neuraminidase
/5-50u/ml/ in 0.15 M cacodylate buffer pH 6.3 for 3 hours at 37°C.
Pronase /0.1-0.5 mg/ml/-at room temperature or 37°C for 10-30 min.
and trypsin /0.1-0.5 mg/ml/ for 30 min. at 37°C were used in 0.2 M
Tris-HCl buffer pH 7.6. Lipid extraction was performed with chloro-
form:methanol /2:1, v/v/ 15 min. with shaking or with acetone
/twice 5 min. and once 30 min./.

4. Results

The plasma membrane of P. aurelia appears to be covered with a
continuous, 200 Å thick layer, forming the surface coat. This structure displays the affinity towards RR and posses also the binding
sites for Con A. Its fuzzy appearance which is revealed after RR
is masked completely after the reaction with Con A. In the first

case the electron dense RR positive material is present as well on the tiny filaments forming the coat, as in the outer plasma leaflet, what may be seen on fig. 1 presenting a fragment with a longitudinally sectioned cilium and on fig. 2 with a crossectioned one. The Con A positive material forms a uniform layer /fig. 3/. Both the reactions are significantly affected with a previous treatment of the cells with neuraminidase and some proteases. Staining with RR reveals that the applied enzymes cut off the fuzzy layer-the deeper, the better were the digestion conditions. As to the neuraminidase this is shown on fig. 4 and 5 which present the fragments of cells treated with 10 u/ml of enzyme solution /fig. 4/ and with 50 u/ml /fig. 5/, both for 3 h at $37^{o}C$. Fig. 6 and 7 show the effect of pronase treatment with the same enzyme concentration: 0.5 mg/ml, however the former for 15 min. at room temp. and the latter for 10 min. but at $37^{o}C$. Fig. 8 shows that trypsin, 0.1 mg/ml, applied for 30 min. at $37^{o}C$ does not remove completely the affinity of the plasma membrane to bind RR.

Similar grad ation in the effect of enzyme treatment depending on the above factors is visible also when the reaction with Con A is applied. The specific character of Con A binding is proved by addition of α-methyl-D-mannoside and α-methyl-D-glucoside /0.2M/.

Lipid extraction with chloroform:methanol diminishes the affinity of the surface coat to RR /fig. 9/, but the longer acetone extraction impaires significantly the cell membrane /fig. 10/.

5. Discussion

The results obtained indicate that reactive groups responsible for binding RR and Con A being localized within the surface coat are accesible to different degree to the enzyme applied. This is in accordance with various cell surface models in which glycoproteins are more extended into the environment than glycolipids and can be partially removed /mainly as the sialoglycopeptides/ by enzyme treatment /5, 6, 7, 8/.

References

1. Beale, G.H. and Kacser, H., 1972, Z. Zellforsch. 134, 153.
2. Winzler, R.J., 1970, Int. Rev. Cyt. 29, 77.
3. Luft, J.H., 1971, Anat. Rec. 171, 347.
4. Bernhard, W. and Avrameas, S., 1971, Exptl. Cell Res. 64, 232.

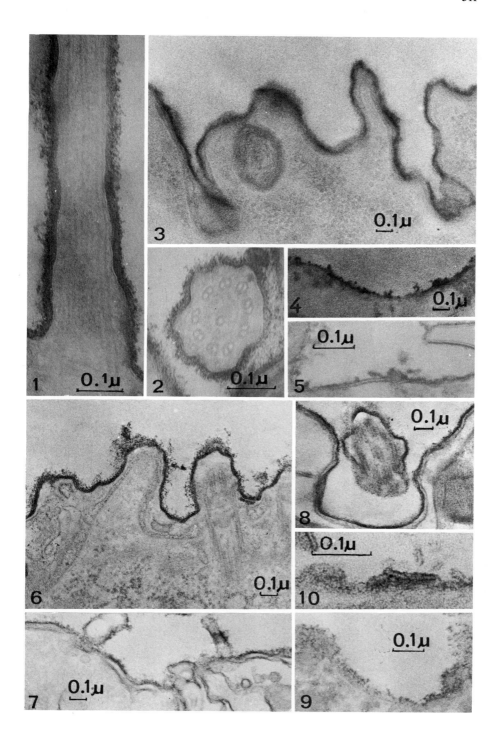

312

5. Ginsburg, V. and Kobata, A., 1971, in: Structure and function of biological membranes, ed. L.I. Rothfield /Academic Press, New York/ p. 439
6. Uhlenbruck, G., Rothe, A. and Pardoe, G.I., 1968, Z. Immunforsch. 136, 79.
7. Langley, O.K. and Ambrose, E.J., 1963, Nature 204, 53.
8. Woo, J. and Carter, D.B., 1972, Bioch. J. 128, 1273.

AUTORADIOGRAPHY

ELECTRON MICROSCOPE AUTORADIOGRAPHY: A PERSONAL ASSESSMENT

Miriam M. Salpeter

Applied Physics and Neurobiology

Cornell University, Ithaca, N.Y. 14850

Electron microscope (EM) autoradiography began with a paper by Liquier-Milward in 1956[1]. It was not until the 1960's, however, that the technique became a practical tool for the electron micros-copist. The work of Caro and collaborators[2,3] stands out among the most significant early advances in the field, in that they were the first to relate theoretical prediction with experimental verifica-tion. Among the basic contributions to the development of the now standard specimen preparation procedures are the introduction of various nuclear emulsions for use in EM autoradiography (Ilford L4, Gevaert 307, Kodak NTE and Sakura-NR-H1) as well as different emul-sion coating and developing procedures. Other early concerns in-volved enhancing specimen contrast and providing for a flat sub-strate. As the technique developed, more studies were directed towards standardization and the determination of the factors which affect sensitivity and resolution as well as with problems involved in analyzing autoradiograms.

The literature in EM autoradiography is by now so vast that I do not accept the responsibility of writing a full review. Several re-views covering the developments outlined above have appeared in the last few years and provide an adequate introduction to the early literature and concepts[4-7]. I would therefore like to use this op-portunity to give a quite personal discussion of the technique and to indicate my views on the current state of the art. I want to first mention, however, several studies not yet adequately covered in the reviews which indicate the current trend in the field and are relevant to the subsequent discussion. These studies deal with clarification of the extent to which various factors actually effect sensitivity and resolution, and also with the application of this information to the analysis of autoradiograms. One such factor is heavy metal staining. It was early recognized that fixation in OsO_4 and staining with uranium and/or lead could theoretically effect sensitivity and resolution. The extent was assumed to be relatively small, and all published calibrations did not include it. Recent studies indicate that this assumption was reasonable for stained tissue sections of thicknesses used in EM autoradiography. Such staining was found to increase self absorption by <10% for tritium[8] and <15% for ^{125}I[9]. The effect on resolution is expected to be in the same order, but its extent is now being assessed experimentally.

Another factor is the energy of radiation of various isotopes. Information on the nature of the grain distribution around radioactive sources, which was previously available only for tritium, ^{14}C, and ^{35}S, has been extended to ^{55}Fe[10] and ^{125}I[11]. Finally a new approach to the analysis of autoradiograms based on the full grain distributions is elegantly illustrated in a recent study by Blackett and Parry.[12]

These studies reflect approaches to EM autoradiography that I have been most concerned with myself recently, i.e. the need for continuing studies on calibration and the application of information regarding resolution to the analysis of autoradiograms. I would like to spend the remainder of this discussion on these aspects. Continuing calibration studies are crucial since EM autoradiography as a technique depends on the photographic process, which approaches more nearly the art of alchemy than any other branch of modern science. Continuing calibration studies have a two-fold purpose. The first is to ascertain whether previous calibrations are still valid. For instance every new batch of emulsion should be tested to determine that the optimum conditions for forming emulsion layers have not been altered by some changes in its manufacture. In addition, the sensitivity should be recalibrated both as an emulsion ages and before applying a new batch of emulsion for quantitative use. We have found the sensitivity of emulsion batches to vary by as much as 25%. The second purpose of recalibration is to ferret out additional factors which affect sensitivity or resolution. Some such studies have already led to adjustments of previously published numerical values[8,13]. Judging from the magnitude of these adjustments and from our current understanding of EM autoradiography, it is safe to say that, although EM autoradiography is not likely to give results with accuracies of 10% or 20%, it has for some time been able to claim accuracies of well within a factor of two. I do not anticipate that any as yet untested factor is likely to render that claim inoperative.

One must not assume however that the calibration done on one system can be carried over to that of others, and new applications require new assessments. One good example of such a new application is EM autoradiography of negatively stained specimens. Several factors such as geometry, self absorption (i.e. the stopping of an electron within a specimen before it reaches the emulsion) and chemography (i.e. the chemical effects of specimen on the emulsion) may be markedly different in such specimens than in thin sections. These factors can affect either resolution or sensitivity or both. A priori I would predict that resolution with a negatively stained specimen should be better than with a thin section. This is due not only to its more favorable geometry (since the negatively stained specimen tends to be thinner than the tissue section) but also to the other factors mentioned above which I will call the differential effects of both self absorption and chemography. Briefly, I am referring to the unique nature of a negatively stained specimen in which the radioactive structure (the biological material) has much lower density than the surrounding area. Since self absorption is greater in higher density material, this will produce a more rapid fall-off of developed grains with distance from the source than is the case in sections in which the dense material is distributed throughout. A similar differential effect could result from chemography, since heavy metal can be demonstrated to lower emulsion sensitivity. If the stain is dense enough, this effect of lowered sensitivity may not be eliminated even when a carbon layer is interposed between the stain and the emulsion, a device used quite effectively with sections. In negatively stained specimens, a lowered grain yield outside the source relative to that over the source would again produce a more rapid fall-off of developed grains with distance from the source, and thus produce better resolution. Thus chemography, which is a nuisance in autoradiography with tissue sections, where the stain is randomly distributed, may prove to be a benefit with negatively stained specimens. Since chemography is not well understood, however, one would have to demonstrate that the extent of its effect is reproducible and controllable before one can use negatively stained specimens for quantitation. This is only one example of the need for independent calibrations for all different types of specimens. Other new and interesting applications of EM autoradiography which would need careful calibration are, for instance, DNA:RNA hybrids when shadowed with heavy metals, or frozen specimens, either frozen sectioned or freeze cleaved.

The topic of greatest interest at the moment is that of analysis of autoradiograms. I therefore would like to direct some comments to that subject. Autoradiography is frequently used to determine (1) the distribution of radioactivity in the sample, (2) the absolute amount of radioactivity present in any one location, or (3) the movement of radioactivity with time from one compartment in a tissue to another. As for determining the distribution of radioactivity in tissue, the basic problem is that no answers obtained from any method of analysis of EM autoradiograms can uniquely describe the true distribution of radioactivity in the specimen (i.e. alternative distributions of radioactivity can be conceived which would result in identical grain distributions). This circumstance stems from the fact that there are not enough grains on any single autoradiogram to get statistically valid information of the full grain distribution independently of underlying structures, which could then subsequently be related to these structures. In practice, in order to get a statistically valid sample, one needs to accumulate information from numerous autoradiograms and, in the process, one loses any possibility of using an independent frame of reference (i.e. independent of the underlying structure). Forced to rely on such structures for reference, grains must be tabulated in relation to identifiable structures in one form or another in order to allow a subsequent pooling of the data. This necessitates a decision regarding which structures are to be tabulated and immediately introduces human judgment and, in effect, hypothesis testing. I have heard people naively argue that they did not approach their autoradiograms with any preconceived hypothesis since they tabulated all structures. But what about the structures they cannot see: i.e. zones of radioactivity that do not exactly coincide with the recognized boundaries of visualized structures? When an autoradiographer counts, by whatever means, the number of developed grains associated with some structure, for example mitochondria, he is testing the hypothesis that the radioactivity is in, on, or near the mitochondria. If he finds a peak of developed grains associated with this organelle, he may be satisfied with this answer, or he may wish to delve further to determine whether the radioactivity is all inside the mitochondria, how it is distributed, or by what factor it is greater than in some other organelle. These questions again have to be answered by testing a series of hypotheses.

The current trend in the literature appears to be a search for a

single best way to analyze autoradiograms. I would like in this discussion to argue against this trend and to emphasize that no one
method is optimum for all problems. The choice of analysis must depend on the questions asked. Furthermore there may be more than one
way to get the answer. Frequently it is best to perform the analysis
in two or even more steps, each step being designed to answer a
specific question. Basic to all analyses is an understanding of the
expected distribution of developed grains around radioactive sources
as a function of various parameters. We will call the normalized
distribution around a point source the Universal Curve[14]. From the
Universal Curve for a point source one can extrapolate to expected
grain distributions of a variety of sources and thus it is the basis
for all analytic methods. It allows the comparison of experimental
values with expected (predicted) values in a variety of ways; the
most recent example is illustrated in the analytic method described
by Blackett and Parry[12]. I cannot here enumerate all the ways in
which the Universal Curves can be used. I want to emphasize that the
Universal Curve contains all the information regarding grains associated with a point source. Any single valid method of analysis uses
some features of the Universal Curve, to a greater or lesser extent.
Therefore, after one method of analysis has been applied, the curve
can be used, for instance, merely to define the limits of accuracy of
the data by indicating whether the value is an upper or lower limit
(e.g. as used in Salpeter and Faeder[15]) or to suggest additional
hypotheses to be tested.

Universal Curves are now available for tritium[14], ^{14}C and ^{35}S[16],
^{55}Fe[10] and ^{125}I[11]. These all have basic similarities which allow
them to be used interchangeably for most purposes. It must be emphasized however that there are differences, especially in the tails of
the distributions (e.g. the higher energies have larger and flatter
tails). Therefore to draw detailed conclusions from one's autoradio-

grams, for instance regarding low levels of grain density even considerable distances from radioactive regions, these tail distributions and their unique properties cannot be ignored.

In a recent review by Salpeter and McHenry[17] the various analytical methods now in use were discussed (except that of Blackett and Parry[12] which was not available at that time). There is no need to repeat our discussion here. Some additional comments and a brief summary may be useful, however. In that discussion we tried to emphasize that all the available methods can provide meaningful answers but that each method is based on assumptions which must be recognized. Most useful therefore is a stepwise sequence of analyses where one method is used to determine the likely locations of radioactivity and subsequent additional analyses are performed to define more precisely the nature of the distribution of radioactivity within given regions of interest. Specifically we discussed first the "simple grain density" method introduced by Ross and Benditt[18]. This method gives average grain density for any designated organelle much as one would get an averaged specific activity by scintillation counting of purified cellular fractions containing these organelles. One precaution required for the autoradiography is to correct for radiation spread which gives rise to developed grains in regions outside an organelle and which is relatively greater for small than for large structures. We also discussed a second method, that of the probability circle (or extended density), derived from the definition of resolution for EM autoradiograms of Bachmann and Salpeter[19,20], and used in various ways by Salpeter[21], Williams[22] and Nadler[23]. This method is most useful for locating very small radioactive structures, or for making a rough estimation regarding the contribution of neighboring structures to grain density at the interface between the two. Inherent in such an estimation is the assumption that each of the two neighboring structures has a uniform distribution of its label. Such an

assumption is not always justified and should be tested. The most
recent contribution to the analysis of autoradiograms is the method
of Blackett and Parry[12] where a series of hypotheses as to the relative amount of radioactivity in different structures are tested
sequentially. It is gratifying to be able to conclude this aspect of
my discussion by mentioning this most elegant study which in my opinion most fully uses the information contained in the universal curves.
The additional labor in its application is well worthwhile if one
wishes to get accurate quantitative data from autoradiograms containing irregularly shaped interlacing sources.

   For the second application of EM autoradiography, that of computing
absolute amounts of radioactivity or specific activity within a given
structure, two conditions must be met: the sensitivity or efficiency
of the method must be calibrated and the location of the radioactivity
defined. Parameters controlling the sensitivity have been described
in a variety of publications, most recently by Vrensen[24] and Salpeter
and Szabo[13] and calibration for sensitivity is not a problem. From
such calibrations one can convert data on developed grains in the
autoradiograms to numbers of radioactive decays that have occurred in
the tissue during the exposure time. However the accuracy of any
conversion to specific activity (decays/unit volume of tissue) depends not only on the reliability of the sensitivity value but also
on how precisely the source has been identified. The main advantage
of EM autoradiography over fractionation and scintillation counting
procedures lies in the greater ability of the former to segregate and
assay only the radioactive structures, giving values of specific
activity with less dilution from nonradioactive tissue. This involves
all the aspects of analysis of grain distribution already discussed.

   Another problem with absolute quantitation in EM autoradiography
is correcting for radiation spread. Not only grains over a structure
but also those which fall outside must be identified and collected

before making the conversion to specific activity. Several ways of
accomplishing this are possible. One may do a grain density distribution outside the structure (leaving out occasional regions of cross
scatter due to neighboring sources, discussed in Budd and Salpeter[27]).
If this distribution fits the expected spread from the source,
then all the scattered grains can easily be collected and included in
the conversion to specific activity. If there are considerable regions of cross scatter from neighboring sources, the procedure of
Blackett and Parry[12] is again most useful for making the corrections.

For the third application of EM autoradiography mentioned above,
that of determining the movement of radioactivity with time, unequivocal answers are in many ways most difficult to obtain. One can
tabulate grain densities for each of a group of compartments, at different times after exposure to radioactivity, and plot the rise and
decay of activity for each compartment. Before discussing what information can and cannot be gleaned from such curves, I must emphasize that for this purpose it is very dangerous to plot these curves
from data on relative density (i.e. % grains/% area) rather than on
absolute density (grains/area). This precaution applies whether one
is using grain centers or circled grains for the tabulation. I am
not arguing against the usefulness of information on percentage of
total grains or on relative density as ways of assessing overall
label for any single time period. However, when used to judge fluxes
of radioactivity between compartments, such data can be misleading
(see example in reference 17). Obviously whenever the radioactivity
in only one compartment is altered, it appears from relative density
tabulations as if all other compartments have changed as well. Even
when the more meaningful data (i.e. absolute density) are plotted, it
is sometimes hard to take into consideration in one's interpretation
such factors as multiple pathways, differential rates of turnover for
different compartments, and dilution effects as radioactivity moves

from smaller to larger compartments. Therefore it is useful to consider several aspects of the data to establish the most likely pathways. In addition to the sequential appearance of peaks in the grain
density curves with time, another helpful clue comes from the slopes
of the rising phase of these curves as discussed by Salpeter[21]. This
paper is not clearly written, is little known and is as often mis-
quoted as quoted. Yet it contains information relevant to this discussion. I will therefore use this opportunity to clarify and sum-
marize its basic thesis. It is based on the fact that the slopes
tell us whether a compartment is first, second, or third order, i.e.
a first order compartment is fed directly from the blood, a second
compartment is fed by a first, etc. Thus although one still cannot
give unequivocal answers regarding the exact sequence of compartments
in the pathway of any given product, one introduces additional information. Using this type of analysis we investigated the turnover of
^{3}H-proline in developing cartilage. We found that both the extracellular matrix and the Golgi complex were primarily second order
compartments whereas the endoplasmic reticulum and ground cytoplasm
were both first order. The conclusions drawn were that, since the
transport has to go from a first to a second order compartment, the
bulk of the collagen precursor leaving the cell does not follow the
pathway ER → Golgi → extracellular matrix, as was suggested for some
secretory cells. Much of it must bypass the Golgi. The most likely
pathway then is ER (a first order compartment) → extracellular matrix
(a second order compartment). This conclusion is consistent with
that of Ross and Benditt[18], which they reached by different means for
the fibroblast, but is at variance with that of Revel and Hay[25] and
more recently Hay and Dodson[26]. We however also indicated that the
ER is not the only first order compartment. The cytoplasmic ground
substance also qualifies on the basis of the logic of our analysis as
a source of extracellular label, and cannot be excluded on the basis

of EM autoradiography alone. I believe that this last statement with its implied heresy may have been partly instrumental in discrediting this method of analysis. I feel however that, although still not unequivocal, conclusions drawn from the peaks of the grain densities plus the slopes of the grain density curves are at present the best approach to the kinetic analysis of autoradiograms. Application of this approach to a "cleaner" system than the cartilage cells would be useful to illustrate its basic strength.

The examples given here by no means exhaust the possible applications of EM autoradiography. Unfortunately EM autoradiography is much simpler to apply technically than its results are to interpret. The difficulties in interpretation have fostered a distrust within the scientific community of information derived from EM autoradiography. To overcome this skepticism, we who use the technique must discipline ourselves to justify for our readers what we are doing whenever we analyze any autoradiograms. We should state why a given method is most appropriate for the question at hand, as well as pointing out the inherent limitations in the analysis. This will go a long way towards establishing the legitimacy of any conclusions drawn from EM autoradiographic studies and will allow an intelligent separation of the valid from the less valid conclusions. Only then will EM autoradiography truly win its place as the quantitative assay of intracellular radioactivity that its unique capacities warrant.

References

1. Liquier-Milward, J., 1956, Nature 177, 619.

2. Caro, L.G., 1962, J. Cell Biol. 15, 189.

3. Caro, L.G. and Van Tubergen, R.P., 1962, J. Cell Biol. 15, 173.

4. Budd, G.C., 1971, Int. Rev. Cytol. 31, 21.

5. Jacob, J., 1971, Int. Rev. Cytol. 30, 91.

326

6. Salpeter, M.M. and Bachmann L., 1972, in: Principles and techniques of electron microscopy, vol. 2, ed. M.A. Hayat (Van Nostrand Reinhold, New York) p. 221.

7. Rogers, A.W., 1973, Techniques of autoradiography (Elsevier, New York), 2nd edition.

8. Salpeter, M.M., 1973, J. Histochem. Cytochem. 21, in press.

9. Fertuck, H.C. and Salpeter, M.M., 1973, J. Histochem. Cytochem., in press.

10. Parry, D.M. and Blackett, N.M., 1973, J. Cell Biol. 53, 16.

11. Salpeter, M.M. and Fertuck, H.C., Resolution in electron microscope autoradiography. III. I^{125}, in preparation.

12. Blackett, N.M. and Parry, D.M., 1973, J. Cell Biol. 53, 9.

13. Salpeter, M.M. and Szabo, M., 1972, J. Histochem. Cytochem. 20, 425.

14. Salpeter, M.M., Bachmann, L. and Salpeter, E.E., 1969, J. Cell Biol. 41, 1.

15. Salpeter, M.M. and Faeder, I.R., 1971, Prog. in Brain Res. 34, 103.

16. Salpeter, M.M. and Salpeter, E.E., 1971, J. Cell Biol. 50, 324.

17. Salpeter, M.M. and McHenry, F.A., 1973, in: Advanced techniques in biological electron microscop , ed. J.K.Koehler (Springer-Verlag, Berlin).

18. Ross, R. and Benditt, E.P., 1965, J. Cell Biol. 27, 83.

19. Bachmann, L. and Salpeter, M.M., 1965, Lab. Invest. 14, 1041.

20. Bachmann, L., Salpeter, M.M. and Salpeter, E.E., 1968, Histochemie. 15, 234.

21. Salpeter, M.M., 1968, J. Morph. 124, 387.

22. Williams, M.A., 1969, in: Advances in optical and electron microscopy, vol. 3, eds. R. Baker and V.E. Cosslett (Academic Press, London, New York) p. 219.

23. Nadler, N.J., 1971, J. Cell Biol. 49, 877.

24. Vrensen, G.F.J.M., 1970, J. Histochem. Cytochem. 18, 278.

25. Revel, J.P. and Hay, E.D., 1963, Z. Zellforsch. U. Mikroskop. Anat. 61, 110.

26. Hay, E.D. and Dodson, J.W., 1973, J. Cell Biol. 57, 190.

27. Budd, G.C. and Salpeter, M.M., 1969, J. Cell Biol. 41, 21.

Electron microscopy and Cytochemistry, eds. E. Wisse, W.Th. Daems, I. Molenaar and P. van Duijn.
© 1973, North-Holland Publishing Company - Amsterdam, The Netherlands.

PROGRESS IN THE ANALYSIS OF ELECTRON MICROSCOPIC AUTORADIOGRAPHS

M.A. Williams

Department of Human Biology and Anatomy
University of Sheffield

In recent years we have seen a great change in the style and
objectives of autoradiographic experiments. In technical papers the
preoccupation with specimen preparation problems has given way to a
widespread realization that problems of interpretation are now the
main barrier to further progress in autoradiography. Many important
data have been accumulated relating the fixation and preservation of
various biochemicals[1,2], such that we are generally now well aware
of the chemical nature of the specimens from which we make our autoradiographs.

The early studies initiated in Dr. Salpeter's laboratory provided us with a usable definition for the 'resolution' of any particular autoradiographic system - and latterly that same laboratory has
provided us with numerical estimates of this parameter[3,4]. We have
also gained a greater realization of the limits of the technique in
terms of efficiency[5,6,7] and in particular we now have a greater
appreciation of the interrelation between the attainable resolution
and attainable grain numbers.

In papers reporting applications of the technique to biological
problems, we have seen a widening selection of topics[8] ranging over
the synthesis and translocation of proteins, nucleic acids, polysaccharides, and some other biopolymers, and also lipids. In addition a great many studies are being made on the binding sites of
small molecules - especially enzyme inhibitors, transmitter substances and some drugs and hormones. While this greatly increased
volume of published work shows the very real advances that have been
made, much of the work still involves surprisingly crude methods of
analysis. In many cases the workers seem reluctant to use the quite
sophisticated analyses which are now available even though they
would greatly enhance the value of their experiments.

In this paper I wish to review the analytical methods that are
at present available for EM autoradiographs, some of their advantages and limitations. Each of the methods I describe involves the
preparation of a large set of micrographs by a systematic sampling
procedure. Data is then collected from each micrograph and the

whole summed together. The processes have much in common with the
systematic stratified random sampling procedures used in stereological analyses[20]. Occasionally biassed sampling methods have to be
used. All the methods are, therefore, essentially statistical.
Particular organelle types are treated as populations of similar
objects and are described by the mean value for grain density or
other relevant parameter. This line of attack has great advantages
in most experimental situations, although in occasional instances
it is not the best.

I   The use of 'HD' curves for sources of various shapes

Salpeter and her colleagues, and others[3,4,9] have measured the
resolution of EM autoradiographs by means of line sources of radioactivity. They have been able to show that the curves of grain
density per unit area about these lines always have similar shape
whatever the section thickness, emulsion crystal size or developer.
They have thus been able to describe the resolution for any particular set of conditions as an 'half distance' or 'HD' value (the
distance within which half the grains fall). It subsequently proved
possible to use this HD data for line sources to compute 'HD curves'
for radioactive sources of other shapes such as solid and hollow
discs, bands etc. Such curves can be used to analyse autoradiographs of tissues[10,11,12]. Figure 1 illustrates this approach
using silver grain density data taken from a set of autoradiographs
of plasma cells pulse-labelled with ^{3}H-tyrosine 30 minutes before
fixation. It is hypothesized that the Golgi apparatus is a major
source of silver grains in these specimens. For the purpose of the
analysis the Golgi region was assumed to be disc-shaped and the mean
radius of the profiles was 4HD. Superimposed on the histogram of
grain density vs. distance are three curves. The labelled lines are
the curves expected for uniformly labelled discs of the radius
indicated. The uppermost line represents the approximate best fit -
this represents a hybrid of two curves - for uniformly labelled discs
of similar specific activity but different diameters (see fig. 1).
Thus it is suggested that the Golgi apparatus in these circumstances
has a lower concentration of newly synthesized protein near its
periphery than at its interior. It is clear of course from other
considerations that the Golgi apparatus is not the sole source of
radioactivity in these cells.

The 'HD curve' method is an excellent approach provided 1) that
the radioactivity is largely confined to a certain restricted number

of features, 2) that the features concerned are well separated - i.e.
that the silver grain spreads of different profiles do not significantly overlap and 3) that the sources are not of too complex a
shape. These restrictions mean that this approach is useful in many
experiments where the radioactivity is largely confined to a small
number of features of appropriate shape, but that it is less useful
in instances where multiple sources of radioactivity are present -
perhaps assuming a variety of profile shapes and patterns.

II Circle methods

These methods originally grew out of the need to analyse
specimens in which the radioactivity seemed to be dispersed over a
range of organelle types[13,2]. In its original form the method made
use of the calculations by Bachmann & Salpeter[14] of the sizes of
circles to be drawn around point radioactive sources to have 50%
chance of enclosing the silver grain derived from that source.
Since the method was originally derived, experience has suggested a
number of modifications to the procedures and has clarified our ideas
about the objectives and priorities during such analyses.

The steps in the method can be summarized as follows:
1. Division of the tissue into primary items (fig. 3).
2. Circumscription of each silver grain by a circle (usually of
    1.7HD radius).
3. Collection of data on the position of each silver grain.
    This is done by apportioning each to a primary item, the
    junction between two primary items or to a compound item
    (see fig. 3) also table 1.
4. Casting a set of circles (1.7 x HD radius) on to each
    micrograph using a transparent overlay screen or a projec-
    tion sytem and collecting data for each circle as under 3.
5. 'Grouping' of the data from various items as necessary.

Generally, provided the set of micrographs is a valid representative sample from the autoradiographs, the factor most likely to
limit the analysis which can be performed is the number of silver
grains recorded. In pilot experiments about 400 silver grains should
be the aim, although some useful inferences can often be drawn from
as few as 100 grains. Later more sophisticated analyses may require
the collection of 100-400 grains over specific groups of items. The
numbers of circles to be applied can be adjusted to order and will
normally exceed 500 for a particular set of micrographs.

The data is used to make inferences at several levels.

1. Comparison of the real grain distribution with the circle (hypothetical random grain distribution) using Chi^2. (Table 2 - data from posterior pituitary labelled with ^{3}H-cystine).

Table 1

Data derived during grain and circle analysis of micrographs of neurohypophysis

10 hours after labelling with ^{3}H-cystine

Items	Number of circles	Grouping of circles	Silver grains grouped
1. N.S. axon axoplasm	4502	4502	175
2. Synaptic vesicles	12		
3. Empty granules/ mito./axop.	2		
4. Empty granules/ axoplasm	238	270	16
5. E.G./N.S.G./axop.	18		
6. N.S.G./synaptic vesicles	0		
7. N.S.G./axoplasm	1104	1116	292
8. N.S.G./mit./axop.	12		
.			
.			
.		etc.	etc.
.			
28			
Total circles 11,048		Total grains 553	

Fig. 1. Analysis of the average grain density around Golgi apparatus profiles of plasma cells fixed 30 mins. after a pulse label of ^{3}H-tyrosine. Method of Salpeter et al[3].

Fig. 2. Electron microscopic autoradiograph of regranulating rabbit parotid gland acinar cells following a pulse of ^{3}H-tyrosine. Note the two intensely labelled granules among many 'unlabelled' ones.

Fig. 3. Drawing of posterior pituitary tissue showing how it was itemized[18] for analysis by the circle[2] or hypothetical grain distribution methods. Circles = 1,2 - pituicyte, 3,9 - N.S.G./axoplasm (compound item), 4 - mitochondrion/axoplasm (junctional item), 6 - Non-N.S. axon/blood vessel (junctional item), 7,8 - blood vessel.

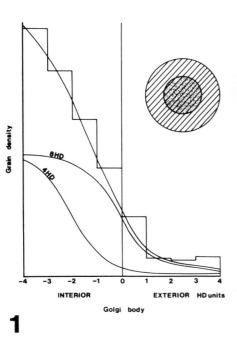

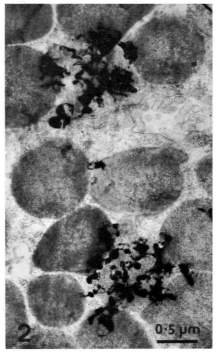

1

2 0·5 µm

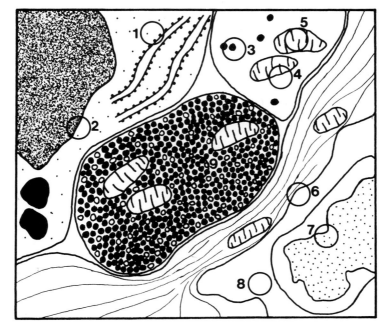

3

2. Expression of the silver grain counts as silver grains/circle for each group of items to give relative specific radioactivity estimates. This may be done at several levels - for example grouping the data into large blocks representing different cell types or smaller blocks representing parts of cells (tables 3,4).

3. The detailed analysis of junctional or compound items. This is achieved by means of a combination of point analysis with circle analysis. This sort of analysis allows one to infer if the silver grain density at an interface represents radioactivity in surface membranes or is merely the resultant of the grain densities of the two items coming into adjacence (see the example in Parry & Blackett[9] who observed a concentration of ^{55}Fe in the junction between the heterochromatin and euchromatin of erythroid cells). An example of ;the analysis of a compound item is shown in Table 5.

4. The relative specific activity data for different items can be used in kinetic experiments in which samples of tissue taken at different times after labelling are compared. For this purpose it is important 1) to have determined if any junctions are labelled and 2) to be quite clear which items give rise to reliable specific activity data as opposed to ones whose data are unreliable due to extensive cross-fire losses and gains. This can be inferred by scrutiny of the mean profile diameters of the features and from their position relative to other labelled items. The grain counts from unlabelled junctions can be particuoned according to the content of the two primary items in the junction (data determined by point counting). This process is simpler and more reliable than the complex series of approximations proposed by Nadler[15] and superior also to the earlier method from the same laboratory[16,17] which necessitates the sacrifice of a large fraction of the grain data.

III Methods of analysis using hypothetical grain distributions

Blackett & Parry[18] have given details of a method in which the real grain distribution is compared with a series of hypothetical distributions (using Chi^2 tests) until a distribution of good fit is found. The method in fact makes considerable allowance for cross-fire effects by the following means.

Table 2

Complete circle and grain analysis of complete
neural lobe tissue

10 hours after labelling with ^{3}H-cystine

Grouped item	Silver grains observed	Silver grains "expected"	$\frac{(OBS-EXP)^2}{EXP}$
Axoplasm of N.S. axons	175	225.3	11.25
Empty granules etc.	16	15.5	0.46
N.S.G./axoplasm etc.	292	55.9	997.0
Mitochondria etc.	14	21.2	2.44
Axon membranes etc.	1	17.3	15.36
Perivascular space	9	36.9	21.10
Perivascular space etc.	7	17.4	6.21
Blood vessels	4	12.3	5.60
Blood vessel contents	2	19.9	16.61
Pituicyte	37	133.3	69.50

$$\chi^2 = 1145.02$$

Table 3

Combined circle and grain analysis
of complete neural lobe tissue

10 hours after labelling with ^{3}H-cystine

Grouped item	% circles	% silver grains	Relative specific activity
Neurosecretory axons	57.2	89.7	1.57
Pituicytes	24.1	6.67	0.28
Perivascular space	8.65	1.81	0.021
Blood vessels	5.82	1.08	0.19

Table 4

Combined circle and grain analysis of
neurosecretory axons
10 hours after labelling with ^{3}H-cystine

Grouped item	% circles	% silver grains	Relative specific activity
Axoplasm	71.3	35.2	0.49
Empty granules/ axoplasm	4.30	5.20	0.74
N.S.G./axoplasm	17.7	58.8	3.33
Mitochondria	6.70	2.80	0.42

Table 5

Calculation of the relative specific radioactivity of
dense-cored granules in untreated rats

		Relative specific activity	Volume % in item	
			Axoplasm	N.S.G.
10 hr after ^{3}H-cystine	axoplasm	0.49	100.0	0
	N.S.G./axoplasm	3.33	86.0	14.0

Relative specific activity
N.S.G. = $\dfrac{3.33 - (0.86 \times 0.49)}{0.14}$ = 20.78

		Relative specific activity	Volume % in item	
			Axoplasm	N.S.G.
19 hr after ^{3}H-cystine	Axoplasm	0.52	100.0	0
	N.S.G./axoplasm	2.02	78.3	21.7

Relative specific activity
N.S.G. = $\dfrac{2.02 - (0.783 \times 0.52)}{0.217}$ = 7.43

Table 6

Posterior pituitary tissue labelled with ^{3}H-cystine.
Analysis by the use of hypothetical grain distribution
and the circle method

Grouped primary item	Relative specific activity	Total activity %	Relative specific activity (circle and point method)
Pituicyte	0.3	2	0.18
N.S.G.	5.50	50	19.91
Axoplasm	1.00	30	1.00
Mitochondria	4.50	14	0.84
Non N.S. axons	1.60	2	1.32
Blood vessels	0.20	2	0.16

Chi^2 (7° Freedom) = 1.2   p = 90%
Total grouped items including
junctional and compound = 10

1. The grain density-distance curve for a point source is mathematically simulated and stored in a computer.
2. A lattice of points (hypothetical 'point sources') is applied to each micrograph.
3. The stored curve is randomly sampled for distance and a random direction also chosen (by computer).
4. The distance and direction are applied to each point and the datum for each collected as 'items' (see above) by circumscribing the end of the line with a circle of radius 1.7 HD, giving a set of 'hypothetical grains' (fig. 4).
5. The hypothetical grain distribution expressed in terms of primary, junctional and compound items, is compared to the real grain distribution by Chi^2.
6. A series of further hypothetical grain distributions (representing various ratios of radioisotope concentration in the different primary items) is tested against the real distribution until the best fit is attained (i.e. a minimal Chi^2 value).

The method is illustrated by an analysis made by Miss Parry and myself on the same labelled posterior pituitary tissue discussed above and in tables 1-5.

Table 6 shows the results. 500 hypothetical grains were applied and data were collected into 12 items which were finally put into 10 groups for computer analysis. Relative specific activity and radioactivity proportion data were obtained for six tissue components. Table 6 includes relative specific activity data from the same tissue sample obtained by circle and point analysis. In both analyses the neurosecretory granules have both the highest concentration and the highest proportion of the tissue radioactivity. The most notable difference in the results obtained by the two methods lies in the mitochondrial specific activity. The circle method clearly underestimates this activity, probably due to the close correspondence between the chosen circle size and the mitochondrial profile size. There is some suggestion, therefore, that the experimenters attitude to the choice of circle size should not be inflexible.

The best fit values from hypothetical grain distribution method were chosen after careful testing of a large range of distributions. The goodness of fit of the distributions (expressed as P values from Chi^2 tables) can be plotted graphically (see figures 5,6,7). It is noticeable that each curve has a region indicating a range of values which would represent a 'satisfactory fit' (any P value $> 20\%$ in fact). Approximate standard error values can be attached to each best-fitted activity. For neurosecretory granules the mean and standard error for proportion of tissue radioactivity was $50\%^{+8}_{-7}$. For lower specific activity components the standard errors are proportionally larger, e.g. Axoplasm 30^{+11}_{-7}.

Some difficulties of the statistical approach

While there is a very strong case for the types of method described here, which average the results from large samples of cells or more particularly parts of cells and organelles, there are some difficulties in certain types of experiment. In most experiments, dealing with the 'average organelle' almost certainly conceals genuine differences within the populations, but this is generally quite justified and indeed necessary as a first treatment. In some experiments, however, differences within a population of organelles can be quite massive and hence data for the 'mean' organelle quite unrepresentative. An example is seen in studies on the genesis of zymogen granules in the regranulating parotid acinar cell. In this case short exposure times and the statistical treatment indicate only that zymogen granules are being synthesized. On the other hand long exposure times make it clear that only a few granules out of a large population are in fact being labelled (fig. 2). Some care must be taken to avoid overlooking situations of this kind. (Cf. the results of Bergeron & Droz[19] on mitochondrial protein synthesis).

Discussion and Conclusions

Methods for the analysis of EM autoradiographs are still evolving. The methods we have owe a great deal to Salpeters calculations and measurements on autoradiograph resolution. All three methods depend to some extent on the basic resolution data and all owe something to the other methods for their development. None of the methods are exclusive and it is possible to visualise situations where all three methods might contribute useful information.

The full capabilities of these techniques have not yet been revealed and it is only by experience of their use in a variety of situations will their full potential be realized. However, it is

Fig. 4. Electron microscopic autoradiograph of posterior pituitary tissue labelled by intraventricular injection of ^{3}H-cystine 10 hr previously. Six points are shown illustrating the hypothetical grain distribution method[18]. Three of the points have distances and directions indicated for placement of circles by computer sampling of the 'HD curve' of a point source.

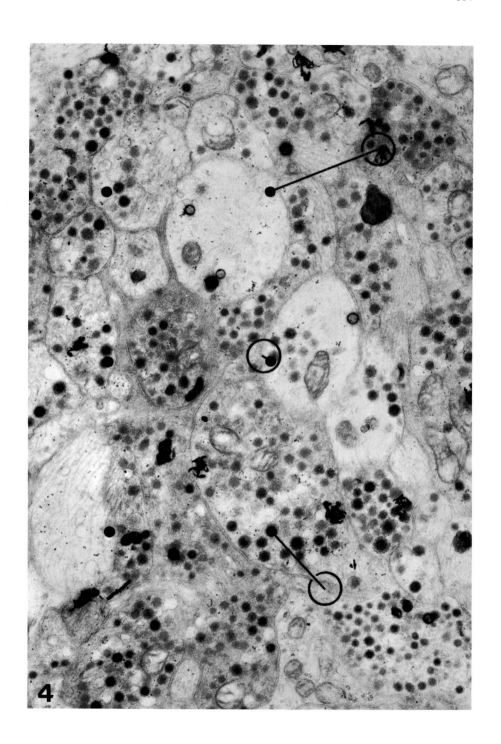

4

338

already clear that the shape and pattern of organelles in the cells
concerned is of crucial importance to the success or otherwise of a
particular analytical approach. Further refinement of these analytical methods will provide answers to numerous questions in cellular
biology many of which will be far more sophisticated than those at
present being posed.

References

1. Stein, O. and Stein, Y., 1971, Adv. Lipid Res. 9, 1.

2. Williams, M.A., 1969, Adv. Opt. Electr. Microsc. 3, 219.

3. Salpeter, M.M., Bachmann, L. and Salpeter, E.E., 1969, J. Cell
   Biol. 41, 1.

4. Salpeter, M.M. and Salpeter, E.E., 1971, J. Cell Biol. 50, 324.

5. Bachmann, L. and Salpeter, M.M., 1967, J. Cell Biol. 33, 299.

6. Caro, L. and Schnös, M., 1965, Science N.Y. 149, 60.

7. Vrensen, G.F.J.M., 1970, J. Histochem. Cytochem. 18, 278.

8. Williams, M.A., 1973, Techn. in Protein Biosyn. 3, 125.

9. Parry, D.M. and Blackett, N.M., 1973, J. Cell Biol. 57, 16.

10. Budd, G.C. and Salpeter, M.M., 1969, J. Cell Biol. 41, 21.

11. Gambetti, P., Autilio-Gambetti, A., Gonatas, N.K. and Schafer, B.,
    1972, J. Cell Biol. 52, 526.

12. Lentz, T.L., 1972, J. Cell Biol. 52, 719.

13. Williams, M.A. and Baba, W.I., 1967, J. Endocr. 39, 543.

14. Bachmann, L. and Salpeter, M.M., 1965, Lab. Invest. 14, 1041.

15. Nadler, N.J., 1971, Appendix to paper by Haddad et al. J. Cell
    Biol. 49, 877.

16. Whur, P., Herscovics, A. and Leblond, C.P., 1969, J. Cell Biol.
    43, 289.

17. Nakagami, K., Warshawsky, H. and Leblond, C.P., 1971, J. Cell
    Biol. 51, 596.

18. Blackett, N.M. and Parry, D.M., 1973, J. Cell Biol. 57, 9.

Figs. 5, 6, 7. Probability curves for the relative specific activity
of different features in posterior pituitary tissue labelled with
^{3}H-cystine. Data obtained by the hypothetical grain distribution
method.

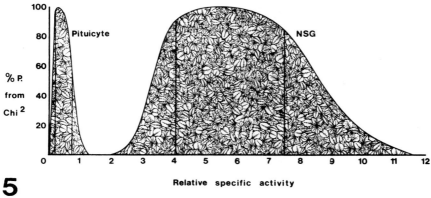

5

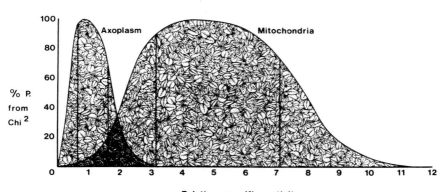

6

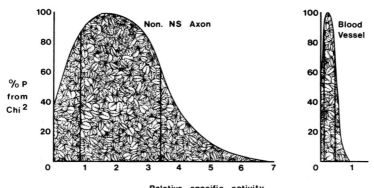

7

340

19. Bergeron, H. and Droz, B., 1969, J. Ultrastruct. Res. 26, 17.

20. Weibel, E.R., 1969, Int. Rev. Cytol. 26, 235.

Addendum

The low estimate of mitochondrial specific activity in the results of circle analysis probably stem not only from the correspon-
dance between mitochondrial profile size and the circle size. The
mitochondrial profiles tend to occur singly or in very small groups.
Hence cross-fired grains originating in mitochondria are likely to
overlie surrounding tissue - usually axoplasma - containing neurosecretory granules (NSG/axoplasm). Thus while the grain count of
mitochondria is depressed NSG/axoplasm is increased. In contrast,
cross-fire losses from neurosecretory granules are not extensive
since the organelles occur in large closely-packed masses. Thus
grains formed by cross-fire from one granule will likely land over-
lying other organelles of the same type. The degree of clustering
of the organelles and their relative patterns in the cell are quite
evidently of considerable importance in determining the feasibility
of a particular analytical method.

Electron microscopy and Cytochemistry, eds. E. Wisse, W.Th. Daems, I. Molenaar and P. van Duijn.
© 1973, North-Holland Publishing Company - Amsterdam, The Netherlands.

QUANTITATIVE ANALYSIS OF ELECTRON MICROSCOPE AUTORADIOGRAPHS

N.M. BLACKETT and D.M. PARRY

Institute of Cancer Research, Belmont, Sutton, Surrey.

A new method for analysing electron microscope autoradiographs
has been devised which combines the resolution considerations reported by Salpeter et al (1) with the circle analysis of Williams (2).

The method offers distinct advantages over previous ones.   It
provides information about cross-scatter between adjacent irregularly shaped sources of differing radioactivity in the section.
Estimates can therefore be obtained for the activity in the different structures which is consistent with the distribution of the
autoradiographic grains, without the need to assume idealised
geometrical shapes for these structures.

Essentially the method involves setting up a theoretical random
distribution of hypothetical grains over an autoradiograph and comparing this with the actual distribution of silver grains found.

The first step is to determine the distribution of grains about a
point source and from this to derive a list of distances chosen at
random such that the frequency of any distance in the list conforms
with the distribution of grains about a point source (3).   In
addition a random direction is generated with each distance (Table 1).

Table 1

Random grains for tritium.   Mag. 10,000 ; HD. 1000 Å

Directions and distances in mm.

N	ESE	SSE	SSE	ESE	ESE	ENE	NNW	SSW	W	SE
2.90	.18	1.60	.36	.58	1.89	.12	11.96	3.95	1.55	.34
NNE	SSW	ENE	SW	NNW	W	NE	ESE	ESE	SW	WSW
4.34	1.45	.23	.51	.66	18.00	1.35	.95	.95	6.68	4.72
SSW	NNE	S	NNW	SSW	NE	SW	ENE	ENE	E	SW
.77	.63	2.37	4.25	4.35	.40	2.74	3.99	.61	2.17	.30

etc.

A grid of points is then superimposed at random over the micrographs
to be analysed, each point being taken to represent a source of
radioactive disintegration.   Using the list of 'directions' and
'distances' the position at which a grain might be produced is

predicted, and the structure lying within a circle surrounding this hypothetical grain is recorded together with the position of the grid point (Table 2). This gives an estimate of the cross-fire

Table 2

Position of grid point (Source of hypothetical grains)	Position of Circle (Site of hypothetical grains)									Total Grains in each source
	ER	ER/ Mit	ER/ Zym	ER/ Golgi	ER/ Memb	Golgi	Golgi /Zym	Zym	Mit	
Endoplasmic reticulum	419	48	40	6	21	3	5	4	2	548
Golgi	3	0	5	4	0	39	16	1	0	68
Zymogen granules	3	1	50	0	0	5	15	26	0	100
Mitochondria	15	22	0	1	2	1	1	0	6	48
Total normalised	75	12	16	2	4	8	6	5	2	131
Real grains	54	3	20	3	1	22	19	9	0	131
χ^2	6.1	6.9	0.8	0	2.2	23.0	25.2	2.6	0	66.8

between neighbouring regions for the particular isotope and experimental conditions in the micrographs being analysed; as shown by the site of the grains originating from the different sources.

The sum of the columns gives the distribution of grains for a uniform distribution of isotope within the section and when compared with the distribution of circles surrounding the real grains (using the χ^2-test of significance) is equivalent to the circle analysis of Williams (2).

It is now possible to allow for differing activities in the various sources, by multiplying the numbers in each row by an appropriate value, until the best agreement between the two distributions is obtained, as shown by the χ^2-test.

It is important to explore the range of activities that can be ascribed to the different sources with the required degree of probability and so determine the accuracy of the best activity values.

An example is shown in Fig. 1 for the distribution of tritiated leucine in the rat pancreas (analysis from micrographs kindly supplied by Dr. Vrensen).

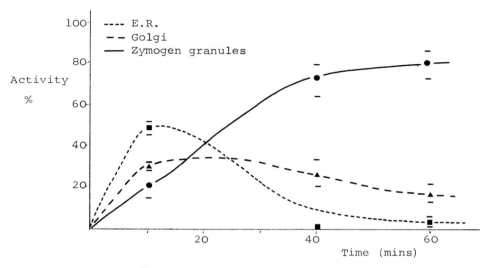

Fig.1. Uptake of ^{3}H-leucine in the rat pancreas

References

1. Salpeter M.M., Bachman, L. and Salpeter, E.E.   J.Cell.Biol.41, 1969.

2. Williams, M.A.   Adv.Opt.& Electron Micros.3, 1969.

3. Blackett, N.M. and Parry D.M.   J.Cell.Biol.57, 1973.

Electron microscopy and Cytochemistry, eds. E. Wisse, W.Th. Daems, I. Molenaar and P. van Duijn.
© 1973, North-Holland Publishing Company - Amsterdam, The Netherlands.

A STUDY OF THE GROWTH OF COLLAGEN FIBRILS IN VITRO USING E.M. AUTORADIOGRAPHY

R.A.Haworth and J.A.Chapman

Rheumatism Research Centre, Clinical Sciences Building,
York Place, Manchester M13 OJJ.

1. Summary

Fibrils of normal collagen precipitated under near-physiological conditions grow at a similar rate at both ends, whereas fibrils of iodinated collagen grow almost wholly from the N-terminal end.   Studies of growth as a function of temperature have allowed values for the activation enthalpy and entropy of precipitation to be calculated for both normal and iodinated collagen.   The autoradiographs produced in this study show a half-distance of 420 Å.

2. Introduction

Factors which affect the formation of collagen fibrils in vitro have been widely investigated[10].   In particular, Bensusan and Scanu[1] have observed a marked acceleration of the precipitation process on iodination of the tyrosine residues, which are located in the terminal 'telopeptide' regions of the collagen molecule.   Furthermore Leibovich and Weiss[6] have found that these telopeptide regions play a major role in the determination of fibril morphology. Here, the growth of normal and iodinated collagen fibrils has been studied using e.m. autoradiography.

3. Materials and Methods

A solution of acetic-acid-soluble collagen was prepared by the method of Jackson and Cleary[5].   Part of this was iodinated at 0°C by adding two equivalents of 10^{-3}M iodine in 0.015M KI to collagen in phosphate buffer, I = 0.2, pH 7.0.   Excess thiosulphate was added after 1 hr, followed by exhaustive dialysis against buffer.   Chromatography of alkaline hydrolysates of collagen iodinated in this way showed that the tyrosine residues were completely di-iodinated, while there was no trace of histidine iodination.

Samples of normal and iodinated collagen were rendered radioactive by iodination with ^{125}I to a specific activity of up to 1 mCi/mg by the method of Greenwood et al.[3].   This corresponds to a degree of iodination of < 1% of the tyrosine residues in the "normal" collagen sample.

Precipitation of the radioactive collagen was initiated by warming for 15 minutes at 27°C (normal) and 21°C (iodinated).   Autoradiographs prepared from precipitates at this stage showed many short uniformly labelled fibrils.   To study relative growth rates, samples were then added to at least 25 volumes of unlabelled normal or iodinated collagen and warmed again to the desired precipitation temperature.   For normal collagen temperatures of 24°, 27° and 30°C were used, while 15°, 18° and 21°C were used for the iodinated collagen.   Drops of solution were removed at various intervals of time and deposited on the "inside" surface of formvar films supported over 4 mm holes in perspex slides[7]. After removal of excess liquid and drying, the other surface was coated with a thin layer of carbon and a monolayer of Ilford L4 emulsion[4].

Following exposure at 4°C, autoradiographs were developed in Microdol X, positively stained with phosphotungstic acid and uranyl acetate, and transferred to copper grids.

4. Results

In many normal fibrils, autoradiography revealed intense activity over a mid-region with the end-regions exhibiting no more activity than background (Fig.2). Using the band pattern of the stained fibrils to establish polarity, the growth

Table 1

Activation enthalpy and entropy values for the precipitation of normal and iodinated collagen

Collagen	$\Delta H^{\ddagger}$ (kcal/mole)	$\Delta S^{\ddagger}$ (cal/mole/deg)
Normal, N-terminal	64.5 ± 3.5	190 ± 12
Normal, C-terminal	52 ± 8	147 ± 27
Iodinated, C-terminal	55 ± 1.7	163 ± 6

Fig. 1. An Aarhenius Plot of fibril growth rate.
A: Normal collagen, C-terminal growth
B: Normal collagen, N-terminal growth
C: Iodinated collagen, N-terminal growth.

rate of the N- and C-terminal ends was measured at each temperature on at least twenty fibrils. It was found that normal collagen grew at up to twice as fast at the N-terminal end as at the C-terminal end, under these conditions. By contrast, growth of iodinated collagen fibrils occurred almost exclusively at the N-terminal end.

Fig. 1 shows this data formulated as an Aarhenius plot. Analysis of precipitation rates in terms of the transition state theory gives values of activation enthalpy and entropy shown in table 1.

5. Discussion

This study illustrates the sensitivity of the fibril formation process to modifications in the telopeptide regions of the collagen molecule. The observed result is difficult to explain in terms of molecular mechanism, since six tyrosine residues occur in the C-terminal telopeptides and five in the N-terminal.

The ability of normal collagen fibrils to grow from both ends may well be of biological significance in the laying down of connective tissues. The half-distance of 420 Å measured on the histogram compares unfavourably with that of 1450 Å found by Salpeter for a similar geometric situation. The measured resolution agrees much better with the predictions of Caro[2], who allowed for the shielding effects of adjacent crystals.

6. References

1. Bensusan H.B. & Scanu, A.W. 1960, J.Am.Chem.Soc. 82, 4990.
2. Caro, L.G. 1962, J.Cell.Biol. 15, 189.
3. Greenwood F.C., Hunter W.M. & Glover J.S., 1963, Biochem.J. 89, 144.
4. Haworth R.A. & Chapman, J.A. 1972, Proc. Fifth European Congress on Electron Microscopy, p.284.
5. Jackson D.S. & Cleary E.G. 1967, Methods of Biochem. Anal. 15, 25.
6. Leibovich S.J. & Weiss J.B., 1970, Biochim.Biophys.Acta, 214, 445.
7. Pelc S.R., Coombes J.D. & Budd G.C., 1961, Exp.Cell.Res. 24, 192.
8. Rauterberg J. & Kuhn K., 1971, Eur.J.Bioch. 19, 398.
9. Salpeter M.M., Bachmann L. & Salpeter E.C., 1969, J.Cell.Biol. 41, 1.
10. Wood G.C. & Keech M.K., 1960, Biochem.J. 75, 588.

Fig. 2. An autoradiograph of two normal collagen fibrils, showing growth at both ends.

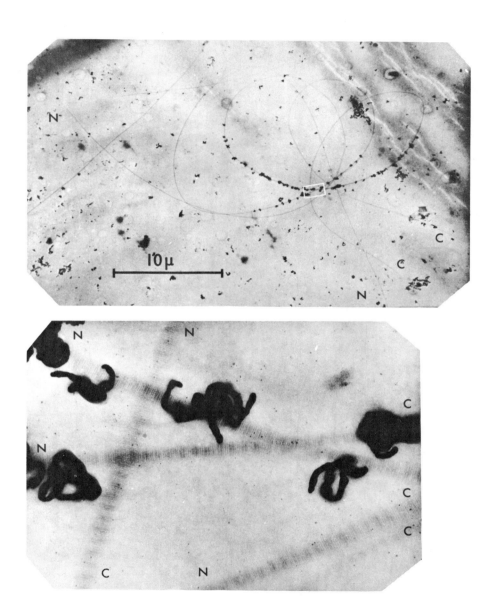

Electron microscopy and Cytochemistry, eds. E. Wisse, W.Th. Daems, I. Molenaar and P. van Duijn.
© 1973, North-Holland Publishing Company - Amsterdam, The Netherlands.

ELECTRON MICROSCOPE AUTORADIOGRAPHY
FOR NEGATIVE STAINED SPECIMENS

N. M. Maraldi, P. Simoni, G. Biagini and R. Laschi

Istituto di Microscopia elettronica clinica

Università di Bologna, 40126 Bologna, Italy

1. Summary

This report gives details of a technique for high resolution autoradiography on negatively stained specimens.

The grain distribution around negatively stained polysomes, using Kodak NTE, indicates that a considerable improvement in resolution power in comparison with usual e. m. autoradiography on sectioned specimens can be achieved. The possibility of further improvements in resolution with Sakura NR-H2 nuclear emulsion is discussed.

2. Introduction

Electron microscope autoradiography has been till now utilized almost exclusively on sectioned specimens. Recently the application of this technique to isolated shadow casted[5] and negatively stained[4] macromolecules has been reported.

Here we describe the experimental procedure and the results of an high resolution e. m. autoradiographic investigation on the ^{3}H-uridine uptake in negatively stained ribosome fractions.

3. Materials and methods

Polysomes have been isolated[2] from Wistar albino rats (weight 150 gr) which received 2 mC of ^{3}H-uridine 5 hr before killing. Ribosome microcrystals have been isolated[4] from 5 days old chick embryos incubated in Eagle MEM containing 10 µC/ml of ^{3}H-uridine for 5 hr before cooling at 5°C for 36 hr. A small drop of the resuspended ribosomes has been deposited on carbon coated Formvar covered grids. The specimens were fixed on grids with 4% acqueous glutaraldehyde before staining with 2% acqueous uranyl acetate (pH 4) and then covered with a thin layer of evaporated carbon; this procedure prevents a loss of stain during the photographic processes. Kodak NTE nuclear emulsion monolayers were prepared according to a previously described method[3] and the autoradiograms, after exposition, were treated for gold latensification, developed in Elon-ascorbic acid for 4 min at 30°C and fixed in Kodak acid fixer for 1 min. Longer photographic treatments may cause a loss of stain.

4. Results

The autoradiograms so obtained display a good preservation of the fine morphology of the negatively stained ribosomes and a very accurate localization of the developed grains (Figs. 1, 2). The grain distribution around the ribosomes indicates that the distance from the source within which 50% of the grain fall (Half Distance)[6] is 363 Å. Using probability circles to denote resolution the radius of a circle of 50% probability is 1.7 HD, i.e. 617 Å.

Preliminary observations indicate that the resolution power could be further improved using the Sakura NR-H2 nuclear emulsion associated with the photographic procedures described by Uchida and Mizuhira[7]. Typical aspects of the developed grains before and after gelatin digestion are shown in Figs. 3, 4.

5. Discussion

The improvement in resolution obtainable with this method respect to sectioned specimens is due to the smallest thickness of the isolated ribosomes (< 300 Å) respect to the mean section thickness. Researches are in progress in order to verify if the use of the Sakura NR-H2 could further improve the ultimate resolution power.

References

1. Barbieri, M., Pettazzoni, P., Bersani, F. and Maraldi, N. M., 1970, J. Mol. Biol. 54, 121.

2. Blobel, G. and Sabatini, D. D., 1970, J. Cell Biol. 45, 130.

3. Fantazzini, A., Serra, U., Maraldi, N. M. and Laschi, R., 1972, J. Submicr. Cytol. 4, 114.

4. Maraldi, N. M., Biagini, G., Simoni, P. and Laschi, R., 1973, Histochemie 35, 67.

5. Niveleau, A., 1971, J. Microscopie 11, 175.

6. Salpeter, M. M., Bachmann, L. and Salpeter, E. E., 1969, J. Cell Biol. 41, 1.

7. Uchida, K. and Mizuhira, V., 1970, Arch. Histol. Jap. 31, 291.

Figs. 1, 2. Kodak NTE autoradiograms. The grains are indicated by the arrows (Fig. 1). A silver grain (arrow) is superimposed on two ribosomes in which the cleft subdividing the subunits is clearly visible (Fig. 2).

Figs. 3, 4. Sakura NR-H2 autoradiograms. Fig. 3 shows the typical "cat's paw" pattern of the developed grain. In Fig. 4 some small "silver specks" which remain after a brief treatment with 0.05 M KOH are visible.

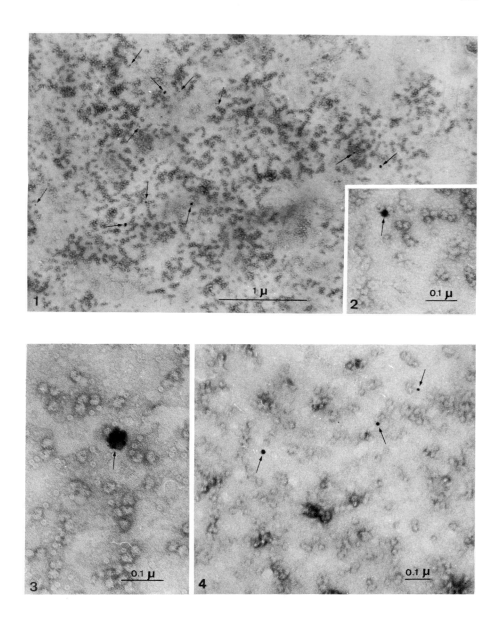

Electron microscopy and Cytochemistry, eds. E. Wisse, W.Th. Daems, I. Molenaar and P. van Duijn.
© 1973, North-Holland Publishing Company - Amsterdam, The Netherlands.

AN ENZYMATIC ASSAY FOR DNA IN TISSUE SECTIONS BY
ELECTRON MICROSCOPE AUTORADIOGRAPHY

S. Fakan and S.P. Modak

Departments of Cell and Molecular Biology
Swiss Institute for Experimental Cancer Research,
1011 Lausanne, Switzerland

1. Summary

Calf thymus terminal deoxynucleotidyl transferase-catalyzed
incorporation of ^{3}H-dAMP in ultrathin sections of fixed tissues is
localized in situ by autoradiography. The product of this reaction
is TCA-insoluble and DNase-sensitive. It allows detection of DNA in
the nucleus, nucleolus, mitotic chromosomes, and mitochondria.

2. Introduction

Exogenous DNA polymerising enzymes can be used to catalyze poly-
deoxynucleotide synthesis using as 'primer' the DNA in paraffin- or
plastic-embedded tissue sections[1-3]. In these conditions the incorporated radioactivity was localized by light microscope autoradiography[1-3]. This method has now been applied[4], using terminal deoxynucleotidyl transferase, to ultrathin sections followed by electron
microscope autoradiography.

3. Materials and methods

Ultrathin sections of tissues or cultured cells fixed in
glutaraldehyde, or glutaraldehyde followed by osmium tetroxide, and
embedded in Epon, araldite, or in glycol methacrylate (GMA) are
picked up with plastic rings in which they are kept during the
entire incubation procedure. To increase the occurrence of 3'-OH
ends, sections can be pretreated with 0.01 N HCl in order to denature DNA. Sections are then incubated for 1 to 3 hours at 37°C in
a reaction mixture containing: 0.1 ml ^{3}H-dATP (spec. act. 4-20 Ci/mM;
1 mCi/ml), 0.2 ml sodium cacodylate (1 M, pH 7.5), 0.1 ml MgCl$_2$
(40 mM), 0.1 ml β-mercaptoethanol (10 mM), 0.01 ml Triton X-100 (1%),
0.4 ml distilled water, and 0.1 ml calf thymus terminal deoxynucleotidyl transferase[5] (1000-20000 units/ml). After incubation, sections
are washed with 0.15 M NaCl followed by 5% trichloroacetic acid (TCA)

with 1% pyrophosphate and by 5% TCA, all at 0ºC. This is followed
by 3 rinsings in distilled water at room temperature. Finally,
sections are placed either on grids or collodion-coated slides and
processed for autoradiography by usual techniques. To test the
specificity of the enzymatic reaction some sections (after the final
wash in distilled water) are incubated in deoxyribonuclease solution
(0.1%), then washed again in TCA-water sequence, and placed on
slides or grids. Some preparations are incubated in a reaction mixture in the absence of enzyme in order to check for non-specific
binding of ^{3}H-dATP to the sections.

4. Results and discussion

The results of one such experiment are shown in the figure 1.
One sees label over the nucleus, nucleolus and mitochondria, and,
possibly, over the endoplasmic reticulum (fig. 1a). In the nucleus
strong labeling is found associated with the condensed chromatin.
As seen in figure 1b, the reaction product of terminal transferase
is sensitive to DNase.

Terminal transferase catalyzes synthesis of deoxynucleotide homo-
polymer using 3'-OH ends of denatured DNA as 'initiator'[5]. The
radioactive product of such synthesis is visualized by the present
autoradiographic method which thus detects DNA. For this purpose,
it is advisable that sections be pretreated to denature DNA which
increases the probability of detection. On the other hand, a comparison of label between the undenatured and the denatured preparations allow determination of the physico-chemical state of the
DNA[1,2]. Such a comparison, however, is not meaningful when the
effect of original fixation on DNA is not well understood as would
be the case for tissues used for electron microscopy.

The present results confirm our earlier studies[4] using the same
method whereby we have detected DNA at similar intracellular sites.[6]

Fig. 1.   Mouse liver cells fixed in glutaraldehyde and embedded in
glycol methycrylate. Autoradiographs coated with Ilford L4 emulsion,
exposed for 38 days and developed in D-19. a) Acid denatured section
incubated with terminal transferase and ^{3}H-dATP for 3 hours.
b) Similar to 'a' but treated subsequently with DNase.

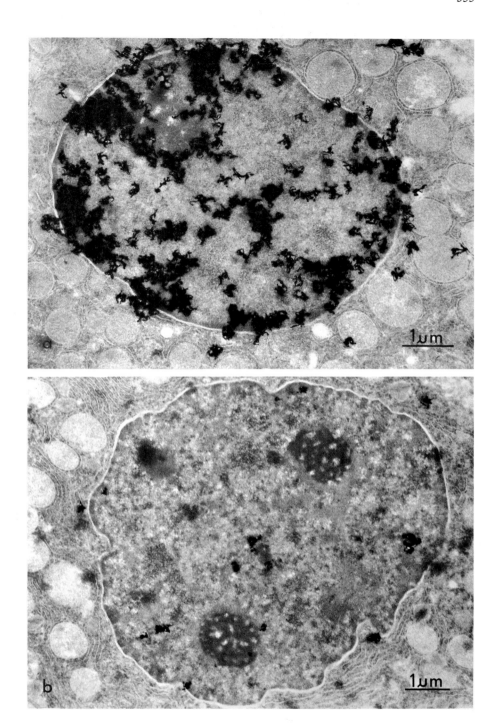

356

References

1. Modak, S.P., von Borstel, R.C. and Bollum, F.J., 1969, Exptl. cell res. 56, 105.

2. Modak, S.P. and Bollum, F.J., 1970, Exptl. cell res. 62, 421.

3. Modak, S.P., Donnelly, G.M., Karasaki, S., Harding, C.V., Reddan, J.R. and Unakar, N., 1973, Exptl. cell res. 76, 218.

4. Fakan, S. and Modak, S.P., 1973, Exptl. cell res. 77, 95.

5. Yoneda, M. and Bollum, F.J., 1965, J. biol. chem. 240, 3385.

6. Research supported by the Centre de Recherches sur les Lymphomes malins, Lausanne, and the Fonds National Suisse de la Recherche Scientifique (3.798.72).

Electron microscopy and Cytochemistry, eds. E. Wisse, W.Th. Daems, I. Molenaar and P. van Duijn.
© 1973, North-Holland Publishing Company - Amsterdam, The Netherlands.

LOCALIZATION OF DNA IN FIXED TISSUES BY RADIOAUTOGRAPHY

Improvement of the specificity of ^{3}H-Actinomycin D binding to DNA in ultrathin
sections of plastic embedded tissues.

M. Geuskens

Institute for Scientific Research on Cancer

94800 Villejuif, France

1. Introduction

The technic of ^{3}H-Actinomycin D binding to fixed or unfixed cytological
preparations has been used in photonic microscopy for detecting radioautographically
small amounts of DNA and   studying the genetic activity of individual cells[2,4,6,7]
The application of this method at the electron microscope level has been achieved
by studying the binding of ^{3}H-Actinomycin D to chromatin in living cells in culture[8], by placing fixed biological material in contact with the tritiated Actinomycin solution before embedding into a plastic[9] or by floating frozen ultrathin
sections on that solution[1,5]. This communication presents a technic which has led
to a successful specific labelling of DNA in ultrathin sections of plastic embedded tissues.

2. Material and methods

Rat lymph node and crab (Cancer pagurus) testis fragments were fixed by cold
1,6 % glutaraldehyde in 0,1 M phosphate buffer, pH 7,2, for 15 minutes and embedded in glycol-methacrylate (GMA). Ultrathin sections were floated for 1 hour on
a 100 µc/ml ^{3}H-Actinomycin D solution (6,1 Ci/mM) in the dark. Sections were then
transferred onto 2 successive distilled water baths and finally harvested onto
formvar coated grids with thick carbon layers on both sides. The grids were then
agitated for 1,5 or 10 minutes in a 50 % pyridine solution in methanol and for
2 minutes in 2 successive baths of methanol. They were covered with a monolayer of
Ilford L_4 emulsion using a loop. The radioautograms were developed after being
exposed for 2 or 3 months a 4°C.

3. Results

Silver grains were found localized over the nuclei of rat lymph node ultrathin sections, particularly concentrated over the condensed chromatin (fig. 1).

358

Silver grains were also seen localized over the nucleolus in some cell sections.

In crab testis sections, silver grains are more numerous over the condensed chromatin in both germinal and somatic cells. In many cases, some chromocenters are more heavily labelled than the neighbouring ones of the same nucleus section (fig. 2). The poly d (G-C) satellite of the crab DNA could be localized in these chromocenters as it is well known that Actinomycin D binds in the minor groove of DNA with a dG-dC dependency. DNA of spermatozoïds, which is not combined with proteins[3], is also strongly labelled. Except in some regions of the grids, the unspecific background is very low.

4. Discussion

The possibility of applying this technic for studying the genetic activity of individual cells at the ultrastructural level will have to be checked. Certain results obtained by applying it at the photonic microscope level show that the quantity of ^{3}H-Actinomycin D bound to the chromatin may reflect the quantity of DNA template available for transcription[7], but this seems to be true only for the ^{3}H-Actinomycin D bound to living cells[7].

However, our technic could be useful to corroborate the DNA nature of some structures as even old blocks of GMA embedded tissues can be used.

References

1. Bernier, R., Iglesias, R. and Simard, R., 1972, J. Cell Biol. 53, 798.
2. Brachet, J. and Ficq, A., 1965, Exptl. Cell Res. 38, 153.
3. Chevaillier, Ph., 1969, C.R. Acad. Sci. Paris 269, 2251.
4. Ebstein, B.S., 1967, J. Cell Biol. 35, 709.
5. Geuskens, M., 1972, J. Microscopie 13, 153.
6. Plessmann Camargo, E. and Plaut, W., 1967, J. Cell Biol. 35, 713.

7. Ringertz, N.R., Darzynkiewiez, Z. and Bolund, L., 1969, Exptl. Cell Res. 56, 411.
8. Simard, R., 1967, J. Cell Biol. 35, 716

Fig. 1. Ultrathin section of GMA embedded rat lymph node : binding of ^{3}H-Actinomycin to the chromatin. Kodak D19 developer.
Fig. 2. Ultrathin section of GMA embedded crab testis : binding of [3]H- Actinomycin to the chromatin in germinal cells : some chromocenters are more heavily labelled (⟶). Phenidon containing developer.

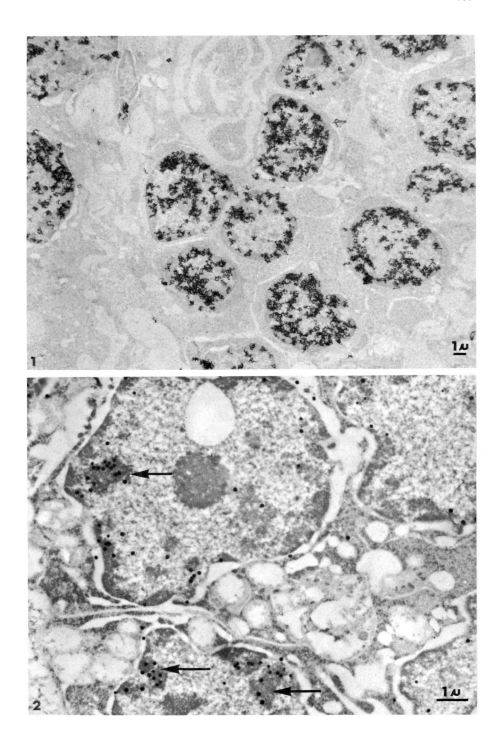

9. Steinert, G. and Van Gansen, P., 1971, Exptl. Cell Res. 64, 355.

The author is a qualified research worker of the belgian national Fund for scientific research.

Electron microscopy and Cytochemistry, eds. E. Wisse, W.Th. Daems, I. Molenaar and P. van Duijn.
© 1973, North-Holland Publishing Company - Amsterdam, The Netherlands.

PROTEIN TRANSPORT THROUGH THE GOLGI SYSTEM IN THE FROG EXOCRINE PANCREAS CELL

J.W. Slot and J.J. Geuze

Centre for Electron Microscopy, Faculty of Medicine,
University of Utrecht, The Netherlands

1. Summary.

Morphological and autoradiographic studies on the frog exocrine pancreas, both "in vivo" and "in vitro", revealed an intracellular transport route for secretory proteins from the RER through the Golgi system to the secretory granules. Some details of the way in which they pass the Golgi system are discussed.

2. Introduction.

The RER-Golgi system-secretory granule route for protein transport has been established broadly in many serous cell types. Nevertheless, in detail there are some discrepancies between reports pertaining to different tissues. So, the transport of protein from the RER may be performed by small vesicles[1], or by the transformation of certain parts of the RER into Golgi cisternae[2], while sometimes the endoplasmic reticulum is described to be continuous with Golgi elements[3]. Furthermore, there is no consistency with respect to the involvement of Golgi cisternae in the transport process[1,4].

Some elements from our work with the frog exocrine pancreas will be presented to demonstrate the way in which in this particular cell type the Golgi system is involved in protein transport.

3. Materials and methods.

Pancreas tissue of fed frogs was fixed in 1% OsO_4 (pH 7.4) for 1 h at 0° C, or impregnated with 2% OsO_4 (unbuffered) during 24 hrs at 37° C[5]. The structure of the Golgi system was studied in ultrathin and, 0,2 μm thick Epon sections.

For autoradiography, tissue was labeled with ^{3}H-leucine "in vivo" after intravenous injection, or "in vitro" by a 5 min pulse, followed by a chase in [1]H-leucine supplied medium. All incubations were done at 22° C. Silver-gold sections of OsO_4-fixed and Epon-embedded tissue were transferred to collodion-coated slides, stained with lead and covered with Ilford L_4 emulsion by the dipping method. After exposure at 4° C, the autoradiograms were developed in phenidone[6].

4. Results.

In the vicinity of smooth transitional elements of the RER at the periphery of the Golgi system, clusters of smooth vesicles and tubules are present, which show

blackening after osmium impregnation (fig. 1). The tubules are shown to be continuous with the outermost Golgi cisternae, which, in thick sections, seem to have
the shape of a flat tubular meshwork (fig. 2).

Quantitation of autoradiographs demonstrates the RER-Golgi system-secretory
granule transport route for proteins in the frog exocrine pancreas cell. (table I)
The incorporation into proteins of label administered "in vivo", continues over

Table 1.

Distribution of silver grains over cell structures, expressed as percentages
of the total number of grains counted at each particular time interval after
injection of ^{3}H-leucine.

Time after injection - - - - -	10min	20min	60min	120min	240min
R E R- - - - - - - - - - - - -	80.1	55.4	37.3	29.3	25.2
Golgi system - - - - - - - - -	4.3	35.0	50.4	19.1	19.8
cisternae - - - - - - - - -	(0.5)	(13.9)	(13.7)	(4.6)	(4.4)
condensing vacuoles - - - -	(0.5)	(1.0)	(17.7)	(5.6)	(7.1)
Zymogen granules (immature)- -	0.0	0.2	3.9	19.3	8.3
Zymogen granules (mature)- - -	0.5	2.4	1.9	26.3	39.3
Nucleus- - - - - - - - - - - -	7.0	2.7	4.0	4.5	3.6
Mitochondria - - - - - - - - -	8.0	4.3	2.5	1.3	3.4

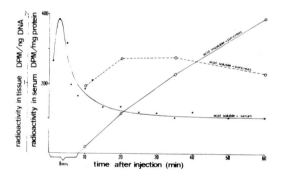

Fig. 3. Radioactivity in serum and pancreas tissue of the frog during the
first h after injection of ^{3}H-leucine (4 µCi/gm body weight). The serum
values are obtained from blood samples of one animal. Those concerning
activity in tissues are means of results from 2-4 animals.

Fig. 1. Golgi system in frog exocrine pancreas cell. Osmium impregnation,
thin section. Osmium deposit can be seen in peripheral Golgi vesicles and
tubules and in the outermost Golgi cisterna.
Fig. 2. Same tissue, 0.2 µm section Face view of a Golgi cisterna, filled
with osmium precipitate.
Fig. 4. Electron microscope autoradiograph of tissue, which was pulse-labeled
"in vitro" during 5 min, followed by a 15 min chase period.

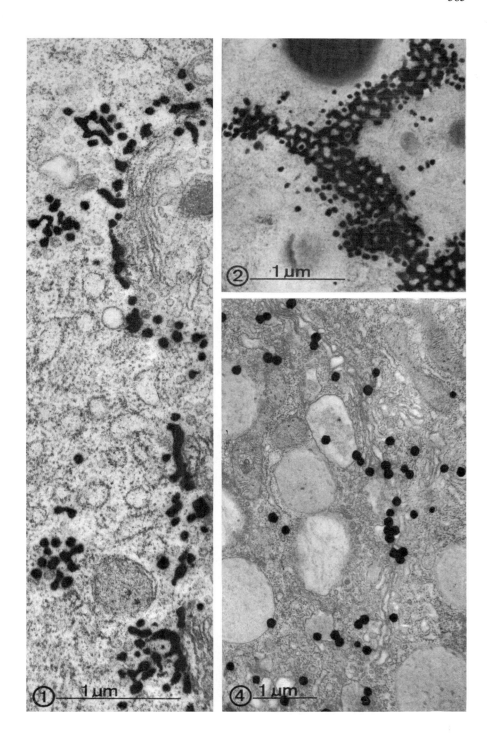

① 1 μm

② 1 μm

④ 1 μm

a rather long period (fig. 3), so that radioactivity is spread over a long
stretch of the transport route[7].

Nevertheless, the shift in labeling intensity of the Golgi cisternae during
the first hrs after ^{3}H-leucine injection, suggests that these cisternae participate in protein transport (table I). After "in vitro" pulse labeling, this is
even more clearly seen (fig. 4).

5. Concluding remarks.

The autoradiographic results clearly demonstrate that in the frog exocrine
pancreas secretory proteins pass through the Golgi cisternae. In view of the
morphological observations it seems most likely that they are transported to
these structures by the small outer Golgi vesicles, which yield tubules by fusion.
The tubules coalesce, forming a flat meshwork, which is the outermost Golgi cisterna. This cisterna moves inwards, while the transported proteins are concentrated in dilated sites, which leave the cisternal stack as condensing vacuoles.

References.

1. Jamieson, J.D. and Palade, G.E., 1967, J. Cell Biol. 34, 577.
2. Shin, W.Y., Ma, M., Quintana, N. and Novikoff, A.B. 1970, Abstracts 7th Int.
   Congr. Electr. Micr. Grenoble III, 79.
3. Claude, A., 1970, J. Cell Biol. 47, 745.
4. Castle, J.D., Jamieson, J.D. and Palade, G.E., 1972, J. Cell Biol. 53, 290.
5. Friend, D.S., 1967, J. Cell Biol. 35, 357.
6. Lettré, H. und Paweletz, N., 1966, Naturwissensch. 53, 268.
7. Slot, J.W. and Geuze, J.J., 1973, Abstracts 2nd Int. Symp. Electr. Micr. and
   Cytochem. (Demonstration).

PREPARATIVE TECHNIQUES

Electron microscopy and Cytochemistry, eds. E. Wisse, W.Th. Daems, I. Molenaar and P. van Duijn.
© 1973, North-Holland Publishing Company - Amsterdam, The Netherlands.

RECENT ADVANCES IN FIXATION OF TISSUES

David Hopwood

Department of Pathology, University of Dundee

Introduction

Tissue fixation lies at the beginning of most morphological studies. The practical aspects of the procedure are well-established and have been described by Pearse (1968), Hayat (1971), Pease (1964). The earlier work was reviewed by Baker (1960). Since then, several reviews have appeared on various aspects of this topic, notably by Wolman (1955), Lojda (1965), Pearse (1968), Lillie (1965), Hopwood (1969a, 1972a), Hayat (1971) and Riemersma (1970). The reader is also directed towards the proceedings of a symposium on fixation organized by the Royal Microscopical Society where fixation for the following aspects were discussed: electron histochemistry of phosphatases – Davies and Garrett (1972), Brunk and Ericsson (1972); immunohistochemistry Miller (1972), Avrameas (1972). There were also discussions of the fixation of fatty acids (Jones, 1972), the osmolarity of fixatives (Bone and Ryan, 1972), and the problem for pathologists (Dawson, 1972) and the use of glutaraldehyde (Hopwood, 1972b). The chemical processes underlying fixation are not so well understood.

Most of this work on mechanisms fixation has involved the reaction of fixatives with proteins and Baker (1960) has emphasised the importance of this. The reactions of fixatives with nucleic acids, lipids and muco substances are not well understood at all. The biochemists and virologists are continuing to produce information on the reactions between aldehydes and nucleic acids. This process was given impetus by the publication of the Watson-Crick model for DNA. Knowledge of fixation of lipids and muco substances is even scantier. The difficulty in fixing muco substances is reflected in the large number of substances which have been proposed for this function.

The present review is concerned with some of the more recent advances in fixation of proteins and nucleic acids, with emphasis on the underlying chemical mechanisms.

Artefacts

A number of artefacts are produced by the action of fixatives on the tissues and these are well-documented. In the case of glutaraldehyde there are some specific artefacts. It is known to produce carbonyl groups in the tissue which will stain with Schiff's reagent (Hopwood, 1967). False localisation and excess binding of radioactive amino acids occurs $[^3H]$ leucine - Peters and Ashley (1967); $[^3H]$ tyrosine - Hodson and Marshall (1967); $[^3H]$ mannose $[^3H]$ thymidine, $[^3H]$ uridine Vanha-Perttula and Grimley (1970). Also diffusion artefacts may occur with glutaraldehyde, even using relatively small tissue blocks (2 x 2 x 5 mm). This has been established in the case of $[^{59}Fe]$ haemo-globin by Reale et al (1970). Glutaraldehyde is also known to cause shrinkage (Hopwood, 1967) but Diers and Schieren (1972), working with Elodea chloroplasts, have shown that post fixation in osmium tetroxide returns the dimensions almost to those in the living cells. Araldite pro-duced no change in volume.

The tonicity of the fixative solution may also play an important part in the image finally seen. Earlier observations with glutaraldehyde were discussed by Hopwood (1972b). Davey (1973) has investigated the structure of frog striated muscle after fixation in acrolein-osmium tetroxide at various tonicities. He concluded that tissues exposed to modifications of normal physiological solutions should use the same modifications as fixative vehicles.

The vehicle for the fixative can produce artefacts in other ways. Pratt and Napolitano (1969) found that there was a glycocalyx surrounding intestinal microvilli which was visualised by glutaraldehyde followed by osmium tetroxide in phosphate buffer. If the post-osmication took place with osmium tetroxide dissolved in carbon tetrachloride then an electron lucent zone resulted. This, obviously, raises the general question of interpretation of electron dense and electron lucent areas in electron micrographs.

Kilburn et al (1973) introduced a new vehicle to overcome an artefact produced by fixation with glutaraldehyde. These workers were interest-ed in the recruitment of leukocytes through airway walls induced by vegetable products. Glutaraldehyde washed these white cells off. When fixation was performed with osmium tetroxide in the fluorocarbon FC-80 the recruited cells were kept in place. The Fc-80 is a solvent for oxygen and does not itself appear to produce artefacts.

Artefacts produced by one fixative may at times be revealed by comparing the results with those obtained from other fixatives and by varying the vehicle composition. The results of such an exercise were reported by Busson-Mabrillot (1971).

The chemical nature of dialdehydes

The dialdehydes, glutaraldehydes and α-hydroxyadipaldehyde, are much used in electron microscopy. There has been considerable discussion over their chemical nature (see Hopwood, 1972b). Recent work by Hardy, Nicholls and Rydon (1972) using [1]H n.m.r. spectroscopy has shown that the glutaraldehyde consists mainly of the cyclic monohydrate. At higher temperatures the free aldehyde content of the mixture increases; α-hydroxydipaldehyde forms only the acyclic dihydrate. These workers showed that aqueous solutions at neutral pH were stable over a long period although the undiluted aldehydes polymerized readily. Korn, Feairheller and Filachione (1972) used n.m.r. spectroscopy to investigate glutaraldehyde from six commercial sources. They showed that 15% of free aldehyde existed at room temperature in aqueous solutions. This proportion increased with temperature reaching 75% at 90°C. The major variation these workers found in the glutaraldehyde from various sources was in their methanol content. The material which absorbs at 235 nm may represent some impurity. Gillet and Gull (1972) used absorbance at this wavelength to assess the stability of glutaraldehydes from various sources stored under differing conditions following vacuum distillation or treatment with charcoal. Storage at higher temperatures increased the rate at which this material, absorbing at 235 nm, reappears. Methods for purification of various dialdehydes are given in detail by Hardy et al. (1972).

Reaction of glutaraldehyde with substances of biological interest

Glutaraldehyde has been shown to react with 10 or more amino acids. It also reacts with biogenic amines, in solution or the vapour phase.

This forms the basis for various histochemical methods - see Hopwood (1972b).

There have now been a considerable number of studies on the reaction of glutaraldehyde with proteins investigating a number of changes in the properties of the proteins and the kinetics of the reactions.

Fixatives decrease enzyme activity. It has been suggested that this effect may be due to limitations in substrate diffusion or the formation of crosslinks between protein molecules. This problem has been investigated quantitatively by numerous authors, in the case of glutaraldehyde using small blocks of tissue, perfused tissue homogenates and crystaline enzymes (see Hopwood, 1972b). Arborgh et al. (1971) pointed out that some of the anomalies were due to the slow rate of penetration of glutaraldehyde into tissue blocks. These workers found that washing to fixed blocks of tissue increased the enzyme activity. There is also evidence that antigen-antibody reactions are impaired by fixation. Hopwood (1969b) investigated the effects of several fixatives on a BSA-anti BSA system. Glutaraldehyde produced the greatest decrease in the reaction.

The mechanisms whereby glutaraldehyde produce fixation have been investigated by a number of techniques. There have been a number of reports on amino acid analyses of proteins, including BSA, various enzymes and wool, which have been reacted with glutaraldehyde. The amino acid most affected is lysine which decreases by 50-60% of the initial content. Glutaraldehyde reacts with its ϵ-amino group (see Hopwood, 1972b). There is some evidence that aromatic amino acids also react with glutaraldehyde (Hopwood, Allen and McCabe, 1970).

When glutaraldehyde reacts with proteins inter- and intramolecular crosslinks are formed (Bowes and Cater, 1968; Hopwood, 1969b). Using polyacrylamide gel electrophoresis and gel chromatography the last author showed that glutaraldehyde was very much more efficient at crosslinking proteins than formaldehyde or α-hydroxyadipaldehyde. Tomimatsu et al. (1971) investigated the reaction between α-chymotrypsin and glutaraldehyde by light-scattering methods. They found an initial rapid reaction of the ϵ-amino groups of lysine on the surface of the protein with crosslinking. In the second phase which occurred a matter of hours, later there was further polymerisation of the initially formed protein oligomers to form long linear polymers.

The efficiency of glutaraldehyde fixation at the tissue level has been shown by Hopwood (1969b) using 1 mm slices of liver treated with various fixatives and then subjected to starchgel electrophoresis. At the ultrastructural level Ellar et al. (1971) investigated the effect of glutaraldehyde on membrane-bound enzymes in <u>Micrococcus lysodeiktus.</u>

The chemical kinetics of the reaction between proteins and glutaraldehyde have been studied in some detail by Hopwood et al. (1970) and were found to be pseudo-first order. The reactions proceeded more rapidly at higher pH and temperature. Determination of the apparent energies of activation for the reaction suggested that little denaturation had taken place, in agreement with the ORD and CD investigations of Lenard and Singer (1968). On the other hand Bagdasar'yan et al. (1971) found that human serum albumen was denatured by a series of aldehydes. Tissue fixation with glutaraldehyde

Recently, Peracchia and Mittler (1972a) have introduced a new technique for tissue fixation, which they claim gives better results than glutaraldehyde alone. These workers used a glutaraldehyde-hydrogen peroxide mixture. Difficult plant and animal specimens are apparently well fixed. The effective products are probably epoxides which are generated by reaction of the hydrogen peroxide with the unsaturated groups in glutaraldehyde polymers. The fixative mixture had a high reactivity with hydroxyl groups which may explain the improved preservation of muco-substances they observed. This mixture appears to have potential for electron histochemistry. Goldfischer et al. (1971) found that it penetrated rat liver more rapidly than glutaraldehyde and that nucleoside diphosphatase and thiamine pyrophosphatase activities were well-preserved. They could give no explanation for this.

Peracchia and Mittler (1972b) also investigated the effect of temperature on fixation with glutaraldehyde. They found that after an initial fixation at room temperature, if the temperature was then raised to 45°C for two hours then the resulting morphology was improved. The mechanism for this, they suggested, was that increasing the temperature caused the glutaraldehyde to polymerize. This gave different lengths of active molecules which could span different distances between organelles, proteins and other cellular contents, thus fixing them. However, Hardy et al. (1972) and Korn et al. (1972) showed that increasing the temperature increases the amount of free aldehyde present in solution.

Effects of glutaraldehyde on cell membrane systems

Shrager et al. (1969) investigated the electrical effects of chemical modification of crayfish axons by protein crosslinking aldehydes. Non-crosslinking aldehydes had no effect or caused a gradual decline in resting potential and produce no widening of the spike. Strongly crosslinking aldehydes, such as acrolein, crotonal and glutaraldehyde widen the action potential and reduce its amplitude, perhaps by altering sodium conductance. The changes may be due to conformational change in axon protein. Similar electrical results were reported by Denoit-Mazet and Vassort (1971). Hubbard and Laskowski (1972) investigated the effects of glutaraldehyde on rat phrenic nerve diaphragm preparations. They found that the miniature and plate potentials frequency and amplitude were significantly reduced after 45 seconds of exposure as were the post-synaptic potentials.

Hertz and Kaplan (1972) investigated the effects of glutaraldehyde on erythrocyte membranes. They found there was increased resistance to mechanical deformation and also an irreversible loss of acetyl cholinesterase. The enzyme loss was dependent on the pH of the reaction. Steck (1972) used various crosslinking agents to study the topological distribution of major protein fractions in the erythrocyte membranes assessing the results by gel electrophoresis. Some proteins were distinctly unreactive whereas others crosslinked. He suggested that these proteins may exist in oligomeric associations.

Use of aldehydes as chemical probes

Glutaraldehyde is beginning to be used by biochemists in investigating ribosomes. Subramanian (1972) pointed out that free ribosomes dissociated when centrifuged on gradients with a low concentration of magnesium ions. Fixation of the isolated ribosomes by glutaraldehyde could prevent this. Any tendency for the ribosomes to be crosslinked could be overcome by the addition of bovine serum albumin. Kahan and Kaltschmidt

(1972) have used glutaraldehyde as a probe of chemical structure in sibo-somes.   The topography of large and small subunits of <u>Escherichia coli</u> ribosomes were studied after exposure to glutaraldehyde and separation by two dimensional polyacrylamide gel electrophoresis.   They concluded that most of the reactive proteins are situated externally in the ribosome.

Glutaraldehyde has also been used to separate two distinct forms of collagen in chick cartilage.   (Trelstad et al. 1970). Marini and Martin (1971) used formaldehyde as a chemical probe to investigate the active centre of chymotrypsin.   They also pointed out that crosslinking was slow. Roux (1972) has used glutaraldehyde to investigate conformatinal differences between the red and far-red absorbing forms of oat phytochrome.   More lysine residues were available in the red form whose photoreversibility was more inhibited than the far red form.

We may expect to see increasing use of glutaraldehyde as a probe by biochemists.   This should furnish important information on the mechanisms of fixation.

Metal containing fixatives

A number of metals are in common use as fixatives.   Mercuric chloride is used in the author's laboratory for most surgical specimens and as a second fixation for necropsy material.   Its use has recently been reviewed (Hopwood, 1972a).   Passow (1970) has also recently reviewed the penetration of distribution and toxic actions of various heavy metals on red blood cells.   Hoogeven (1970) has investigated the effects of metal ions on the stability of model membranes.   Uranyl ions had the greatest affinity for the system, mercuric ions the least.   Ferrous and calcium ions promoted the structural properties of the model.   The uses of osmium tetroxide, potassium dichromate and potassium permanganate in fixation have been reviewed recently at length by Riermersma (1970). Also, Litman and Barrnett (1972) have investigated the mechanism whereby osmium tetroxide fixes and stains tissue.   They suggested that Os (VIII) was responsible for much of the membrane density, probably by hydrogen bonds to protein and aliphatic side chains.

Nucleic acids

A better understanding of the fixation of nucleic acids can be obtained if their chemical structure is known.   Detailed accounts of these are now available and the reader is referred to these; Watson (1965). The helical nature of DNA is well-known. More recently RNA has been shown to consist of clover leaf pattern molecules. These probably undergo further folding in the presence of divalent cations to give them a tertiary structure (Levitt, 1972).

The different structures of DNA and RNA are important in determining the way in which they will react with fixatives. These structures are reflected in their thermal transitional profiles (Doty et al. 1959). DNA melts relatively rapidly at about $90^{\circ}C$, when the hydrogen bonds between the bases are destroyed. On the other hand, RNA melts more gradually over a larger temperature range. Römer et al. (1970) have shown that overall picture is due to the melting of individual loops of clover leaf secondary structure. The presence of cations is important in determining the melting temperature for a nucleic acid.

A further factor of importance in understanding the fixation of nucleic acids, is the chemical constituents of nuclei and ribosomes. Nuclei are known to be surrounded by the double nuclear membrane. Besides the nucleic acids there are also other proteins present including enzymes concerned with nucleic acid synthesis. There are also some residual proteins (Leveson and Peacocke, 1966). The other important groups of substances are the histones which are rich in basic amino acids. Lysine is the amino acid which glutaraldehyde reacts with chiefly. (Hopwood, 1972b).

Reactions with formaldehyde

The reactions between formaldehyde and nucleic acids are probably the best understood. These have been known for some length of time since Fraenkel-Conrat described an increase in absorbance and a shift to the blue during the reaction of formaldehyde with tobacco mosaic virus. Marciello and Zubray (1964) investigated the effect of temperature on the reaction of formaldehyde with adapter RNA. They found reactions take place in 3 phases in 3 different temperature ranges. More generally Boedtker (1967) pointed out that most RNAs were unstable at temperatures above $60^{\circ}C$.

In the case of DNA, the effects of temperature on the reaction with formaldehyde were studied by Grossman, Levine and Allison (1961) and Berns and Thomas (1961). Up to 45°C there was no reaction. Above 70°C the reaction was rapid and apparently first order. Clearly, reaction does not occur until some hydrogen bonds of the DNA have been broken, after which it proceeds readily.

The nature of the reaction between the nucleic acids and formaldehyde has been investigated by a number of workers. Penniston and Doty (1963) reported detailed investigations in the case of soluble RNA. They suggested that either a Schiff base or a methylol group was formed, both adducts being labile. Later work has favoured the latter (Utiyama and Doty, 1971). They concluded the first effect of the formaldehyde was to denature the RNA and then react with the revealed bases. The spectral effects are the sum of these two processes. By making measurements of the hyperchromicity at different wavelengths it is possible to separate the extent of denaturation and reaction of RNA with formaldehyde (Utiyama et al. 1971). The reactions of formaldehyde with RNA follows pseudo first-order kinetics, occurring more readily at pH 7.0 than 4.6. Eyring and Ofengand (1967) investigated the reaction of formaldehyde with the heterocyclicimino nitrogen of purine and pyrimidine nucleotides.

These simple adducts are formed rapidly, are labile and may be destroyed by simple dilution. Feldman (1967) showed that another group of derivatives could be formed between formaldehyde and nucleotides or nucleic acids. These stable derivatives were formed over a period of days at room temperature and were due to the formation of methylene bridges. Their production is favoured by a neutral or alkaline pH and high concentrations of formaldehyde. The crosslinks which formed, however, were mostly intramolecular (Axelrod et al. 1969).

The significance of this temperature effect on the reaction of nucleic acids with formaldehyde is obvious. At the practical tissue level there is a considerable literature on the loss of nucleic acids and nucleotides during preparation for microscopy. Davies (1954) showed that 10-35% of nucleic acids and nucleotides were lost from chick fibroblasts during fixation in formalin or Carnoy. Losses also occur during embedding in paraffin wax (Berenbom, Yokoyama and Stowell, 1952), and during the flattening of sections on a water bath (Jonsson and Lagerstedt, 1958).

The reaction of nucleic acids with glyoxal

The reactions of glyoxal with nucleic acids have been investigated by a number of groups. Nakaya et al. (1968) showed that at low concentrations of glyoxal only guanine and guanylic acid react, but at higher concentration all nucleotide bases react. Glyoxal, however, does not react with native DNA at room temperature although it will do with heat denatured DNA. Broude and Budowsky (1971) made detailed investigations of the kinetics of the reactions between glyoxal and nucleic acids and the nucleotides. The adduct formed with guanine is stable, the rate of reaction increasing with pH.

Reactions between nucleic acids malonaldehyde and glutaraldehyde

Malonaldehyde was shown to react with DNA by Brooks and Klamerth (1968) by the use of transition profiles and chromatographic behaviour.

The reactions of glutaraldehyde with nucleic acids have been investigated by Hopwood (in preparation). At temperatures up to 64°C no reaction occurred between native DNA and glutaraldehyde. At temperatures above 75°C the reaction followed pseudo-first order kinetics. The kinetic studies were corroborated by thermal transition profile studies. There was little evidence for the formation of crosslinks between nucleic acid molecules, even at elevated temperatures, using gel filtration and polyacrylamide gel electrophoresis techniques.

The problem of how nucleic acids are stabilised in the tissue remains. At the temperatures commonly used for fixation, 0°-23°C, for light and electron microscopy, only little reaction will occur. A possible mechanism may be that nucleic acids are physically entrapped in the gel formed by cellular proteins when they react with fixatives. In the case of glutaraldehyde, the lysine rich histones may play an important part in the entrapment. Crosslinking may also occur between histones and DNA. In spite of these theoretical problems in the fixation of nucleic acids the quantitation of nucleic acids in tissue sections is a well-established technique. Swift (1966) has recommended various properties that a fixative for such a purpose should have. Fixation is clearly a compromise between various ideals sought by the investigators.

Metal containing fixatives and nucleic acids

A number of metals in common use as fixatives are known to react with nucleic acids. Beer et al. (1970) reviewed the progress in the understanding of the reactions between DNA and osmium tetroxide. Reaction will only take place with thymidine and that in the denatured nucleic acid. They also reported that reaction mixtures which contained cyanide ions and osmium tetroxide at pH 7.0 gave promising results for staining nucleic acids. Beer and his colleagues also reported that gold chloride will react with adenosine, d-AMP d-GMP and d-CMP to form insoluble products with high molecular weights.

The use of mercuric chloride as a fixative agent for nucleic acids has been reviewed recently by Hopwood (1972a). Yamane and Davidson (1961) have shown that mercuric chloride reacts with the heterocyclic bases of the nucleic acids. The reaction is reversible by excess halide ions. The reactions of mercuric chloride with virus nucleic acids have been reported by various workers (Singer 1964; Dorne and Hirth, 1970, 1971). Reaction is associated with loss of infectivity, which will return with removal of the metal. The presence of mercury in nucleic acids has been shown at the electron microscopic level by Frithz and Lagerholm (1968) and biochemically by Norseth (1968).

Eichorn and his colleagues (1970) have investigated the reactions of nucleic acids with a large number of other metals. Binding was by the ribose hydroxyl groups, the phosphate group or the bases. The metals investigated could be arranged into a series producing decreasing stability of the nucleic acid: magnesium, cobalt, nickel manganese, zinc, cadmium, copper, silver, mercury.

Summary

Our understanding of the mechanisms of fixation is increasing, thanks in part to the use of fixatives as chemical probes by the biochemists. The time may come soon when fixatives for specific purposes may be designed and used.

378

Arborgh, B., Ericsson, J. L. E. and Helminen, H., 1971, J. Histo-chem. Cytochem. 19, 449.

Avrameas, S., 1972, Histochem. J. 4, 321.

Bagdasar'yan, S. N. and Troitskii, G. V., 1971, Biokhimiya 36, 732.

Baker, J. R., 1960, Principles of biological micro-technique (Methuen, London).

Beer, M., Gibson, D. W. and Koller, T., 1970, in: Effects of metals on cells, subcellular particles and macromolecules, eds. J. Maniloff, J. R. Coleman and M. W. Muller (C. C. Thomas, Springfield).

Berns, K. I. and Thomas, C. A., 1961, J. Mol. Biol. 3, 289.

Boedtker, H., 1967, Biochemistry 6, 2718.

Bone, Q. and Ryan, K. P., 1972, Histochem. J. 4, 331.

Bowes, J. H. and Cater, C. W., 1968, Biochim. biophys. Acta, 168, 341.

Brooks, B. R. and Klamerth, O. L., 1968, Europ. J. Biochem. 5, 178.

Broude, N. E. and Budowsky, E. I., 1971, Biochim. biophys. Acta 254, 380.

Brunk, U. T. and Ericsson, J. L. E., 1972, Histochem. J. 4, 349.

Busson-Mabillot, S., 1971, J. Microsc. 12, 317.

Davies, H. G., 1954, Q. J. Microscop. Sci. 95, 433.

Davies, K. and Garrett, J. R., 1972, Histochem. J. 4, 365.

Davey, D. F., 1973, Histochem. J. 5, 87.

Dawson, I. M. P., 1972, Histochem. J. 4, 381.

Denoit-Mazet, M. and Vassort, G., 1971, C.R. Acad. Sci. D, 273,1851.

Diers, L. and Schieren M. T., 1972, Protoplasma, 74, 321.

Dorne, B. and Hirth, L., 1970, Biochemistry 9, 119.

Dorne, B. and Hirth, L., 1971, Biochimie 53, 469.

Doty, P., Boedtker, H., Fresco, J. R. and Haselkorn, M Proc. Nat. Acad. Sci. U.S. 45, 482.

Eichhorn, G. L., Buteow, J. J., Clark, P. and Shin, Y. A., 1970, in: Effects of metals on cells, subcellular particles and macromolecules, eds. J. Maniloff, J. R. Coleman and M. W. Miller (C. C. Thomas, Springfield).

Ellar, D. J., Munoz, E. and Salton, M. R., 1971, Biochem. Biophys. Acta 225, 140.

Eyring, E. J. and Ofengand, J., 1967, Biochemistry 6, 2500.

Feldman, M., 1967, Biochim. biophys. Acta 149, 20.

Frithz, A. and Lagerholm, B., 1968, Acta derm- venereol 48, 403.

Gillet, R. and Gull, K., 1972, Histochemie 30, 162.

Goldfischer, S., Essner, E. and Schiller, B., 1971, J. Histochem. Cytochem. 19, 319.

Grossman, L., Levine, S. S. and Allison, W. S., 1961, J. Mol. Biol. 3, 47.

Hardy, P. M., Nicholls, A. C. and Rydon, H. N., 1972, J. Chem. Soc. 2270.

Hayat, M. A., 1970, Principles and techniques of electron microscopy. Biological applications (Van Nostrand Reinhold Co., London).

Hertz, F. and Kaplan, E., 1972, Proc. Soc. exp. biol. Med. 140, 720.

Hodson, S. and Marshall, J., 1967, J. cell. biol. 35, 722.

Hoogeven, J. T., 1970, in: Effects of metals on cells, subcellular elements and macromolecules, eds. J. Maniloff, J. R. Coleman and M. W. Miller (C. C. Thomas, Springfield).

Hopwood, D., 1967, J. Anat. 101, 83.

Hopwood, D., 1969, Histochem. J. 1, 323.

Hopwood, D., 1969b, Histochemie 17, 151.

Hopwood, D., 1972a, Prog. Histochem. Cytochem. 4, 193.

Hopwood, D., 1972b, Histochem. J. 4, 267.

Hopwood, D., Allen, C. R., and McCabe, M., 1970, Histochem. J. 2, 137.

Hubbard, J. I. and Laskowski, M. B., 1972, Life Sci. II, 781.

Jones, D., 1972, Histochem. J. 4, 421.

Jonsson, N. and Lagerstedt, 1958, Experientia 14, 157.

Kahan, L. and Kaltschmidt, E., 1972, Biochemistry II, 2691.

Kilburn, K. H., Lynn, W. S., Tres, L. L. and McKenzie, W. N., 1973, Lab. Invest. 24, 55.

Korn, A. H., Feairheller, S. H. and Filachione, E. M., 1972, J. mol. Biol. 65, 525.

Lenard, J. and Singer, S. J., 1968, J. cell. Biol. 37, 117.

Leveson, J. E. and Peacocke, A. R., 1966, Biochim. biophys. Acta 123, 329.

Levitt, M., 1972, in: Polymerization in biological systems 147, eds. G. E. W. Wolstenholme and M. O'Connor (Elsevier-Excerpta Medica – North Holland, Amsterdam).

Lillie, R. D., 1965, Histopathologic technique and practical histochemistry (McGraw-Hill, London).

380

Litman, R. B. and Barrnett, R. J., 1972, J. ultrastruct. Res. 38, 63.

Lojda, Z., 1965, Folia morph. 13, 65.

Marciello, R. and Zubay, G., 1964, Biochem. biophys. res. Commun. 14, 272.

Marini, M. A. and Martin, C. J., 1971, Eur. J. Biochem. 19, 153.

Miller, H. R. P., 1972, Histochem. J. 4, 305.

Nakaya, K., Takenaka, O., Horinishi, H. and Shibata, K., 1968, Biochim. biophys. Acta 161, 23.

Norseth, T., 1968, Biochem. Pharmacol. 17, 581.

Passow, H., 1970, in: Effects of metals on cells, subcellular elements and macromolecules, eds. J. Maniloff, J. R. Coleman and M. W. Miller (C. C. Thomas, Springfield).

Pearse, A. G. E., 1968, Histochemistry Theoretical and Applied, 3rd Edn., vol. I (Churchill, London).

Pease, D. C., 1964, Histological techniques for electron microscopy (Academic Press, London).

Penniston, J. T. and Doty, P., 1963, Biopolymers 1, 145.

Peracchia, C. and Mittler, B. S., 1972, J. cell. Biol. 53, 234.

Peracchia, C. and Mittler, B. S., 1972b, J. ultrastruct. Res. 39, 57.

Peters, T. and Ashley, C. A., 1967, J. cell. Biol. 33, 53.

Pratt, S. A. and Napolitano, L., 1969, Anat. Rec. 165, 197.

Reale, E. and Luciano, L., 1970, Histochemie 23, 144.

Riemersuma, J. C., 1970, in: Some biological techniques in electron microscopy, eds. D. F. Parsons (Academic Press, London).

Römer, R., Riesner, D., Coutts, S. M. and Maass, G., 1970, Evn. J. Biochem. 15, 77.

Roux, S. J., 1972, Biochemistry 11, 1930.

Shrager, P. S., Strickholm, A. and Macey, R. I., 1969, J. Cell. Physiol. 74, 91.

Singer, B., 1964, Biochim. biophys. Acta 80, 137.

Steck, T. L., 1972, J. mol. Biol. 66, 295.

Subramanian, A. R., 1972, Biochemistry 11, 2710.

Swift, H., 1966, in: Introduction to quantitative cytochemistry, ed. G. L. Wied (Academic Press, London).

Tomimatsu, Y., Jansen, E. F., Gaffield, W. and Olsen, A. C., 1971, J. colloid interface science 36, 57.

Trelstad, R. L., Kang, A. H., Igarashi, S. and Gross, J., 1970, Biochemistry 9, 4993.

Utiyama, H. and Doty, P., 1971, Biochemistry 10, 1254.

Vanha-Perttula, T., and Grimley, P. M., 1970, J. Histochem. Cytochem 18, 565.

Watson, J. D., 1965, Molecular biology of the gene (Benjamin, New York).

Wolman, M., 1955, Int. Rev. Cytol. 4, 79.

Yamane, T. and Davidson, N., 1961, J. Amer. chem. soc. 83, 2599.

Electron microscopy and Cytochemistry, eds. E. Wisse, W.Th. Daems, I. Molenaar and P. van Duijn.
© 1973, North-Holland Publishing Company - Amsterdam, The Netherlands.

ULTRAHISTOCHEMICAL DEMONSTRATION OF FREE INORGANIC IONS

G. Geyer

Anatomisches Institut,

Friedrich-Schiller-Universität Jena, DDR.

SUMMARY.

In an aqueous trapping medium the histochemical detection of free
inorganic ions suffers from dislocation. More precise localization may
be achieved by a freeze-substitution method consisting of a) rapid
immobilization of fluids by quick freezing of the tissue, and b) trapping
of ions during the dissolution of cellular and interstitial ice. This principle
proved valid for the demonstration of inorganic phosphate, calcium, and
chloride ions.

1. Introduction

Solubility renders topochemical detection of free inorganic ions
most difficult. During an incubation of fresh tissue samples dislocation of ions might be caused by several factors, as for
instance: inhibition of ion pumps in cytomembranes, restricted
permeability of trapping agents, low concentration and shortage of
reagent at the site of precipitation, outflow of ions from superficial layers of the tissue, building up of artificial ion
gradients in the sample, affinity to ions and reaction products
respectively of some tissue components (1).
Transitory binding of ions to organic cellular and interstitial
constituents seems to be strong enough to prevent translocation
within a certain range (2). Nevertheless, histochemical demonstration of soluble ions should be done with minimum risk of
dislocation.

2. Ion trapping by freeze-substitution

Therefore, the procedure should start with the rapid immobilization of the tissue fluid and its solutes, i.e. quick freezing of
the tissue in situ or shortly after dissection. Unless the detection of radioactive elements is intended which could be initiated

by freeze-drying of tissue samples, ion trapping by freeze-substitution will be the next step.

Freeze-substitution in this case includes

a) substitution of tissue water, and

b) trapping of inorganic ions at the moment they are set free by the dissolution of ice.

It should result in opaque precipitates of high specificity and true location. The procedure requires reagents which are

a) soluble in organic solvents such as methanol, ethanol, or acetone,

b) retaining their activity at low temperature, and

c) precipitating instantly.

As the substitution proceeds slowly it will continue for some days, unless ultrathin frozen sections of unfixed samples can be prepared, which should not be soaked with a cryo-protecting medium.

In most cases the formation of ice crystals will be a severe drawback invalidating large areas of tissue. On the other hand, microcrystallization causes a great number of sites of elevated ion concentration. Therefore, microcrystallization might be considered a prerequisite for ultrahistochemical detectability of ions, at least of those distributed homogeneously in a subthreshold concentration,for instance,in the cytoplasm.

3. Recent results in ultrahistochemical localization of ions by freeze-substitution

Inorganic phosphate, chloride, and calcium ions were demonstrated by freeze-substitution methods (studies done in collaboration with Drs. K.-J. Halbhuber, J.H. Stibenz, and A. Benser). In liver cells of normal mice Pi-precipitation occured predominantly in the cytoplasm and in the mitochondria (3). A small number of granular products has been observed in nuclei. CCl_4-poisoned mice showed enhanced precipitates in mitochondria (4). However, no nucleolar phosphate pool (5) was detectable by this method (fig.1).

Fig. 1. Mouse liver cell. Freeze-substitution with lead acetate for the detection of free inorganic phosphate.

Fig. 2. Human erythrocyte, fixed with glutaraldehyde and soaked with 1% $CaCl_2$. Freeze-substitution with oxalic acid for the demonstration of calcium ions.

Fig. 3. Human red blood cell, fixed with glutaraldehyde and soaked with 1% NaCl. Freeze-substitution with silver nitrate for the demonstration of chloride ions.

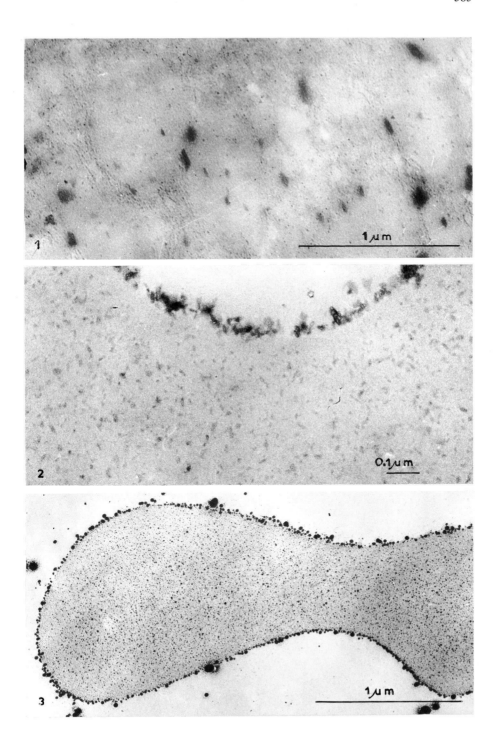

386

Preliminary experiments were carried out with both aldehyde fixed red blood cells or blood serum previously soaked with calcium and chloride ions respectively. Freeze-substitution of these model substances resulted in electron dense precipitates (fig. 2 and 3). Ultraautoradiographic demonstration of ^{45}Ca proved the trapping of calcium ions. In this procedure a total loss $< 0.1\%$ radioactive calcium was found with a liquid scintillation counter.

References

1. Geyer, G., 1973, Ultrahistochemie. 2. Aufl. (Fischer, Jena, Stuttgart).

   Geyer, G., 1973, Wiss. Z. Universität Jena (in press).

2. Klein, R.L., Yen, S.-S. and Thureson-Klein, A., 1972, J. Histochem. Cytochem. 20, 65.

   Komnick, H. and Bierther, M., 1969, Histochemie 18, 337.

   Komnick, H., Rhees, R.W. and Abel, J.H., 1972, Cytobiologie 5, 65.

3. Halbhuber, K.-J. and Geyer, G., 1972, Acta histochem. 43, 21.

4. Halbhuber, K.-J. and Geyer, G., 1972, Acta histochem. 44, 226.

5. Tandler, C.J., 1960, Biochim. Biophys. Acta 44, 536.

   Tandler, C.J. and Solari, A.J., 1969, J. Cell Biol. 41, 91.

   Halbhuber, K.-J., Stibenz, J.H. and Geyer, G., 1971, Acta histochem. 39, 32.

Electron microscopy and Cytochemistry, eds. E. Wisse, W.Th. Daems, I. Molenaar and P. van Duijn.
© 1973, North-Holland Publishing Company - Amsterdam, The Netherlands.

CRYO-ULTRAMICROTOMY OF MUSCLES IN DEFINED FUNCTIONAL STATES

Methodological aspects

M. Sjöström, R. Johansson and L.E. Thornell

Department of Anatomy, University of Umeå, S-901 87 Umeå, Sweden

Introduction. The advantages of cryo-ultramicrotomy in the preparation of tissue for electron microscopy include the possibility of eliminating chemical fixation, dehydration with organic solvents and plastic embedding. The naturally occurring ions can be identified in situ in the ultrathin freeze-dried sections of the chemically untreated and frozen tissue by means of X-ray microanalysis 1-5. However, phenomena such as postmortem diffusion and diffusion during freezing, cutting, drying, storage and examination may cause distortion, these phenomena have not been completely investigated yet.

In studying the ultrastructure and elemental constituents of muscle tissue we have applied the technique of cryo-ultramicrotomy. To elucidate whether the sections obtained by the dry cutting mode of the technique are suitable for microanalysis we have studied in more detail different steps of the specimen preparation. The following describes a method permitting momentary freezing of chemically untreated skeletal muscle tissue of physiologically defined length and functional state. The aspects on other steps of both the dry and wet cutting mode of the preparation procedure have been discussed elsewhere 3, 6, 7.

Materials and methods. Fresh toe extensor muscles (∅ about 0.5 mm) of frog (Rana temporaria) were carefully dissected out and immediately after removal they were suspended in an oxygenated bath (4-10° C) containing a modified Ringer solution 8. The muscles were mounted horizontally between a force transducer (Sanborn Co., FTA-10-1) and a micrometer screwe. Electrical stimulation with alternating positive and negative square pulses of 2 msec duration was achieved by two parallel plates of platinum mounted so as to cover the two sides of the bath. If desirable, the muscle could be stimulated instead through its nerve supply. Tension recording was performed either by paper recording (Devices Instr. Ltd., M4) or by photographing the oscilloscope trace (Tektronix Inc., Type 564). A length-tension diagram was obtained for each muscle by measuring total tension during smooth tetanus and passive tension at different muscle lengths. The physiological parameters could be calculated from these diagrams.

The freezing apparatus included two pneumatically controlled chilled hammers with five copper specimen holders mounted on one of them (fig. 1). With the muscle either stimulated or not and its length physiologically defined the chilled hammers were instantaneously brought together and at the same time the bath was drawn away. The muscle was frozen between the hammers and simultaneously attached to the specimen holders. These were lowered into the hammer to avoid breakage of the muscle by excessive pressure. The temperature of the holders was measured continuously with a nickel resistance thermometer. The rapid movement of the hammers arose from their mechanical connection with a pneumatic cylinder (Mechman 1300 ev 1/20) electronically operated by a magnetic valve. The movement of the bath was driven by the same pneumatic cylinder as the hammers but delayed to minimize the time of air exposure for the muscle. The bath did not begin to move downwards, therefore, until the hammers had accelerated and reached the edge of the bath. Adhesion between muscle and bathing

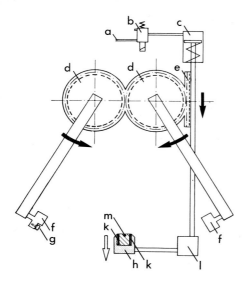

a Air feed pipe
b Magnetic valve
c Pneumatic cylinder
d Gear
e Rack
f Hammer (chilled)
g Specimen holder
h Trough (containing
   oxygenated Ringer)
k Platinum electrode
l Delay unit
m Muscle (cross-section)

Fig. 1. Freezing apparatus. For details, see Materials and methods.

solution was minimized by mounting the muscle just below the solution surface. After freezing, the hammer with the muscle attached to the holders was immersed into liquid nitrogen where the muscle was cut off between each holder. The specimens were then stored in liquid nitrogen for cryo-sectioning subsequently.

The cryo-sectioning was performed with an LKB CryoKit equipped with dry glass knives. Specimen and knife temperatures were -70° C and -50° C respectively. Flattened sections were pressed against cooled formvar-coated grids, freeze-dried for two hours in gelatin capsules in the cryo chamber and then transferred in cold atmosphere to an exsiccator and allowed to reach room temperature. The sections were examined in a Siemens Elmiskop 1A and the elemental analyses were performed in a Philips 301 combined with an energy dispersive X-ray spectrometer (EDAX).

Results. The muscles responded as would be expected during stimulation (fig. 2). When the bath was drawn away and the muscle passed the surface of the bathing solution a small tension peak was detectable, measuring about 10-20 per cent of the total tension developed during tetanus. The time the muscle was exposed to air and thus with-

Fig. 2. Tension (A) with corresponding stimulation (B) recorded during single twitches and smooth tetanus. The a-b-c-complex was recorded in connection with freezing. (a) tension change at bath movement, (b) tension change while the hammers reached the muscle, (c) mechanical oscillations in the system after freezing.

Fig. 3. Dry cut, unstained and freeze-dried section of chemically untreated skeletal muscle, frozen in tetanus (see fig 2) when stretched 10% compared to its rest length.

Fig. 4. Elements detected in a part of the sections in fig. 3. Numerous peaks are detectable indicating that the ions have been remained in the section throughout the preparative procedure. Outside the section (*) no elements was found.

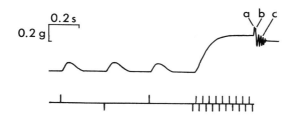

Fig. 2

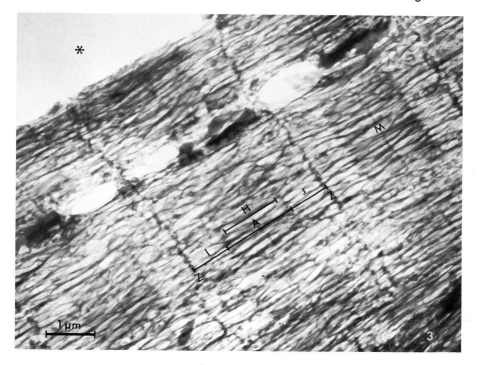

Fig. 4

out possibility for stimulation was less than 10 msec when studied on photographs of the oscilloscope trace. This is a shorter time than the interval between electrical stimuli leading to tetanus even if the time from the previous stimulus to the movement of the bath was encountered. The best attachment of the muscle to the specimen holders was obtained when these had a temperature of -70° - -75° C. The other hammer was kept at liquid nitrogen temperature.

The myofibrillar cross-striations (Z-, A-, I-, H-, pseudo-H-bands and overlap regions) were easily distinguishable, which facilitated topographic orientation during analytical electron microscopy (fig. 3). Fine folding occurred often, however, and was probably due to too rapid freeze-drying eventually combined with ice crystal damage and shrinkage in the electron beam. Freeze-drying in the cryo chamber and warming up in vacuo were decisive steps since thawing was eliminated thereby and diffusable substances were presumed to remain in the sections. X-ray microanalyses of such sections revealed numerous peaks (fig. 4) while no peaks were detectable on the supporting film outside the section 3.

Discussion. There were two main reasons for bringing this first study carefully through. The first reason was purely procedural in nature due to the difficulties in freezing an unfixed muscle when both in in vivo length and attached to specimen holders. If not mounted the muscle shortens maximally when cut free from its bony attachments, and this is morphologically unacceptable. For analysis of the contractile mechanism this rapid freezing technique affords new possibilities, as well, as practically every phase of a contraction can be "frozen".

The other reason was an attempt to make microanalysis more meaningful. In that some of the naturally occurring ions are supposed to move between morphologically defined structures during the contraction cycle 9, it should be of value to examine the presence of these in different compartments involved in exitation - contraction coupling at rest and during a contraction. It is too early yet to appraise the possibilities for a meaningful elemental analysis of sections obtained with this rapid freezing technique with subsequent dry cutting and freeze-drying. There are many factors in addition to uncontrolled postmortem events occurring during specimen preparation. These additional factors may cause distortion and have to be better understood. The freezing speed should be rapid enough theoretically for at least the superficial parts of the muscle fibres 10. Ice crystal formation during freezing should be overcome, if serious, by stimulating the muscle in solutions containing either low percent (e.g. 15-20 %) of glycerol or dimethyl sulfoxide (DMSO) 10-15 %. Diffusion during cutting is no problem at least within the limits in resolution of the present electron microscopic analytical instrumentation 11. It is essential after freeze-drying that the sections are kept in vacuo or in dry atmosphere since the dry sections are extremely hygroscopic. The moisture in room air will cause in other cases rapid melting of the sections (within minutes, perhaps seconds). The sensitivity and resolution of the X-ray spectrometers and the influence of the electron beam on the section during examination must be documented further.

(Supported by Swedish Medical Research Council, No 12X-3934 and the Medical Faculty of Umeå University).

References

1. Appleton, T.C., 1972, Micron 3, 101.
2. Marshall, J., 1972, Micron 3, 99.
3. Sjöström, M., 1973, J. Ultrastruct. Res. (in press).
4. Weavers, B.A., 1972, Micron 3, 107.
5. Weavers, B.A., 1973, J. Microscopy 97, 331.

6. Sjöström, M., Thornell, L.E., and Cedergren, E., 1973.
    J. Microscopy (in press).
7. Thornell, L.E., Sjöström, M., and Cedergren, E., 1973
    J. Ultrastruct. Res. (in press).
8. Huxley, H.E., Brown, W., 1967, J. Mol. Biol. 30, 383.
9. Adrian, R.H., Chandler, W.K., Hodgkin, A.L., 1969,
    J. Physiol. 204, 207.
10. Moore, H., 1971, Phil. Trans. Roy. Soc. Lond. B. 261, 121.
11. Appleton, T.C., 1972, in: Proc. fifth european cong. on electron
    micr., EMCON 72, Manchester, England p.56.

Electron microscopy and Cytochemistry, eds. E. Wisse, W.Th. Daems, I. Molenaar and P. van Duijn.
© 1973, North-Holland Publishing Company - Amsterdam, The Netherlands.

IMPROVED CONTRAST IN DRY ULTRATHIN FROZEN SECTIONS

G. Werner, E. Morgenstern, and K. Neumann

Medizinische Biologie, Fachbereich Theoretische Medizin der Universität
des Saarlandes, D-6650 Homburg/Saar

1. Summary

Ultrathin frozen sections of native tissues have been cut with a dry knife and stained, after thawing, by osmium tetroxide vapour. After this treatment cell organelles become clearly visible. As the sections have not come into contact with a liquid, the staining by osmium tetroxide vapour should not reduce their usefulness for microanalysis.

2. Introduction

Ultrathin frozen sections of native material offer the best conditions for microanalysis[1]. Such sections have to be cut with a dry knife and must not come into contact with a liquid even after sectioning. The main difficulties are encountered with observing rather than with preparing these sections. Although in some favourable cases organelles may be recognized by transmission electron microscopy, finer structural details remain obscure due to lack of differences in contrast. One way to get better information would be to stain these sections after thawing in the dry state with osmiumtetroxide vapour.

3. Materials and Methods

Fresh liver and testis of an albino rat were cut into 1mm thick slices and simultanously frozen to the temperature of liquid nitrogen with the quick-freeze pliers of Reichert. Ultrathin frozen sections were made with the freezing attachment FC150 in connection with an ultramicrotome OmU 3 of Reichert using dry glass knives[2]. Specimen and knife were maintained at temperatures from -70 to -90°C. Sections from large surfaces cut with thermal advance and minimal speed were mounted on formvar coated grids and flattened by the method of Christensen[3]. After enclosing in a BEEM capsule, the grids were transferred from the freezing chamber to an incubator at 50 to 60°C. After drying, the grids were placed into small glass vials containing crystals of osmium tetroxide and exposed to the osmium vapour for 1/2 to 1 hour.

4. Results

In very thin untreated dry sections practically no details can be recognized. In thicker sections one can discern the nucleus; dense organelles, which in size and shape resemble mitochondria can also be seen. However, an unequivocal identification of cell structures is possible only after treatment with osmium tetroxide vapour.

At low magnification of a liver section (fig.1) the polygonal cells are demarcated from each other. Bile capillaries become evident as light areas between the cells. The sinuses some of which contain erythrocytes show a marginal irregular band which corresponds to Disse´s space. The nucleus is round, though its contour is somewhat wrinkled. Darker-stained chromatin may be distinguished from a round, pale area which apparently represents the nucleolus. In the cytoplasm the mitochondria stand out as organelles with very high contrast. At higher magnification (fig.2) their typical structure can be recognized. Surrounding membranes and cristae show even more contrast than the matrix. Areas of the granular endoplasmic reticulum are identified by the characteristic parallel arrangement of their cisternae. The light fields occuring close by correspond to the agranular reticulum. Interspersed glycogen is not stained by osmium.

Though large coherent sections could not be obtained from testis, sections through spermatozoa were found in sufficient number. In the tail (fig. 3) the surrounding membrane and the longitudinally cut microtubules of the flagellum are clearly visible.

5. Discussion

The lack of contrast in ultrathin frozen sections of native material renders impossible the unambiguous interpretation of the electron microscopic image. On the other hand, if microanalysis is intended, the sections must not come into contact with a liquid, as this inevitably causes dislocation and even loss of substances thereby precluding a study of the distribution of ions and molecules in the natural position. The staining of dry sections by vapour of osmium tetroxide may be regarded as a way to avoid this difficulty. The obtainable contrast allows the identification of fine structural details, as can be seen best in the mitochondria. The osmium treatment should not interfere with microanalysis, as the osmium peak of an x ray spectrum is sufficiently far apart from the peaks of most of the interesting elements in biological material.

Fig. 1. Liver section at low magnification. AER agranular endoplasmic reticulum, BC bile capillary, E erythrocyte, GER granular endoplasmic reticulum, M mitochondrion, S sinus. Arrows indicate the cell boundaries. - Fig. 2. Liver mito - chondria at high magnification. - Fig. 3. Longitudinal sections through sperm tails. CM cell membrane, MT microtubules of the flagellum.

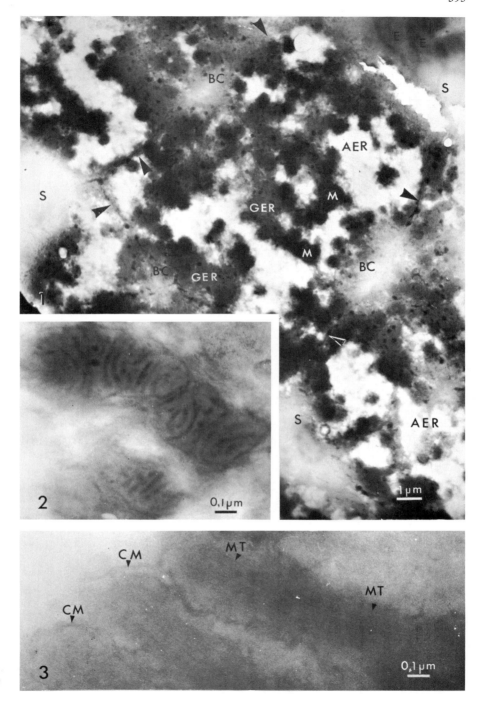

6. References

1. Appleton, T. C. , 1972, Micron 3, 101.
2. Werner, G. , Neumann, K. , and Morgenstern, E. , 1973, Mikroskopie 29, 27.
3. Christensen, A. K. , 1971, J. Cell Biol. 51, 772.

AUTHOR INDEX

SUBJECT INDEX